Machine Learning-Based Modelling in Atomic Layer Deposition Processes

While thin film technology has benefited greatly from artificial intelligence (AI) and machine learning (ML) techniques, there is still much to be learned from a full-scale exploration of these technologies in atomic layer deposition (ALD). This book provides in-depth information regarding the application of ML-based modeling techniques in thin film technology as a standalone approach integrated with the classical simulation and modeling methods. It is the first of its kind to present detailed information regarding approaches in ML-based modeling, optimization, and prediction of the behaviors and characteristics of ALD for improved process quality control and discovery of new materials. As such, this book fills significant knowledge gaps in the existing resources as it provides extensive information on ML and its applications in thin film technology.

- Offers an in-depth overview of the fundamentals of thin film technology, state-of-the-art computational simulation approaches in ALD, ML techniques, algorithms, applications, and challenges.
- Establishes the need for and significance of ML applications in ALD while introducing integration approaches for ML techniques with computation simulation approaches.
- Explores the application of key techniques in ML, such as predictive analysis, classification techniques, feature engineering, image processing capability, and microstructural analysis of deep learning algorithms and generative model benefits in ALD.
- Helps readers gain a holistic understanding of the exciting applications of ML-based solutions to ALD problems and apply them to real-world issues.

Aimed at materials scientists and engineers, this book fills significant knowledge gaps in existing resources as it provides extensive information on ML and its applications in film thin technology. It also opens space for future intensive research and intriguing opportunities for ML-enhanced ALD processes, which scale from academic to industrial applications.

Emerging Materials and Technologies

Series Editor:
Boris I. Kharissov

The *Emerging Materials and Technologies* series is devoted to highlighting publications centered on emerging advanced materials and novel technologies. Attention is paid to those newly discovered or applied materials with potential to solve pressing societal problems and improve quality of life, corresponding to environmental protection, medicine, communications, energy, transportation, advanced manufacturing, and related areas.

The series takes into account that, under present strong demands for energy, material, and cost savings, as well as heavy contamination problems and worldwide pandemic conditions, the area of emerging materials and related scalable technologies is a highly interdisciplinary field, with the need for researchers, professionals, and academics across the spectrum of engineering and technological disciplines. The main objective of this book series is to attract more attention to these materials and technologies and invite conversation among the international R&D community.

Polymer Processing
Design, Printing and Applications of Multi-Dimensional Techniques
Abhijit Bandyopadhyay and Rahul Chatterjee

Nanomaterials for Energy Applications
Edited by L. Syam Sundar, Shaik Feroz, and Faramarz Djavanroodi

Wastewater Treatment with the Fenton Process
Principles and Applications
Dominika Bury, Piotr Marcinowski, Jan Bogacki, Michal Jakubczak, and Agnieszka Jastrzebska

Mechanical Behavior of Advanced Materials: Modeling and Simulation
Edited by Jia Li and Qihong Fang

Shape Memory Polymer Composites
Characterization and Modeling
Nilesh Tiwari and Kanif M. Markad

Impedance Spectroscopy and its Application in Biological Detection
Edited by Geeta Bhatt, Manoj Bhatt and Shantanu Bhattacharya

For more information about this series, please visit: www.routledge.com/Emerging-Materials-and-Technologies/book-series/CRCEMT

Machine Learning-Based Modelling in Atomic Layer Deposition Processes

Oluwatobi Adeleke, Sina Karimzadeh,
and Tien-Chien Jen

CRC Press
Taylor & Francis Group
Boca Raton London New York

CRC Press is an imprint of the
Taylor & Francis Group, an **informa** business

MATLAB® is a trademark of The MathWorks, Inc. and is used with permission. The MathWorks does not warrant the accuracy of the text or exercises in this book. This book's use or discussion of MATLAB® software or related products does not constitute endorsement or sponsorship by The MathWorks of a particular pedagogical approach or particular use of the MATLAB® software.

First edition published 2024
by CRC Press
2385 Executive Center Drive, Suite 320, Boca Raton, FL 33431

and by CRC Press
4 Park Square, Milton Park, Abingdon, Oxon, OX14 4RN

CRC Press is an imprint of Taylor & Francis Group, LLC

© 2024 Oluwatobi Adeleke, Tien-Chien Jen, and Sina Karimzadeh

ISBN: 978-1-032-38670-6 (hbk)
ISBN: 978-1-032-38673-7 (pbk)
ISBN: 978-1-003-34623-4 (ebk)

DOI: 10.1201/9781003346234

Typeset in Times
by codeMantra

This book is dedicated to God, who is the source
of all inspiration, wisdom and knowledge

Contents

PART I   Introduction to Atomic Layer Deposition

PART II   Machine Learning Techniques

PART III   Machine Learning Applications in Atomic Layer Deposition

Preface

Machine learning has gained traction in recent years as a tool for understanding complex systems. Their ability to understand complex systems has revolutionized science and technology, making them more necessary than ever in this era of big data. Thin film technology is one area that has benefited greatly from these techniques, and there is still vast potential for full exploration of their capabilities in atomic layer deposition (ALD). This book titled *Machine Learning-Based Modelling in Atomic Layer Deposition Processes* provides a comprehensive overview of the application of machine-learning-based modeling techniques in thin film technology. This innovative approach can be used either as a standalone method or integrated with classical simulation and modeling techniques to optimize and predict the behavior and characteristics of ALD, leading to improved process quality control. This book is the first of its kind to provide comprehensive information on the application of machine learning in ALD, addressing important knowledge gaps in the literature. Based on analyses of literature-based case studies, it provides a thorough analysis of current uses and potential future applications for ALD procedures.

This book covers the current research in these two emerging spaces. It targets researchers, engineers, and students interested in ALD modeling machine learning algorithm creation and implementation. There are three sections in the book and 15 chapters in all. The first section introduces the fundamentals of ALD and thin film technology. The chapters in this section address ALD and thin film overview, state-of-the-art modeling and simulation approaches in ALD, ALD characterization methods, and industry 4.0 in thin film technology. The second section discusses the fundamentals of machine learning. The chapters in this section discuss machine learning techniques and algorithms, applications, challenges and limitations, and methods of optimal model development. Specifically, Chapters 6–8 discuss supervised learning, unsupervised learning, and deep learning, respectively, while Chapter 9 examines the hard and soft computing techniques.

The third section explores various applications of machine learning-based ALD modeling, ranging from simple predictive analysis and classification tasks in machine to complex deep learning-based applications such as microstructural image analysis, image blurring, 3D characterization, structural zone diagram analysis, and dimensionality reduction. Chapter 10 introduces this section by discussing the need and significance of machine learning-based modeling in ALD, while presenting the opportunities for the integration of numerical computational approaches and machine learning models. Chapter 10 further discusses some of the potential sources of data such as experimental and simulated data and the required pre-processing techniques for ALD data. Chapters 11–13 explore the application of prominent machine learning techniques, namely, predictive analysis, classification task, and deep learning for thin film properties prediction, material classification, microstructural image processing, and materials discovery. Chapter 14 discusses feature engineering, a pertinent machine learning approach that determines the quality of other machine learning model development and implementation in ALD and thin films.

Chapter 15 concludes the chapter by presenting some of the limitations of this book, challenges in machine learning application scaling up to industrial applications, and future directions.

Numerous examples and case studies that highlight the practical value of machine learning-based ALD modeling are provided by the authors throughout the entire book. In addition, the authors emphasize how this strategy has the potential to revolutionize the ALD industry and open the door for the creation of new products with improved functionality and performance. We anticipate that this book will be an invaluable tool for academics and industry professionals working in the disciplines of machine learning and ALD, and that it will spark more investigation and advancement in this fascinating field.

Oluwatobi Adeleke
Mechanical Engineering Science
Faculty of Engineering and Built Environment
University of Johannesburg

Sina Karimzadeh
Mechanical Engineering Science
Faculty of Engineering and Built Environment
University of Johannesburg

Tien-Chien Jen
Mechanical Engineering Science
Faculty of Engineering and Built Environment
University of Johannesburg

Acknowledgments

The authors of this book thank God who is the source of inspiration, wisdom, and strength to start and finish this book.

The authors are grateful to the JENANO research group of the Department of Mechanical Engineering Science, University of Johannesburg and several researches who have inspired the authors on this project. Dr. Oluwatobi Adeleke thanks his wife, Mrs. Grace Adeleke, for her unwavering love, encouragement, and understanding throughout this journey.

The authors thank the University of Johannesburg for the research funds provided through the University Research Committee (URC), GES 4.0 scholarship funding and the Faculty of Engineering and Built Environment funding under which the book was developed. Professor Tien-Chien Jen appreciates the financial support from National Research Foundation.

Author Biographies

Oluwatobi Adeleke is a researcher at the Department of Mechanical Engineering Science, University of Johannesburg. He joined the University of Johannesburg in 2019 as a PhD researcher and completed his PhD research in 2022. He also holds Masters' and Bachelor's degrees in Mechanical Engineering from the University of Ibadan and Ladoke Akintola University of Technology, respectively. His research interest is in artificial intelligence, soft computing techniques and machine learning, renewable energy, bio-energy, solar cells, waste-to-energy, and systems modeling. He has also made extensive contribution to researches in atomic layer deposition, material science, corrosion inhibition, waste management modeling and optimization, and life cycle assessment. He has published articles in reputable journals such as *Neural Computing and Applications, Energy Reports, Fuel, Renewable Energy, Biotechnology Reports, Journal of Material Research and Technology*, and *Engineered Science*. He has published a number of book chapters and conference papers in these fields. He is a reviewer for several reputable journals in the field of waste management, soft computing, and renewable energy. He was awarded the best PhD researcher in the Department of Mechanical Science, under the Faculty of Engineering and Built Environment (FEBE) award of academic excellence, 2021. He is a member of the Council for Regulation of Engineering in Nigeria (COREN), Nigeria Society of Engineers (NSE), and American Society of Mechanical Engineers (ASME).

Sina Karimzadeh is a PhD research candidate at the University of Johannesburg. He holds an MSc degree in Mechanical Engineering from the University of Johannesburg in 2020. He has been selected as one of the prospective candidates for the Chancellor's Medal for the Most Meritorious Master's Study for 2020. His current research interest focuses on the development of Li-ion battery active components and interface engineering by using the atomic layer deposition (ALD) technique. He has also been involved in a number of projects including Hydrogen Storage, Hydrogen Generation, Thin Films and Nanotechnology, Drug Delivery, Heat Transfer, Water Purification Membrane, and Computational Modeling and Simulation. He has published numerous journal articles in *Journal of Electrochemical Energy Reviews, Journal of Energy Storage, Journal of Water Process Engineering, International Journal of Heat and Mass Transfer, Journal of Molecular Liquids*, and *ASME International Mechanical Engineering Congress* and also attended conferences. He has served as the reviewer for journals of Elsevier, Springer, etc. He is currently the Head of the ALD and innovation sub-research group and the Lead experimentalist at the ALD facility.

Tien-Chien Jen is a full professor and the Head of Department, Mechanical Engineering, University of Johannesburg. Earlier, Prof Jen was a faculty member at the University of Wisconsin, Milwaukee. Prof Jen received his PhD in Mechanical and Aerospace Engineering from UCLA, specializing in thermal aspects of grinding. He has received several competitive grants for his research, including those from the US National Science Foundation, the US Department of Energy, and the EPA. He is also the Director of Manufacturing Research Centre of the University of Johannesburg. Meanwhile, SA National Research Foundation has awarded Prof Jen a NNEP grant (National Nano Equipment Program) worth USD 1.5 million to acquire two state-of-the-art Atomic Layer Deposition (ALD) tools to be housed in a 220 m² 10,000 level (ISO 7) clean room facility for ultra-thin film coating. These two ALD research facilities will be the first in South Africa and possibly the first in the African continent. He has made extensive contributions to the field of mechanical engineering, specifically in the area of machining processes, ALD, cold gas dynamics spraying, fuel cells and hydrogen technology, batteries, and material processing. He has published several journal articles, books, and book chapters on these topics in reputable journals, and presented in several conferences.

Part I

*Introduction to Atomic
Layer Deposition*

1 Overview of Atomic Layer Deposition and Thin Film Technology

1.1 INTRODUCTION

Thin films are widespread in modern technology, with uses ranging from surface coatings to cutting-edge nanoelectronics. Thin films are broadly used in an array of materials, procedures, and applications, particularly in terms of electronics. They are essential to advancing the development of ever-more-powerful computing, data storage, communication, energy storage, energy harvesting, and sensing systems. Most of this technology needs high-precision and high-quality film materials of thickness less 100 nm which is achievable by chemical vapor deposition methods. Moreover, the compatibility of process condition, such as temperature and deposition techniques, plays a vital role in the success of these technologies. Atomic layer deposition (ALD) is an effective coating technique for fabrication of high quality and uniform thin films and precise material growths at a nanometric scale. ALD is a special variation of chemical vapor deposition (CVD) technique where gaseous reactants (precursors), each containing various elements, are injected to the reaction chamber to generate the desired thickness of film through chemical vapor interactions. These interactions are in the base of chemisorption which provide strong adsorption of precursor element with surface atoms. ALD properties known as sequential, self-limiting deposition allow controllable film growth, pin hole free, and conformality.

The first use of ALD was the production of ZnS thin films for electroluminescent displays, but subsequently it was expanded to produce ternary and quadric composites [1,2]. For the same purpose, the technique was also employed to deposit amorphous Al_2O_3 dielectric thin films. However, until the 1990s, Atomic layer epitaxy (ALE) was the main industrial use for electroluminescent displays. Further, Suntola et al. [3] pioneered the ALD process for commercial application in Finland in the mid-1980s. Currently, ALD is developed for synthesis of different compounds such as metal oxides, metals, carbides, nitrides, fluorides, sulfurs, and selenides [4]. Also, there are a lot of universities, research institutions, and industries that perform ALD-related research. Meanwhile, by increasing the interest in amorphous and polycrystalline film compounds, the importance of high insulating oxide films with low-leakage high-k dielectrics for use in dynamic random-access memory (DRAM) and complementary metal oxide semiconductors (CMOS) was felt more. Thus, ALD has a key role in continuous miniaturization of semiconductor devices.

This chapter first outlines the fundamentals of ALD process and features, and highlights a brief survey on the current status of ALD. Some chemical examples

DOI: 10.1201/9781003346234-2

3

accomplished by ALD are summarized to present a different surface chemistry. Various ALD reactor types are defined. Advantages and disadvantages of ALD are discussed, and a comparison of thin film deposition methods are explained. The possible fast-growing application of ALD in various fields is described to manifest its use in practise. Finally, the possible challenge and perspective of ALD are discussed.

1.2   PRINCIPLE OF ATOMIC LAYER DEPOSITION

1.2.1   PROCESS AND METHODS

In this section, the basic of an ideal ALD process and some important concepts that are helpful for understanding ALD process and intrinsic features is highlighted. In the ALD process, thin films are generated by repeating deposition cycles through expositing different gaseous species on material surface maintained in a fixed temperature. ALD films are very highly uniform and conformal. ALD is typically regarded as a bottom-up framework, and ideally, a monolayer is formed per ALD cycle. Each ALD cycle is split into two individual half cycles, which consist of four stages as follows:

    i. introducing the precursor gas on substrate surface through chemisorption process
    ii. purging away the excess decomposed residual generated by precursor and surface reactions
    iii. introducing second precursor (co-reactant) to terminate the surface with functional groups
    iv. second purging

Therefore, the film growth achieved by ALD is controlled and self-limiting, which prohibits spontaneous gas-phase reactions. However, certain similar cycles can be repeated to deposit the desired thickness of film. As an example, Figure 1.1 depicts the steps in the ALD process for the deposition of aluminum oxide utilizing trimethylaluminum (TMA) and water (H_2O) as the precursor gases. In the process, TMA undergoes various chemisorptions on the surface as described in Equations 1.1 and 1.2 when $(1 \leq X \leq 3)$:

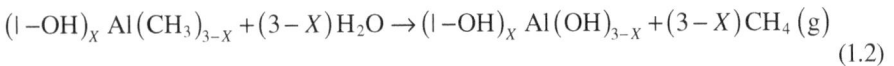

$$X \mid -OH + Al(CH_3)_3 (g) \rightarrow (\mid -O)_X Al(CH_3)_{3-X} + XCH_4 (g) \tag{1.1}$$

$$(\mid -OH)_X Al(CH_3)_{3-X} + (3-X)H_2O \rightarrow (\mid -OH)_X Al(OH)_{3-X} + (3-X)CH_4 (g) \tag{1.2}$$

TMA molecules contain three methyl groups, and each of them can undergo the dissociation and association process via multi exchange reaction (X). Similarly, ALD Al_2O_3 film consists of two half cycles. The initial substrate surface terminated by OH functional groups was presumed. At first, the Precursor A (TMA) was exposed to the surface to react with OH surface groups, and it led to Al-O bond formation through chemisorption and generation of CH_4 residuals. However, the exposure continues to saturate the surface and terminate it with -CH_3 groups which are poor in reactivity,

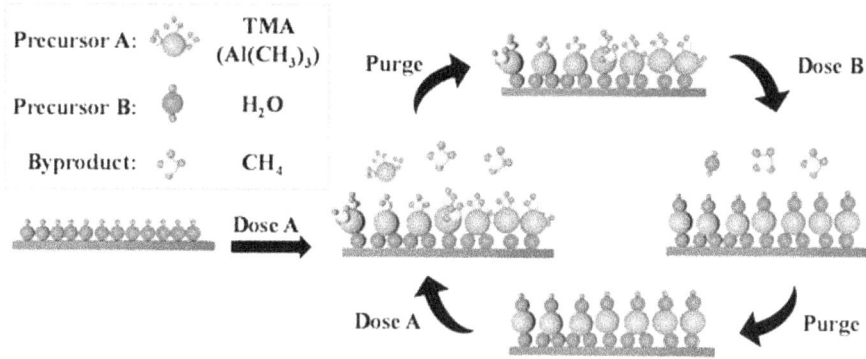

FIGURE 1.1 Schematic demonstration of atomic layer deposition (ALD) Al$_2$O$_3$ film growth mechanism by using precursors trimethylaluminum (TMA) and H$_2$O [5].

meaning the limitation of further precursor reactions. Then, the excessive products were purged away from the chamber zone by inert N$_2$ or argon gases. During the second half of the cycle, water vapour is introduced to the surface, where it undergoes a reaction with -CH$_3$ groups, resulting in the production of residual CH$_4$ and termination of the surface with -OH groups, thereby preparing it for the subsequent deposition cycle. Then, the second purging process removed residuals from the chamber. The physical separation of the precursors and co-reactants during the purge phase following each reaction step is essential for maintaining the self-limiting character of ALD reactions and ensuring that no reactions take place. A targeted material with a specified thickness and composition can be formed in a precisely controlled manner by repeating the cycle of these stages.

The film growth rate of ALD is highly dependent on the precursors type, process condition including temperature and pressure, and process parameters such as gas flows rate, pulsing, and purging times [6]. However, the cycle time can be modified from seconds to minutes, depending on the precursors' chemistry and reactivity, and substrates' geometry. Planar or simple substrates, such as silicon substrates, often need fewer cycles to produce high-quality films than complicated substrates such as biological substrate [7]. Moreover, each ALD reaction needs a certain amount of activation energy to trigger precursor reactions with reactive sites of surface. This activation energy can be provided by heat in thermal ALD, plasma power in plasma-enhanced ALD, and UV light in UV-assisted ALD. Growth per cycle (GPC) is an important factor to analyze an ALD process. Figure 1.2 presents the ALD temperature window which indicates the temperature range when the growth is constant. The failed ALD deposition or poor GPC is frequently caused by temperature outside of the window that can be affected by fast desorption and decomposition of precursor at high temperature or weak reaction kinetic and precursor condensation at low temperature. However, the temperature windows cover most of the precursors; therefore, slight changes in precursor input or temperature have little impact on the GPC.

Generally, ALD process can be carried out by two methods: thermal ALD and plasma enhanced ALD (PEALD). In thermal ALD, the reactions between precursor

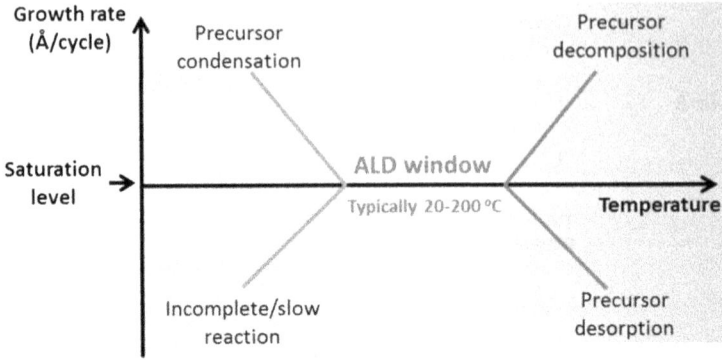

FIGURE 1.2   Schematic representation of a general atomic layer deposition (ALD) temperature window [8].

TABLE 1.1
List of Materials Deposited by ALD [10]

Elemental	Nitrides	Sulfides	Oxides	Other compounds
C, Al, Si, Ti, Fe, Co, Ni, Cu, Zn, Ga, Ge, Mo, Ru, Rh, Pd, Ag, Ta, W, Os, Ir, Pt	B, Al, Si, Ti, Cu, Ga, Zr, Nb, Mo, In, Hf, Ta, W	Ca, Ti, Mn, Cu, Zn, Sr, Y, Cd, In, Sn, Sb, Ba, La, W	Li, Be, B, Mg, Al, Si, P, Ca, Sc, Ti, V, Cr, Mn, Fe, Co, Ni, Cu, Zn, Ga, Ge, Sr, Y, Zr, Nb, Ru, Rh, Pd, In, Sn, Sb, Ba, La, Ce, Pr, Nd, Sm, Eu, Gd, Tb, Dy, Ho, Er, Tm, Yb, Lu, Hf, W, Ir, Pt, Pb, Bi	Li, B, Mg, Al, Si, P, Ca, Ti, Cr, Mn, Co, Cu, Zn, Ga, Ge, As, Sr, Y, Cd, In, Sb, Te, Ba, La, Pr, Nd, Lu, Hf, Ta, W, Bi

ALD, atomic layer deposition.

and co-reactants and surface occur in a thermodynamically favorable manner, and heat plays a major role during the reactions between precursors and surfaces and provides the required energy for possible reactions. Typically, thermal ALD is mostly used for depositing metal oxide films, and the temperature range for successful chemisorption between precursor and surface in thermal ALD is 150°C–350°C. The advantage of this method is the precise control of thickness, which is independent of reactor design and surface morphology.

PEALD is an advanced version of conventional thermal ALD, which is equipped with plasma power and beneficial for performing ALD process in lower temperature (even in room temperature), and it provides wider selection for substrate materials such as organics, polymers, and drug powders. During chemical reactions in the chamber, plasma generates energetic species such as ions and electrons to facilitate the mechanism of reactions. Moreover, PEALD allows the use of various types of co-reactants such as O_3, O_2, H_2, N_2, H_2S, and NH_3, which enables deposition of metal, metal nitride, metal sulfide, metal selenide, metal phosphide, etc. ALD offers a wide range of elements to create various compounds. Table 1.1 and Figure 1.3 present the overview of different elements utilized for ALD materials [9] and thin films. However, besides the advantages, the PEALD has some drawbacks including causing

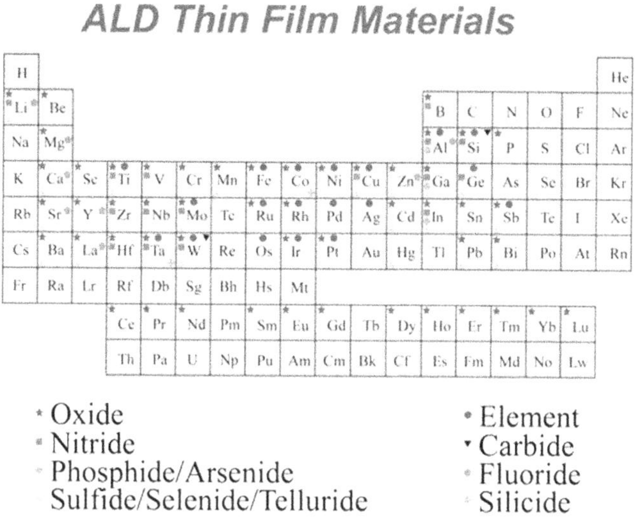

ALD Thin Film Materials

FIGURE 1.3 A list of the elements utilized in atomic layer deposition (ALD) material. Adopted from reference [9].

undesirable reactions such as nitridation and oxidation, bombardment of energetic ions, which causes bond breaking, atom destabilization, and propagation of charge in the dielectric layer, which may damage the film conformality. Also, the industrial scale-up approach and equipment development are challenging. Although the ALD can deposit a variety of materials, it is still not feasible to create every material as seen in the periodic table of elements in Figure 1.3.

1.3 ADVANTAGES AND DISADVANTAGES OF ALD

1.3.1 ADVANTAGES

Table 1.2 summarizes the advantages of ALD in the base of the self-limiting and surface-controlled properties which may support choosing a coating process for a particular application. ALD is a reliable strategy to offer the film growth solution when one or more of these benefits are essential for purposes ranging from research to industrial application. Some of these advantages are discussed in the following.

1.3.1.1 Uniformity and Conformality

The thin films generated by the ALD technique are extremely uniform and conformal over 3D materials surface such as porous materials, deep trenches, and vias, which is one of the unique advantages of ALD compared to other coating methods. Conformal coating follows the contours of the surface and is independent of surface defects or roughness. ALD coating has the same thickness over the whole surface. Figure 1.4 presents the properties of different thin film coating methods. As seen, ALD film has the best uniformity and conformality compared to the sol-gel, CVD, and PVD methods. Thus, ALD is a prominent technique for use in trench coating of semiconductor memory device and trench filling deposition of diffractive optics parts as illustrated in Figure 1.5a and b, respectively.

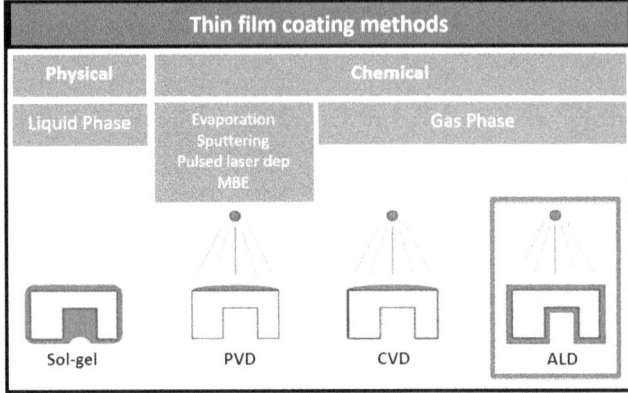

FIGURE 1.4   Schematic comparison of different thin film coating methods.

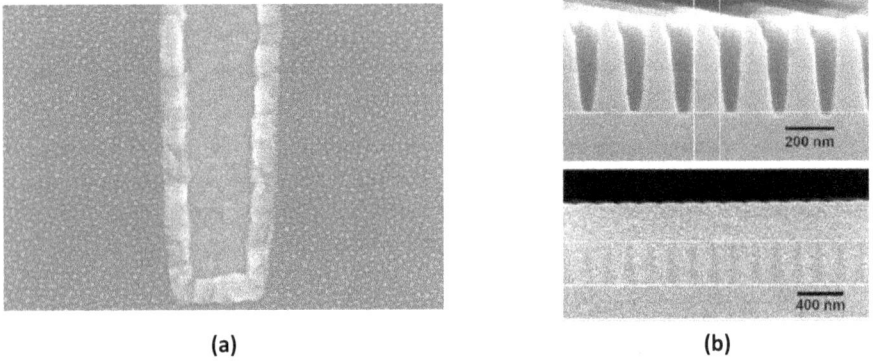

(a)   (b)

FIGURE 1.5   (a) A cross-section Scanning Electron Microscopy (SEM) images of a deep trench with 20 nm HfO_2 coated by a $(CpMe)2HfMe_2/O_3$ precursor at 450°C and (b) atomic layer deposition (ALD) filling of optical parts trench. Adopted from references [11,12].

1.3.1.2   Pinhole-Free

Due to the bottom-up film growth scheme, the ALD thin films are inherently pinhole-free. For instance, the properties mentioned above offer advantages in a variety of applications involving barriers and passivation film. For example, 10 nm thick barriers film are critical to keep oxygen and moisture away from electronic devices and organic materials [13]. Additionally, thin-film electroluminescent (TFEL) screens are classified as high electrical field devices, and the implementation of ALD technology facilitates the production of a high-quality insulating layer for these screens. This insulating layer plays a crucial role in ensuring the reliability of the product [14].

1.3.1.3   Controllable Film Thickness

By virtue of ALD, film growth can be achieved layer by layer and the thickness can be controlled precisely. Generally, the GPC for different ALD films are often less than 2Å/cycle, which is depending on the deposition condition and the material type [15].

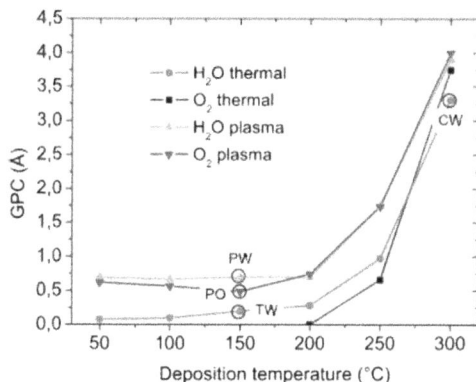

FIGURE 1.6 Temperature-dependent growth per cycle (GPC) for thermal and plasma atomic layer deposition (ALD) with oxygen and water.

ALD is more effective than other thin film deposition methods with thickness control at the nanoscale level, such as CVD and PVD. For example, recent experimental study showed the GPC of V2O5 film (i.e., 0.7 Å/cycle) by using PEALD and vanadium pentoxide (VTIP) and O_2 as precursors [16]. Figure 1.6 presents the comparison ALD windows of VTIP and H_2O and O_2 precursors in two different methods, namely, thermal and plasma.

1.3.1.4 Controllable Film Composition

Another prominent advantage of ALD is the ability to control the composition of a material accurately. In an ALD process, the deposited film's stoichiometry is often near to its theoretical result, but it can be affected by deposition temperature and type of precursors. ALD can tailor the composition by supercycle which consists of different ALD processes [10]. For example, the conductivity and optical features of the film can be modified by tailoring the zinc tin oxide composition by ALD [17].

1.3.1.5 Low-Temperature Deposition

Low-temperature operation of ALD is another advantage that distinguishes it from other conventional methods like CVD. In addition, different types of materials can be deposited using ALD under 100°C or even in ambient temperature [18]. This characteristic of ALD enables it to be used on temperature-sensitive materials such as polymers and organic compounds. Thus, ALD allows wider material selection for deposition. So far, various metal and metal oxides including TiO_2 [19], SiO_2 [20], SnO_2 [21], Al_2O_3 [22], ZnO [23], Pd [24], and Cu [25] are coated by ALD at close to ambient temperature.

1.3.1.6 Artificial Material

ALD is a particularly potent supportive technique in cutting-edge nanotechnology science and surface functionalization. The capacity of ALD to mix two or more materials at nanoscale level allows for the production of new "artificial" materials with distinctive properties, which is perhaps its most prominent advantage. Accordingly,

TABLE 1.2

A Brief Comparison of Different Thin Film Deposition Methods in Terms of Advantages and Disadvantages

Coating Method	Advantages	Disadvantages
ALD	High-density film, conformality, uniformity, great stoichiometry, accurate thickness control	Costly, high waste rate, slow deposition rate, complicated preparation process
CVD	Precursor availability, easy operation, reproducibility, great deposition rate	Low reactive precursors, high-temperature operation
PVD	Low toxicity, cheap, low temperature, safer	Needs annealing time, poor stoichiometry, low deposition rate, bad uniformity
PLD	Simple, great deposition rate, flexible, short test time, low temperature	Limited surface area deposition, impurities, low deposition rate
Spin coating	Simple setup, fast operation, low cost	Low accuracy, nonuniform, lack of material efficiency, limited surface area
Electroplating		
Sol gel	Adhesive strength, simple, low cost, low temperature	Nonuniform, large volume shrinkage and cracking, uncontrollable film thickness and composition
RF sputtering	Low deposition rate, film thickness accuracy, great purity, complicated	Expensive, cracks, ion loss and low conductivity
Hydrothermal	Low cost, simple, flexible, low toxicity	Very slow, difficult condition

ALD, atomic layer deposition; CVD, chemical vapor deposition; PVD, physical vapor deposition; Pulsed laser deposition (PLD); Radio-frequency (RF).

when these functional materials are incorporated with high scalability and repeatability of ALD, it allows for production of novel nanomaterials.

1.3.2 DISADVANTAGES OF ALD

Although ALD has multiple advantages, it suffers from some issues including high cost, high material waste, complicated preparation process, slow deposition rates, and limited precursor availability compared to CVD. However, these issues can be mitigated by utilizing appropriate precursors and optimal reactor design. For comparison, Table 1.2 summarized some specific advantages and disadvantages, while Table 1.3 compares different thin film coating methods based on different properties.

1.4 ALD REACTORS

At the heart of an ALD reactor is a set of valves that control the entry and outflow of different gases to a substrate that has to be coated. Represented in Figure 1.7 is a typical laboratory scale reactor with typical ALD reactor components. Although there are many kinds of ALD reactors, they all contain several fundamental parts

TABLE 1.3

Comparison of Different Thin Film Coating Method Properties [26]

Properties	Coating Method					
	ALD	CVD	PLD	Sputtering	Evaporation	MBE
Film density	Good	Good	Good	Good	Fair	Good
Uniformity	Good	Good	Fair	Good	Fair	Fair
Deposition rate	Poor	Good	Good	Good	Good	Fair
Pinhole-free	Good	Good	Fair	Fair	Fair	Good
Low temperature	Good	Varies	Good	Good	Good	Good
No plasma damage	Good	Varies	Fair	Poor	Good	Good
Sharp dopant profile	Good	Fair	Varies	Poor	Good	Good
Step coverage	Good	Varies	Poor	Poor	Poor	Poor
Smooth interface	Good	Varies	Varies	Varies	Good	Good

ALD, atomic layer deposition; CVD, chemical vapor deposition; Pulsed laser deposition (PLD).

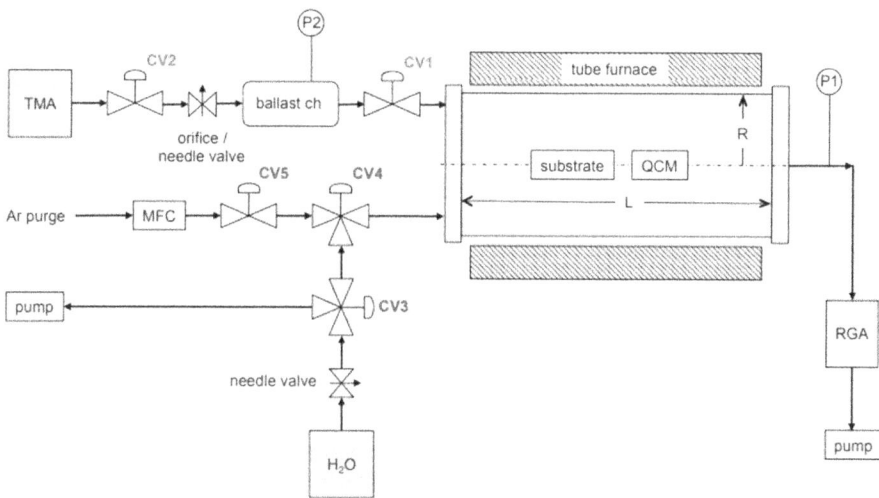

FIGURE 1.7 Schematic illustration of typical atomic layer deposition (ALD) reactor components [27].

that enable it to carry out ALD functions. Some of the important parts of the ALD reactor are the following:

i. **Reaction chamber**: This is the area where the ALD process takes place and substrates are kept. For example, in the thermal ALD, the reaction chamber is frequently heated to stimulate reactions.

ii. **Precursor sources**: Chemicals in the form of liquid, solid, and gas which all incorporate in the ALD reaction.

iii. **Purging gas**: Typically uses inert gases such as argon and nitrogen to purge away the by-product from reaction chamber.

iv. **Vacuum pump**: It provides low-pressure condition inside the chamber and also facilitates the way to exhaust by-products and excess precursors. Typically, reactions take place at 1–10 mTorr.

v. **Valving:** The flow of precursors is controlled by a variety of valves (solenoid, pneumatic, and gate), which are also utilized to isolate different reactor components.

vi. **A flow tube reactor**: The simplest type of ALD reactor that operates by continuously circulating inert gas. Individual precursor sources are dosed into the stream of inert gas via valves, which transports the reactant to the substrate. Then, the valve shuts off the source when a suitable dosage is attained, and the inert gas then transports the reaction by-products and extra precursor to the exhaust. The second precursor is injected in the same way as the first after a purge to assure the removal of unwanted CVD reactions.

1.4.1 TYPES OF ALD REACTORS

ALD reactors can be set up in a variety of ways depending on the substrate type, deposited materials, the ability to do in situ characterization, and many other parameters. In general, reactors may be distinguished by the kind of ALD processes performed, thermal or plasma, and the manner in which precursors are delivered to the substrate, temporally or spatially. However, the ALD reactors can be classified as thermal ALD (only heat drives both reactions), plasma ALD (plasma is incorporated to assist the surface to become activated for one or more reactions), temporal ALD (precursors are injected separately, and the substrate is fixed), and spatial ALD (precursor flows continuously through zones where the substrate is moving). The different ALD reactor designs are depicted in Figure 1.8. Two single-wafer, temporal reactors for thermal ALD are shown in the top half, one on the left with a flow-type reactor and the other on the right with a showerhead reactor. Batch, energy-enhanced, and spatial alternative types are displayed in the bottom half, starting at the bottom left. Moreover, the Computational fluid dynamics (CFD) tool is a useful simulation technique to model and design an optimal configuration of an ALD reactor and further analyze the capability of flows and reactions inside the chamber. Figure 1.9 represents the 3D design of the Picosun R-200 ALD reactor.

1.5 ALD RECIPE

The ALD recipe typically displays the details of a specific ALD process. An ALD recipe consists of the complete setting of an ALD process including pulsing and purging times, process temperature, flow rate of species, and the number and type of deposition cycles. For example, Figure 1.10 indicates the various model systems with different types of ALD cycle described as AB cycle with one precursor and one co-reactant. Regular AB cycle can be used to deposit Li_2O, LiOH, and LiF. A multistep ABC cycle or supercycle technique can be employed by ALD to create doped and ternary materials. For example, Li_2CO_3 ALD requires a multistep process. However, in the supercycle method, various AB processes are merged together to create an ALD deposition. In other words, by repeating the AB cycle m times and the CD cycle

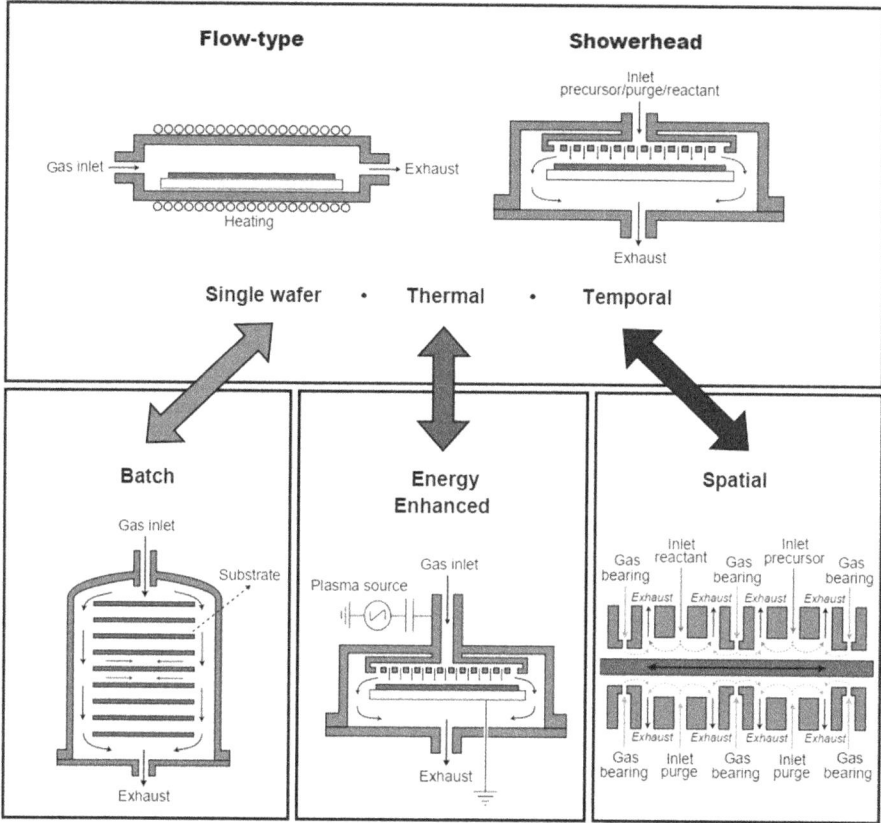

FIGURE 1.8 The illustration of the different atomic layer deposition (ALD) reactor designs [28].

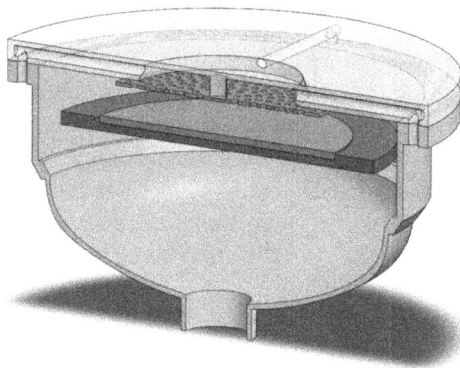

FIGURE 1.9 Schematic of CFD 3D designed R-200 atomic layer deposition (ALD) reactor.

Regular (AB)$_m$ Multistep (ABC)$_m$ Supercycle ((AB)$_m$(CD)$_n$)$_f$

← m →   ← m →   ← f →
                                                        ← m → ← n →

[A] [B] [A] [B]   [A] [B] [C] [A] [B] [C]   [A] [B] [C] [B] [A] [B] [C] [B]

Li$_2$O, LiOH, LiF   Li$_2$CO$_3$ AlP$_x$O$_y$

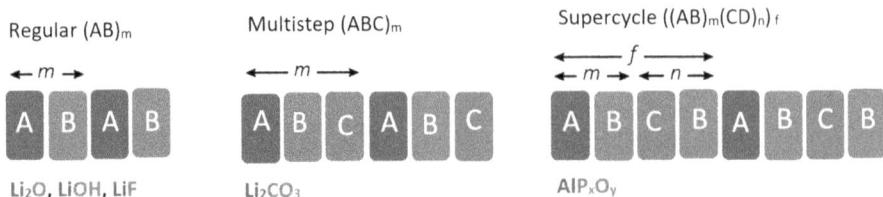

FIGURE 1.10   Various cycle types of an atomic layer deposition (ALD) process. In the regular AB, A denotes the precursor and B denotes the co-reactant. The multiple model ABC refers to the deposition of three different elements. The supercycle (AB)$_m$(CD)$_n$ is characterized by m repeats of the AB cycle followed by n repetitions of the CD cycle. This process is then repeated f times.

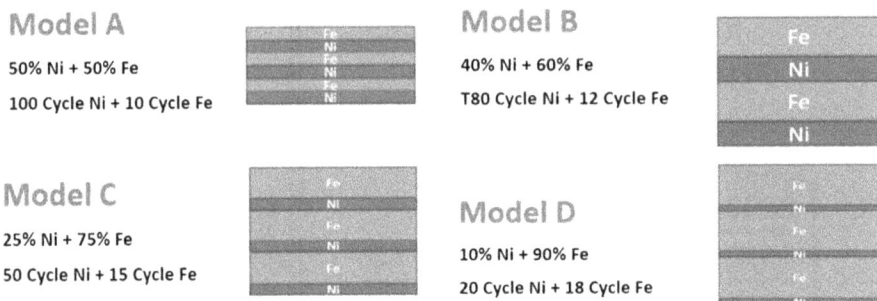

Model A

50% Ni + 50% Fe

100 Cycle Ni + 10 Cycle Fe

Model B

40% Ni + 60% Fe

T80 Cycle Ni + 12 Cycle Fe

Model C

25% Ni + 75% Fe

50 Cycle Ni + 15 Cycle Fe

Model D

10% Ni + 90% Fe

20 Cycle Ni + 18 Cycle Fe

FIGURE 1.11   An atomic layer deposition (ALD) recipe approximation of NiFe film composition with different material dosages (four models) by considering the growth rate of nickel (Ni) and iron (Fe).

n times, the supercycle (AB)m(CD)n is created. Then, the desired thickness can be achieved by repeating the f supercycle. For instance, a supercycle strategy was used in the development of the AlP$_x$O$_y$ process. Here, the supercycle consists of one Al$_2$O$_3$ cycle and n POx cycles [29]. Furthermore, Figure 1.11 demonstrates an example of ALD recipe approximation for fabrication of NiFe compound with different material dosages (Four Models) by considering obtained GPC of each material. For example, model A presents the NiFe film with the same element concentration. Therefore, for attaining this type of film, we need to predict the GPC ratio of each element which may be available in research literature or need to be found experimentally. However, by considering the GPC ratio of Fe and Ni, which is about 1/10, the number of Ni cycles needs to be 10 times more than Fe cycle to achieve the same coating thickness of each layer. The other models show the different GPC ratios of NiFe composition.

1.6   ALD PRECURSORS

In general, ALD precursors are chemical compounds constructed by metal atoms in the center which are surrounded by organic ligands and can be found in the forms of liquid, gas, and solid. Specific bubblers are designed for ALD tools to keep the precursor chemicals and can be converted to the gas phase through vaporization or sublimation phenomenon. The temperature of these containers can be regulated in terms of chemical properties to have an optimal phase transition. In comparison to other gas-phase chemical techniques, ALD has different general criteria since all gas-phase reactions must be avoided and only the reaction with the surface should occur.

Nowadays, varied precursors have been developed for ALD which are highly reactive compared to the CVD precursors. However, some of the CVD precursors can be utilized for ALD. ALD precursors must have a great thermal stability in the gas phase and during contact with material surface in the high temperature (150°C–200°C) to prevent unwanted reactions. Stability is a critical factor especially for industrial application which needs long process time. Due to the self-limiting nature of ALD, just a little quantity of the precursor is needed to load into the surface throughout one pulse, and any more will be evacuated by the inert gas. ALD is known as the gas-phase process; therefore, its precursors must be volatile enough under certain pressure and temperature. An excellent precursor must be highly reactive and completely decompose when reacting with the substrate's active sites at lower temperatures which affect the growth progress. However, the following are a few standard specifications for ALD precursors:

   i. Highly stable and no self-decomposition at high temperature
  ii. High volatility at process temperature
 iii. Nontoxicity and safe operating
  iv. Availability and low cost
   v. Sufficient reactivity with the surface site at low temperature

Figure 1.12 provides an overview of some of the aforementioned precursor properties with regard to volatility. An ideal ALD precursor must be stable at low temperatures and should not simply decompose. This is shown by volatilization at a low temperature and is characterized by a rapid mass decrease to 0%. However, many compounds do not meet these criteria of TG curve, where volatilization occurs slowly over a large temperature range that is greater than that required by the majority of ALD procedures. Accordingly, a substantial residual mass and the loss of ligands, which are shown by several plateaus and mass decreases in the TG curve, are two indicators that the mass loss in this case is most likely an occurrence of decomposition.

Notably, intermolecular forces, such as hydrogen bonds or electrostatic interactions (primarily van der Waals' forces), have an impact on a compound's volatility, which can be affected by molecular weight and coherence of precursor molecule. Thermogravimetric analysis (TG) curve presents weight loss of a compound as a function of temperature. However, the vapor pressure of a precursor can be determined. In general, highly asymmetric, lighter compounds are relatively volatile, whereas heavier, symmetrical molecules tend to possess lower volatilities. However, there are some exceptions. Using a heteroleptic precursor (metal centered with two or multiple ligands) and asymmetry in the compound structure is a common strategy for enhancing volatility. Moreover, the ligands themselves have an impact on the molecular weight and intermolecular forces.

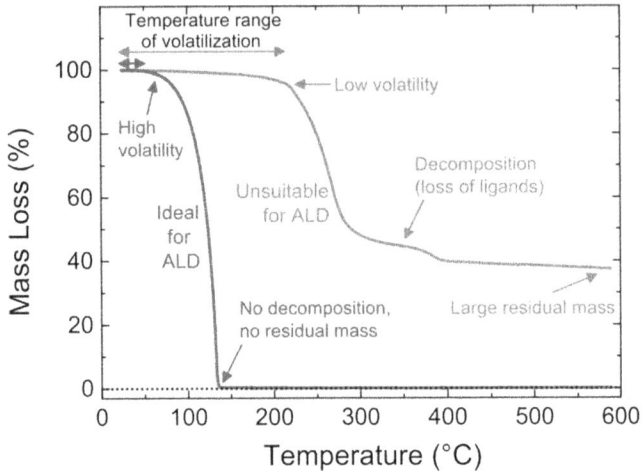

FIGURE 1.12 Thermogravimetric characterization of an ideal and nonideal atomic layer deposition (ALD) precursor (the mass loss of a precursor as a function of temperature at a fixed heating rate, typically 10 C/min).

TABLE 1.4
Brief Details of Various ALD Precursors for Metals

ALD Precursors	Species
Pure elements	-
Metal hydrides	-
Metal halides	Bromides, chlorides, fluorides, iodides
Metal carbon bonds	Alkyls, cyclopentadienyls
Metal oxygen bonds	Alkoxides, beta-diketonates
Metal nitrogen bonds	Amides, imides, amidinates

ALD, atomic layer deposition.

Accordingly, to specify an appropriate precursor for an ALD process, some critical factors should be taken into account including material of interest, deposition condition, desired application, reactivity to other chemical species, and required film properties such as conductivity, dielectric constant, current loos, antibacterial activity, permeability, adsorption ability, and photochemical activity.

Five major groups can be applied to classify the most common volatile metal-containing ALD precursors: -diketonate compounds, halides, alkoxides, N-coordinated complexes (amidinates, amides), and organometallics, which include cyclopentadienyl-type and metal alkyls groups. The detailed types of ALD precursors for metals are summarized in Table 1.4.

However, other species, such as carboxylates, metal nitrates, and isocyanates, have often been utilized as ALD precursors [30,31]. Moreover, a variety of metal halide precursors with O_2 and H_2O as co-reactants have been utilized for ALD. The deposition rate of metal halide precursors is acceptable and affordable for use in industrial applications. Nevertheless, this type of precursors can contaminate the surface at low process temperatures and may cause the etching or corrosion of film through generation of species such as HCl, HF, HBr, and HI [32,33].

Organometallic precursors are often classified as very reactive precursors in ALD process, since their molecule contains metal to carbon bonding. ALD of Al_2O_3 is the most frequently investigated film in thin film technology. For instance, the TMA $(Al(CH_3)_3)$ precursor has been utilized with oxygen, ozone, and water as co-reactants in the ALD to form Al_2O_3 film [34–36]. Moreover, Alumina has also been obtained from other metal alkyls precursors, such as $3Al$ (CH_3CH_2), $AlCl(CH_3)_2$, or $AlH(CH_3)_2$ [37,38]. The ALD of TMA/water can be performed in a wide temperature range of 100°C–500°C with a GPC of 1.2 Å/cycle. However, impurities such as OH can arise at low deposition temperature; thus, the plasma ALD with oxygen source offers solution to mitigate the impurities content which may enhance the electrical features.

Cyclopentadienyl compounds (metallocenes) are another type of ALD precursors, and their chemicals have at least one metal-carbon bonded to C_5H_5 ligand. These precursors are very volatile and stable in high temperatures, and their high reactivity can be regulated by sequentially pulsing the precursor. Despite the existence of a wide variety of diverse ligands based on C_5H_5, only a few of the alkylated cyclopentadiene complexes have been utilized as ALD precursors. Metal oxides such as MgO [39], In_2O_3 [40], and Sc_2O_3 [41] have been deposited using metallocene precursors. Figure 1.13 represents the variety of metal precursors for ALD and only highlights examples of homoleptic precursors. Furthermore, Table 1.5 lists ALD precursors for different species.

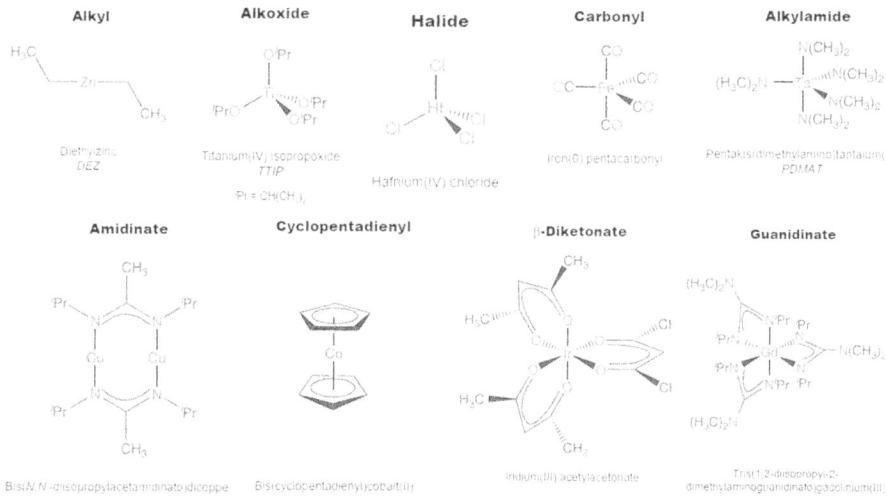

FIGURE 1.13   Variety of metal precursors for atomic layer deposition (ALD). The words in bold refer to the category of the precursor. The name in below indicates the example of the compound.

TABLE 1.5
List of ALD Precursors for Different Species

Species	ALD Precursors
Oxygen	Water vapor, Hydrogen peroxide (H_2O_2), Methanol, Ethanol, Ozone (O_3), Oxygen (O_2), Nitrogen dioxide (NO_2), ROH
Nitrogen	Ammonia (NH_3), Hydrazine (N_2H_4), Dinitrogen (N_2), Nitric oxide (NO)
Carbon	Acetylene gas, Formic acid vapor, Carbon contained metal compound
Fluorine	Hydrogen fluoride (HF), Fluorine contained metal compound (WF_6)
Sulfur	Sulfur vapor (S_n), Hydrogen sulfide (H_2S)
Selenium	Bis(triethylsilyl)selenium, $(Et_3Si)_2Se$
Tellurium	Bis(triethylsilyl)tellurium, $(Et_3Si)_2Te$
Phosphorous	Phosphine gas (PH_3)
Arsenic	Arsine gas, (AsH_3)
Antinomy	Antimony trichloride ($SbCl_3$), tris(dimethylamido)antimony
Pure metal	H_2, H_2N_2, NH_3, O_2N_2, O_2, Si_xH_y, formalin

ALD, atomic layer deposition.

1.7   APPLICATIONS OF ALD

In many fields of science and technology, advancements have been made as a result of the revolutionary use of ALD in the creation of novel materials and gadgets. In this section, we will examine some of the most fascinating and promising ALD applications and how they are influencing future developments in science and technology.

1.7.1   MICROELECTRONICS AND SEMICONDUCTORS

Devices have been driven to scale down to the nano and atomic scales with increasingly highly dispersed structures as technology development. A rising trend involves creating highly organized 3D structures that offer a larger surface area and, as a result, increase device performance. Consequently, the significance of the ALD process has enhanced, and it will eventually be used to manufacture semiconductor devices. Recently, research and development efforts have focused mostly on using ALD to produce uniform films with highly adjustable thickness, high dielectric constant, and pinhole-free structure. ALD is frequently utilized in the production of TFMHs for disk drives and DRAM stack capacitors. ALD development is presently focused on the deposition of films for metal gates and gate dielectrics. The device scalability to 45 nm architecture and beyond is enabled by this method.

Moreover, numerous studies on the fabrication of transistors using ALD process have been reported [42–45]. Liu et al. [46] carried out ALD process to coat zinc oxide on graphene for thin film transistor. In another work, in order to efficiently reduce the hole carrier concentration during the construction of high-performance p-type thin film transistors, Kim et al. employed ALD method on SnO film [47]. Furthermore, high-performance capacitors can be fabricated by ALD on porous alumina membranes [48]. The PEALD was employed by Dustin et al. [49] to create Al_2O_3/SiO_2-based metal-insulator-metal capacitors.

1.7.2 ENERGY STORAGE

Batteries, hydrogen storage, and supercapacitors are just a few examples of the energy storage technologies that have been effective in reducing the use of fossil fuels, addressing environmental concerns and facilitating the growth of the electric vehicle industry. To achieve the highest efficiency for the various energy-related devices, the manufacturing and interface/surface engineering of electrode materials with optimized architectures are essential.

The most often used ALD films are metal oxides, which are also excellent options for Lithium ion battery (LIB) electrode materials. TiO_2 is a desirable anode material for LIBs, which has a great rate capability, chemical stability, minimal volume expansion during cycling, and inherent safety. SnO_2 is another metal oxide that has received a lot of research attention because it has a great specific capacity and energy density than traditional graphite anodes. The use of ALD SnO_2 as an anode material for LIBs was initially presented when it was coated on graphene nanolayers [50]. In another work for the first time, $FePO_4$ compound was deposited via ALD by Gandrud et al. [51] by merging two subcycles of FeO_x and PO_x. Trimethyl phosphate and H_2O/O_3 are utilized as oxygen precursors in this process for deposition of PO_x.

Another significant electrochemical energy storage solution that has a better power density than batteries is the supercapacitor [52]. Nevertheless, supercapacitors have a lower energy density than conventional LIB. To create the electrodes for supercapacitors, transition metal oxides, such as NiO, Co_3O_4, Mn_3O_4, and Fe_2O_3, are coated by ALD on various conductive substrates [53–57].

1.7.3 DESALINATION MEMBRANE

The majority of water filtration facilities across the world use membrane-based separation techniques. Utilizing new membranes or modifying current ones can enhance industrial separation applications, particularly gas and water purification. An innovative method called ALD is being proposed to improve certain membrane types regardless of their chemical structure and shape. All membrane varieties can benefit from the ALD of inorganic materials, primarily metal oxides, on the membrane surface to increase hydrophilicity, permeability, selectivity, and antifouling. However, the use of ALD-based photocatalytic coatings on membranes could represent a viable strategy for the simultaneous isolation and decay of pollutants, with the potential to significantly reduce membrane fouling.

1.7.4 MEDICAL AND BIOLOGICAL APPLICATIONS

ALD offers a number of benefits for the development of thin films that might be used in bioelectronic devices, implanted medical devices, biosensors, drug delivery systems, tissue engineering scaffolds, bioassay systems, and other medical devices. ALD's low processing temperatures make it possible to employ temperature-sensitive materials such as organic materials, polymeric materials, naturally generated materials, and more that are often used in biological and medical applications.

1.7.5 OPTICAL COMPONENTS

To image or change laser light with regard to a particular application, laser systems require nonplanar or 3D-shaped optics. For example, an axicon transforms the point-source laser beam into a ring of light used for optical trapping or laser eye surgery. All of these optical components, from extremely curved lenses to specially constructed aspheres and multi-sided lenses, are designed to shape light. Almost all optical components require thin film coatings to improve their efficiency, such as AR coatings, that lower transmission losses or mirror coatings that reflect beams. The typical method of physical vapor deposition (PVD) exhibits constraints in this application. ALD MgF_2 coatings were successfully developed using a commercially available ALD reactor [58,59]. Al_2O_3, SiO_2, HfO_2, and TiO_2 are other ALD methods that may be used to create optical films [60–63]. New applications for barrier layer and optical coatings are made possible through ALD coating techniques.

1.7.6 FUEL CELL

Proton-exchange membrane (PEM) fuel cells due to their great efficiency in converting hydrogen's chemical energy into electrical power, functioning at close to room temperature, and lack of pollutants are potential alternative power sources for mobility and portable applications [64]. According to reports, platinum is the most effective catalyst for the oxygen reduction reaction (ORR) in PEM [65,66]. However, Pt is classified as a rare and very expensive metal. ALD is a technology that has promise for resolving issues with cost. Aaltonen et al. [67] achieved significant advancements in the synthesis by ALD through depositing Ru metal from $RuCp_2$ and O_2, and the method has now been expanded to include additional metals including Pt, Rh, and Ir [68].

Hydrogen is classified as clean fuel and can be utilized as the energy source in PEM fuel cells. Water electrolysis is one potential environmentally friendly method for producing hydrogen. When water is electrolyzed, one of the half-reactions is the hydrogen evolution reaction (HER), and Pt-based catalysts have demonstrated excellent activity throughout this process. By applying ALD, Pt catalyst can be fabricated in the optimal content which leads to cost reduction. Alshareef et al. [69] used ALD to synthesize monolithic 3D graphitic scaffolds with conformally deposited Pt.

1.7.7 PROTECTIVE FILM

Metal parts can be damaged and fail in industrial applications, particularly in semiconductor manufacturing applications, as a result of exposure to reactive chemicals. Therefore, protecting metals from corrosion is crucial from a technological standpoint, particularly in harsh conditions. The application of protective films or coatings to metal surfaces is a popular form of protection. Numerous protective coatings, including nitrides and metal oxides, have undergone extensive research and testing.

These coatings have excellent levels of corrosion resistance, wear resistance, and mechanical strength.

TiO$_2$ and Al$_2$O$_3$ offer great strength, strong oxidation, and corrosion resistance. They are among the most significant reinforcing materials used as a protective layer. TiO$_2$, in particular, is known to exhibit strong corrosion resistance against aqueous solutions, making it a viable choice to address shortcomings of Al$_2$O$_3$ as a sealing layer against moisture in multilayer coatings. Ta$_2$O$_5$ [70], SiO$_2$ [71], ZrO$_2$ [72], and HfO$_2$ [73] were also created for corrosion protection, either as a single layer or interspersed as a stack similar to TiO$_2$. When traditional metal oxides may not be strong or thin enough to guard against fluorine, chlorine, or bromine plasmas, ALD's anti-corrosion application focuses on techniques for plasma etch-resistant films.

1.8 CHALLENGES AND FUTURE PERSPECTIVES

Despite major advancements in the development of an ALD approach for the synthesis of different materials, the ALD still confronts a number of difficulties. For example, there hasn't been much research done on 2D materials by ALD, especially in the field of energy storage and conversion devices such as graphene, WS$_2$, and MoS$_2$. Moreover, the toxicity of ALD precursors, such as H$_2$S, has severely constrained ALD research. As a result, research into the effective application of ALD for this family of 2D materials is still ongoing.

The high cost is another challenge of ALD technique, since some of the ALD precursors of rare metals such as Au, Pt, Ag, and Pd that are frequently employed for ALD are expensive. Moreover, the substitution of very toxic ALD precursors (such as H$_2$S) has not been well investigated before. In addition, future research must thus concentrate on some other ALD pathways to either use this type of precursors with strong safety standards or replace them with some nontoxic and inexpensive precursors.

In addition, more clarification is needed on the doping process over substrate surface through an ALD process. The key challenge is balancing the dopant and substrate surface in the proper ratio [74,75]. Hence, an effective doping technique must be developed. However, during ALD cycles, it is possible to investigate the simultaneous usage of a number of dopant precursors in a parallel or series arrangement.

While the typical ALD method can produce a polycrystalline or amorphous complex, creating a single crystalline material is extremely desirable. So, further study must focus on designing the morphology and internal framework of the coated materials. In addition, the ALD can be investigated for the synthesis of materials commonly used for high-performance electrochemical energy applications, including (i) alumina-coated polymeric separators, (ii) treated electrode materials for Li-ion batteries (surface treatment via chemical/thermal processes), and (iii) applying a precise surface oxidation level on carbon materials (e.g., oxidized graphene, carbon nanotubes).

Consequently, a successful pilot-scale use in industrial scale of the ALD process must be developed in order to evaluate the variances in operating cost at bigger levels. Thus, it must be possible to optimize the configuration of such a large-scale system for different applications such as electronic devices. Moreover, other ALD methods in the energy field are also required to prepare the foundation for such a wide viewpoint. Nowadays, numerous interesting ALD techniques have been developed in an attempt to improve performance while upholding high standards in terms of excellent uniformity and precisely accurately controlled film thickness, including roll-to-roll ALD, batch ALD, and spatial ALD.

1.9 CONCLUSION

ALD has grown rapidly because of its capacity to build thin films with atomic precision and control. In this chapter, we have covered basics, pros and cons, and reactor types of ALD. We also examined the importance and precursor kinds of ALD recipe. ALD is used in microelectronics, energy storage, biomedical engineering, and nanotechnology, while it has helped develop new materials and gadgets, and advance science and technology. However, challenges still exist in the field of ALD, including the need for more efficient precursors and better control over film properties. ALD technology's continuing development will overcome these hurdles and lead to exciting new advances in the future. Many fascinating opportunities await as ALD technology and applications improve. Atomic-scale film deposition could transform microelectronics, energy storage, and biomedical engineering. ALD helped produce thin film transistors, solar cells, and nanoscale coatings. ALD has several benefits, but optimizing procedures, discovering novel precursors, and understanding its science are still difficult. ALD will continue to improve and develop new applications and breakthroughs with continuing research and development. As we learn more about ALD, new materials and gadgets that could revolutionize science and technology will be developed.

REFERENCES

[1] H. Kim, "Atomic layer deposition of metal and nitride thin films: Current research efforts and applications for semiconductor device processing," *Journal of Vacuum Science & Technology B: Microelectronics and Nanometer Structures Processing, Measurement, and Phenomena*, vol. 21, no. 6, pp. 2231–2261, 2003, doi: 10.1116/1.1622676.
[2] T. Suntola, "Atomic layer epitaxy," *Materials Science Reports*, vol. 4, no. 5, pp. 261–312, 1989, doi: 10.1016/S0920-2307(89)80006-4.
[3] M. Ahonen, M. Pessa, and T. Suntola, "A study of ZnTe films grown on glass substrates using an atomic layer evaporation method," *Thin Solid Films* vol. 65, 301, 1980.
[4] B. S. Lim, A. Rahtu, and R. G. Gordon, "Atomic layer deposition of transition metals," *Nature Materials*, vol. 2, no. 11, pp. 749–754, 2003, doi: 10.1038/nmat1000.
[5] X. Wang, Z. Zhao, C. Zhang, Q. Li, and X. Liang, "Surface modification of catalysts via atomic layer deposition for pollutants elimination," *Catalysts*, vol. 10, no. 11, p. 1298, 2020, doi: 10.3390/CATAL10111298.
[6] R. W. Johnson, A. Hultqvist, and S. F. Bent, "A brief review of atomic layer deposition: From fundamentals to applications," *Materials Today*, vol. 17, no. 5, pp. 236–246, 2014, doi: 10.1016/J.MATTOD.2014.04.026.
[7] S. M. Lee, "Atomic layer deposition on biological matter," Dissertation, Albert-Ludwigs Universität Freiburg.
[8] D. V. Lam, *Atomic Layer Deposition on Carbon-Based Materials*. PhD Dissertation. Daejeon, Korea: University of Science and Technology, 2018. [online]
[9] B. J. Oneill *et al.*, "Catalyst design with atomic layer deposition," *ACS Catalysis*, vol. 5, no. 3, pp. 1804–1825, 2015, doi: 10.1021/CS501862H/ASSET/IMAGES/LARGE/CS-2014-01862H_0013.JPEG.

[10] R. W. Johnson, A. Hultqvist, and S. F. Bent, "A brief review of atomic layer deposition: From fundamentals to applications," *Materials Today*, vol. 17, no. 5, pp. 236–246, 2014, doi: 10.1016/J.MATTOD.2014.04.026.

[11] J. Niinistö *et al.*, "Atomic layer deposition of HfO2 thin films exploiting novel cyclopentadienyl precursors at high temperatures," *Chemistry of Materials*, vol. 19, no. 13, pp. 3319–3324, 2007, doi: 10.1021/CM0626583/ASSET/IMAGES/MEDIUM/CM0626583N00001.GIF.

[12] J. Wang, Y. Zhao, and G. Mao, "Colloidal subwavelength nanostructures for antireflection optical coatings," *Optics Letters*, vol. 30, no. 14, pp. 1885–1887, 2005, doi: 10.1364/OL.30.001885.

[13] P. F. Carcia, R. S. McLean, M. H. Reilly, M. D. Groner, and S. M. George, "Ca test of Al_2O_3 gas diffusion barriers grown by atomic layer deposition on polymers," *Applied Physics Letters*, vol. 89, no. 3, 2006, doi: 10.1063/1.2221912/917233.

[14] J. A. T Suntola, "Atomic layer epitaxy for producing EL-thin films," *Proceedings of SID Symposium Digest of Technical Papers*, pp. 108–109, 1980.

[15] M. Knez, K. Nielsch, and L. Niinistö, "Synthesis and surface engineering of complex nanostructures by atomic layer deposition," *Advanced Materials*, vol. 19, no. 21, pp. 3425–3438, 2007, doi: 10.1002/ADMA.200700079.

[16] J. Musschoot *et al.*, "Comparison of thermal and plasma-enhanced ALD/CVD of vanadium pentoxide," *Journal of the Electrochemical Society*, vol. 156, no. 7, p. P122, 2009, doi: 10.1149/1.3133169/XML.

[17] M. N. Mullings, C. Hägglund, J. T. Tanskanen, Y. Yee, S. Geyer, and S. F. Bent, "Thin film characterization of zinc tin oxide deposited by thermal atomic layer deposition," *Thin Solid Films*, vol. 556, pp. 186–194, 2014, doi: 10.1016/J.TSF.2014.01.068.

[18] S. E. Potts, H. B. Profijt, R. Roelofs, and W. M. M. Kessels, "Room-temperature ALD of metal oxide thin films by energy-enhanced ALD," *ECS Transactions*, vol. 50, no. 13, pp. 93–103, 2013, doi: 10.1149/05013.0093ECST/XML.

[19] M. Knez, A. Kadri, C. Wege, U. Gösele, H. Jeske, and K. Nielsch, "Atomic layer deposition on biological macromolecules: Metal oxide coating of tobacco mosaic virus and ferritin," *Nano Letters*, vol. 6, no. 6, pp. 1172–1177, 2006, doi: 10.1021/NL060413J/SUPPL_FILE/NL060413JSI20060418_032016.PDF.

[20] W. Gasser, Y. Uchida, and M. Matsumura, "Quasi-monolayer deposition of silicon dioxide," *Thin Solid Films*, vol. 250, no. 1–2, pp. 213–218, 1994, doi: 10.1016/0040-6090(94)90188-0.

[21] J. Heo, A. S. Hock, and R. G. Gordon, "Low temperature atomic layer deposition of tin oxide," *Chemistry of Materials*, vol. 22, no. 17, pp. 4964–4973, 2010, doi: 10.1021/CM1011108/ASSET/IMAGES/MEDIUM/CM-2010-011108_0016.GIF.

[22] M. D. Groner, F. H. Fabreguette, J. W. Elam, and S. M. George, "Low-temperature Al_2O_3 atomic layer deposition," *Chemistry of Materials*, vol. 16, no. 4, pp. 639–645, 2004, doi: 10.1021/CM0304546/ASSET/IMAGES/MEDIUM/CM0304546N00001.GIF.

[23] Y. M. Chang, S. R. Jian, H. Y. Lee, C. M. Lin, and J. Y. Juang, "Enhanced visible photoluminescence from ultrathin ZnO films grown on Si-nanowires by atomic layer deposition," *Nanotechnology*, vol. 21, no. 38, p. 385705, 2010, doi: 10.1088/0957-4484/21/38/385705.

[24] G. A. Ten Eyck, S. Pimanpang, H. Bakhru, T. M. Lu, and G. C. Wang, "Atomic layer deposition of Pd on an oxidized metal substrate," *Chemical Vapor Deposition*, vol. 12, no. 5, pp. 290–294, 2006, doi: 10.1002/CVDE.200506456.

[25] B. H. Lee *et al.*, "Low-temperature atomic layer deposition of copper metal thin films: Self-limiting surface reaction of copper dimethylamino-2-propoxide with diethylzinc," *Angewandte Chemie International Edition*, vol. 48, no. 25, pp. 4536–4539, 2009, doi: 10.1002/ANIE.200900414.

[26] J. S. Becker, "Atomic layer deposition of metal oxide and nitride thin films," Thesis, Harvard University, 2003.

[27] C. D. Travis and R. A. Adomaitis, "Dynamic modeling for the design and cyclic operation of an atomic layer deposition (ALD) reactor," *Processes*, vol. 1, no. 2, pp. 128–152, 2013, doi: 10.3390/pr1020128.

[28] H. C. M. Knoops, S. E. Potts, A. A. Bol, and W. M. M. Kessels, "27- Atomic layer deposition," In *Handbook of Crystal Growth* (Second Edition), T. F. Kuech, Ed., Boston: North-Holland, pp. 1101–1134, 2015. doi: https://doi.org/10.1016/B978-0-444-63304-0.00027-5.

[29] Felix Mattelaer, "Atomic layer deposition for lithium-ion batteries," Thesis, Ghent University, Department of Solid state sciences.

[30] K. Kobayashi and S. Okudaira, "Preparation of ZnO films on sapphire (0001) substrates by alternate supply of zinc acetate and H2O," *Chemistry Letters*, vol. 26 no. 6, pp. 511–512, 2004, doi: 10.1246/CL.1997.511.

[31] J. C. Badot *et al.*, "Atomic layer epitaxy of vanadium oxide thin films and electrochemical behavior in presence of lithium ions," *Electrochemical and Solid-State Letters*, vol. 3, no. 10, pp. 485–488, 2000, doi: 10.1149/1.1391187/XML.

[32] Y. Ye *et al.*, Copper etch using HCl and HBr chemistry, U.S. Patent No. 6,008,140, 1997.

[33] W. F. Gtari and B. Tangour, "Interaction of HF, HBr, HCl and HI molecules with carbon nanotubes," *Acta Chimica Slovenica*, vol. 65, no. 2, pp. 289–295, 2018, doi: 10.17344/ACSI.2017.3698.

[34] J. L. van Hemmen *et al.*, "Plasma and thermal ALD of Al2O3 in a commercial 200 mm ALD reactor," *Journal of The Electrochemical Society*, vol. 154, no. 7, p. G165, 2007, doi: 10.1149/1.2737629.

[35] C. W. Jeong, J. S. Lee, and S. K. Joo, "Plasma-assisted atomic layer growth of high-quality aluminum oxide thin films," *Japanese Journal of Applied Physics, Part 1: Regular Papers and Short Notes and Review Papers*, vol. 40, no. 1, pp. 285–289, 2001, doi: 10.1143/JJAP.40.285/XML.

[36] K. Manandhar, J. A. Wollmershauser, and B. N. Feigelson, "Growth mode of alumina atomic layer deposition on nanopowders," *Journal of Vacuum Science & Technology A: Vacuum, Surfaces, and Films*, vol. 35, no. 4, p. 041503, 2017, doi: 10.1116/1.4983445/1006821.

[37] R. Huang and A. H. Kitai, "Preparation and characterization of thin films of MgO, Al_2O_3 and $MgAl_2O_4$ by atomic layer deposition," *Journal of Electronic Materials*, vol. 22, no. 2, pp. 215–220, 1993, doi: 10.1007/BF02665029/METRICS.

[38] K. Kukli, M. Ritala, M. Leskelä, and J. Jokinen, "Atomic layer epitaxy growth of aluminum oxide thin films from a novel Al(CH3)2Cl precursor and H_2O," *Journal of Vacuum Science & Technology A*, vol. 15, no. 4, pp. 2214–2218, 1997, doi: 10.1116/1.580536.

[39] M. Putkoncn, L. S. Johansson, E. Rauhala, and L. Niinisto, "Surface-controlled growth of magnesium oxide thin films by atomic layer epitaxy," *Journal of Materials Chemistry*, vol. 9, no. 10, pp. 2449–2452, 1999, doi: 10.1039/A904315B.

[40] J. W. Elam, A. B. F. Martinson, M. J. Pellin, and J. T. Hupp, "Atomic layer deposition of In_2O_3 using cyclopentadienyl indium: A new synthetic route to transparent conducting oxide films," *Chemistry of Materials*, vol. 18, no. 15, pp. 3571–3578, 2006, doi: 10.1021/CM060754Y/ASSET/IMAGES/MEDIUM/CM060754YN00001.GIF.

[41] M. Putkonen, M. Nieminen, J. Niinistö, L. Niinistö, and T. Sajavaara, "Surface-controlled deposition of Sc_2O_3 thin films by atomic layer epitaxy using β-diketonate and organometallic precursors," *Chemistry of Materials*, vol. 13, no. 12, pp. 4701–4707, 2001, doi: 10.1021/CM011138Z.

[42] A. Tsukazaki, A. Ohtomo, D. Chiba, Y. Ohno, H. Ohno, and M. Kawasaki, "Low-temperature field-effect and magnetotransport properties in a ZnO based heterostructure with atomic-layer-deposited gate dielectric," *Applied Physics Letters*, vol. 93, no. 24, p. 241905, 2008, doi: 10.1063/1.3035844/336176.

[43] A. M. Ma et al., "Schottky barrier source-gated ZnO thin film transistors by low temperature atomic layer deposition," Applied Physics Letters, vol. 103, no. 25, p. 253503, 2013, doi: 10.1063/1.4836955/26069.

[44] B. Y. Oh et al., "High-performance ZnO thin-film transistor fabricated by atomic layer deposition," Semiconductor Science and Technology, vol. 26, no. 8, p. 085007, 2011, doi: 10.1088/0268-1242/26/8/085007.

[45] Y. Y. Lin, C. C. Hsu, M. H. Tseng, J. J. Shyue, and F. Y. Tsai, "Stable and high-performance flexible ZnO thin-film transistors by atomic layer deposition," ACS Applied Materials & Interfaces, vol. 7, no. 40, pp. 22610–22617, 2015, doi: 10.1021/ACSAMI.5B07278.

[46] R. Liu, M. Peng, H. Zhang, X. Wan, and M. Shen, "Atomic layer deposition of ZnO on graphene for thin film transistor," Materials Science in Semiconductor Processing, vol. 56, pp. 324–328, 2016, doi: 10.1016/J.MSSP.2016.09.016.

[47] S. H. Kim et al., "Fabrication of high-performance p-type thin film transistors using atomic-layer-deposited SnO films," Journal of Materials Chemistry C, vol. 5, no. 12, pp. 3139–3145, 2017, doi: 10.1039/C6TC04750E.

[48] L. Assaud, M. Hanbücken, and L. Santinacci, "Atomic layer deposition of TiN/Al_2O_3/TiN nanolaminates for capacitor applications," ECS Transactions, vol. 50, no. 13, pp. 151–157, 2013, doi: 10.1149/05013.0151ECST/XML.

[49] D. Z. Austin, D. Allman, D. Price, S. Hose, and J. F. Conley, "Plasma enhanced atomic layer deposition of Al_2O_3/SiO_2 MIM capacitors," IEEE Electron Device Letters, vol. 36, no. 5, pp. 496–498, 2015, doi: 10.1109/LED.2015.2412685.

[50] X. Li et al., "Tin oxide with controlled morphology and crystallinity by atomic layer deposition onto graphene nanosheets for enhanced lithium storage," Advanced Functional Materials, vol. 22, no. 8, pp. 1647–1654, 2012, doi: 10.1002/ADFM.201101068.

[51] K. B. Gandrud, A. Pettersen, O. Nilsen, and H. Fjellvåg, "High-performing iron phosphate for enhanced lithium ion solid state batteries as grown by atomic layer deposition," Journal of Materials Chemistry A, vol. 1, no. 32, pp. 9054–9059, 2013, doi: 10.1039/C3TA11550J.

[52] C. Zhong, Y. Deng, W. Hu, J. Qiao, L. Zhang, and J. Zhang, "A review of electrolyte materials and compositions for electrochemical supercapacitors," Chemical Society Reviews, vol. 44, no. 21, pp. 7484–7539, 2015, doi: 10.1039/C5CS00303B.

[53] C. Chen et al., "NiO/nanoporous graphene composites with excellent supercapacitive performance produced by atomic layer deposition," Nanotechnology, vol. 25, no. 50, p. 504001, 2014, doi: 10.1088/0957-4484/25/50/504001.

[54] C. Guan et al., "Atomic layer deposition of Co_3O_4 on carbon nanotubes/carbon cloth for high-capacitance and ultrastable supercapacitor electrode," Nanotechnology, vol. 26, no. 9, p. 094001, 2015, doi: 10.1088/0957-4484/26/9/094001.

[55] R. M. Silva et al., "Coating of vertically aligned carbon nanotubes by a novel manganese oxide atomic layer deposition process for binder-free hybrid capacitors," Advanced Materials Interfaces, vol. 3, no. 21, p. 1600313, 2016, doi: 10.1002/ADMI.201600313.

[56] L. Yu et al., "Highly effective synthesis of NiO/CNT nanohybrids by atomic layer deposition for high-rate and long-life supercapacitors," Dalton Transactions, vol. 45, no. 35, pp. 13779–13786, 2016, doi: 10.1039/C6DT01927G.

[57] M. Li and H. He, "Study on electrochemical performance of multi-wall carbon nanotubes coated by iron oxide nanoparticles as advanced electrode materials for supercapacitors," Vacuum, vol. 143, pp. 371–379, 2017, doi: 10.1016/J.VACUUM.2017.06.026.

[58] T. Sato et al., "Atomic layer deposition of magnesium fluoride for optical application," Optical Interference Coatings Conference (OIC) 2022 (2022), Paper TC.5, p. TC.5, 2022, doi: 10.1364/OIC.2022.TC.5.

[59] T. Pilvi et al., "Study of a novel ALD process for depositing MgF_2 thin films," Journal of Materials Chemistry, vol. 17, no. 48, pp. 5077–5083, 2007, doi: 10.1039/B710903B.

[60] M. Szindler, M. M. Szindler, P. Boryło, and T. Jung, "Structure and optical properties of TiO₂ thin films deposited by ALD method," *Open Physics*, vol. 15, no. 1, pp. 1067–1071, 2017, doi: 10.1515/PHYS-2017-0137/MACHINEREADABLECITATION/RIS.

[61] Y. Li, S. Deng, J. Li, G. Li, S. Zhang, and Y. Jin, "Investigation on HfO2 properties grown by ALD using TDMAH as precursor," *Vacuum*, vol. 203, p. 111243, 2022, doi: 10.1016/J.VACUUM.2022.111243.

[62] H. Liu, P. Ma, Y. Pu, and Z. Zhao, "Atomic layer deposition of Al₂O₃ and HfO2 for high power laser application," *Journal of Alloys and Compounds*, vol. 859, p. 157751, 2021, doi: 10.1016/J.JALLCOM.2020.157751.

[63] L. Ristau, D. Tünnermann, and A. Szeghalmi, "Comparative study of ALD SiO₂ thin films for optical applications," *Optical Materials Express*, vol. 6, no. 2, pp. 660–670, 2016, doi: 10.1364/OME.6.000660.

[64] B. C. H. Steele and A. Heinzel, "Materials for fuel-cell technologies," *Nature*, vol. 414, no. 6861, pp. 345–352, 2001, doi: 10.1038/35104620.

[65] L. Wei *et al.*, "Electrochemically shape-controlled synthesis in deep eutectic solvents: Triambic icosahedral platinum nanocrystals with high-index facets and their enhanced catalytic activity," *Chemical Communications*, vol. 49, no. 95, pp. 11152–11154, 2013, doi: 10.1039/C3CC46473C.

[66] L. Zhang, K. Doyle-Davis, and X. Sun, "Pt-based electrocatalysts with high atom utilization efficiency: from nanostructures to single atoms," *Energy & Environmental Science*, vol. 12, no. 2, pp. 492–517, 2019, doi: 10.1039/C8EE02939C.

[67] T. Aaltonen, A. Rahtu, M. Ritala, and M. Leskelä, "Reaction mechanism studies on atomic layer deposition of ruthenium and platinum," *Electrochemical and Solid-State Letters*, vol. 6, no. 9, p. C130, 2003, doi: 10.1149/1.1595312/XML.

[68] T. Aaltonen, M. Ritala, T. Sajavaara, J. Keinonen, and M. Leskelä, "Atomic layer deposition of platinum thin films," *Chemistry of Materials*, vol. 15, no. 9, pp. 1924–1928, 2003, doi: 10.1021/CM021333T/ASSET/IMAGES/MEDIUM/CM021333TN00001.GIF.

[69] P. Nayak, Q. Jiang, N. Kurra, X. Wang, U. Buttner, and H. N. Alshareef, "Monolithic laser scribed graphene scaffolds with atomic layer deposited platinum for the hydrogen evolution reaction," *Journal of Materials Chemistry A*, vol. 5, no. 38, pp. 20422–20427, 2017, doi: 10.1039/C7TA06236B.

[70] C. Li, T. Wang, Z. Luo, D. Zhang, and J. Gong, "Transparent ALD-grown Ta2O5 protective layer for highly stable ZnO photoelectrode in solar water splitting.," *Chemical Communications (Cambridge, England)*, vol. 51, no. 34, pp. 7290–7293, 2015, doi: 10.1039/C5CC01015B.

[71] D. Arl *et al.*, "SiO₂ thin film growth through a pure atomic layer deposition technique at room temperature," *RSC advances*, vol. 10, no. 31, pp. 18073–18081, 2020, doi: 10.1039/d0ra01602k.

[72] M. Dinu *et al.*, "Effects of film thickness of ALD-deposited Al₂O₃, ZrO₂ and HfO2 Nano-layers on the corrosion resistance of Ti(N,O)-coated stainless steel," *Materials*, vol. 16, no. 5, p. 2007, 2023, doi: 10.3390/MA16052007.

[73] J. S. Daubert *et al.*, "Corrosion protection of copper using Al₂O₃, TiO₂, ZnO, HfO2, and ZrO₂ Atomic layer deposition," *ACS Applied Materials and Interfaces*, vol. 9, no. 4, pp. 4192–4201, 2017, doi: 10.1021/ACSAMI.6B13571/ASSET/IMAGES/MEDIUM/AM-2016-13571P_0012.GIF.

[74] O. Zandi, B. M. Klahr, and T. W. Hamann, "Highly photoactive Ti-doped α-Fe₂O₃ thin film electrodes: resurrection of the dead layer," *Energy & Environmental Science*, vol. 6, no. 2, pp. 634–642, 2013, doi: 10.1039/C2EE23620F.

[75] Y. Lin *et al.*, "Growth of p-type hematite by atomic layer deposition and its utilization for improved solar water splitting," *Journal of the American Chemical Society*, vol. 134, no. 12, pp. 5508–5511, 2012, doi: 10.1021/JA300319G/SUPPL_FILE/JA300319G_SI_001.PDF.

2 State of the Art Modeling and Simulation Approaches in ALD

2.1 INTRODUCTION

Atomic layer deposition (ALD) has become a crucial technique for the development of nanostructured materials through the formation of ultra-thin films which play an important role in various applications including energy science, biomedicals, semiconductors, microelectronics, etc. Accordingly, ALD with the specific characteristic of thin film fabrication over surface of substrate such as conformity, pinhole-free, sub-monolayer film growth, and atomic accuracy has garnered more and more interest over the past few decades from both theorists and experimentalists. However, ALD has some limitations; for example, ALD experiments cannot provide a comprehensive understanding of the reaction processes during film generation. Furthermore, different deposition conditions, such as precursor type, temperature, pulse and purge time, flow rate, need to be examined to ensure the reproducibility of the samples and for achieving optimized ALD film growth for the desired application, which is practically very costly and may be impossible. In addition, the most of ALD precursors are hazardous and expensive and their usage should be minimized as much as possible. Therefore, in order to mitigate the limitations of ALD, the incorporation of a theoretical framework is essential and important in validating experimental data. Theoretical methods are appropriate for designing new materials based on existing experimental data and predicting their properties before synthesizing them. Currently, theoretical methods have been implemented by many researchers to simulate, optimize, and estimate various ALD attributes, including reaction mechanism, growth rate, and materials, under different deposition conditions. Thus, this strategy can be effectively validating the experimental data. This chapter highlights the role of theoretical modeling methods and calculation in developing the ALD research and describes how theoretical approaches might improve the effectiveness of experiments. These theoretical modeling methods also provide useful insights into the physics and chemistry of ALD.

2.2 THEORETICAL MODELING METHODS

The complexity of ALD processes and the necessity to optimize deposition parameters have led to theoretical models to better explain and predict ALD system behavior. Theoretical modeling can anticipate and optimize ALD film properties, bringing up interesting new paths for academics and engineers. This section will discuss the

DOI: 10.1201/9781003346234-3

state-of-the-art theoretical modeling such as density function theory, computational fluid dynamics, and molecular dynamics (MD) in ALD and its implications on next-generation materials

2.2.1 Density Functional Theory

Density functional theory (DFT) is known as the first principles (ab initio) method and is an efficient computational approach based on fundamental laws of quantum mechanical for estimating chemical and physical properties of various types of nano-materials including atoms, molecules, crystals, complexes in both gas and liquid phase with high accuracy. DFT can predicts the electronic characteristics, thermo-dynamic properties, energy parameters, atomic forces of materials based on particle interactions without the need for any experimental data. In addition, DFT can determine the reaction mechanism, such as initial surface reactions, reaction pathways, and the precursor chemisorption process, which is useful for the investigation of ALD process.

Nowadays, DFT has been widely used by researchers due to the low computational efforts. Nevertheless, until the 1990s, DFT was not thought to be precise enough for calculations in quantum chemistry and to solve the Schrödinger's equation, Ab initio Hartree-Fock (HF) and second-order MØller–Plesset perturbation theory [1]. The Schrödinger's equation can be expressed in several ways, and a straightforward form is nonrelativistic and time-independent, but in order to effectively interpret it, we need to specify the quantities. The simple Schrödinger's equation is described as follows:

$$H\psi = E\psi \tag{2.1}$$

where H presents the Hamiltonian operator and c refers to the set of solutions, or eigenstates, of the Hamiltonian. ψ_n is correlated with an eigenvalue of E_n, which is a real integer that solves the eigenvalue equation according to Sholl and Steckel [2]. The Schrödinger equation can be accurately solved in various established circumstances, such as the particle confined within a box and the harmonic oscillator, both of which are dependent on the specific physical system. However, in the case that multiple electrons interact with several nuclei, the Schrödinger's equation is defined as follows:

$$\left[\frac{h^2}{2m} \sum_{i=1}^{N} \nabla_i^2 + \sum_{i=1}^{N} V(r_i) + \sum_{i=1}^{N} \sum_{j<i} U(r_i, r_j) \right] \Psi = E\Psi \tag{2.2}$$

where m denotes the electron mass, the first bracket presents the kinetic energy of each electron, and the second bracket shows the interaction energy among electrons and set of atomic nuclei and the next interaction energy between electrons. Symbol ψ indicates the function for electronic wave, which is determined by the spatial coordinates of each of the N electrons. Then, the individual electron wave functions can be used to estimate parameter ψ.

Later, in 1965, Lu Jeu Sham and Walter Kohn [3] proposed a hybrid algorithm of HF to solve the Schrödinger equation, which offers a variational concept for density functionals and enables the many-body electronic ground state to be described using an effective potential and a single-electron equation. In addition, the Hartree approach is a good argument to describe the entire wave function in this way using the sum of one electron wave functions. When compared to conventional techniques, such as the exchange limited Hartree-Fock concept, the computational cost is quite inexpensive due to incorporating the electron correlation. Accordingly, as a product of the various electron functions, Ψ can be determined using the following Hartree equation [4]:

$$\Psi = \Psi_1(r)\Psi_2(r),..., \Psi_N(r) \qquad (2.3)$$

The quantum nature of matter can be accurately described if an accurate exchange-correlation functional is used. However, the Schrödinger equation can only be solved for systems with one electron; therefore, in the late 1980s, the local density corrected approximation (LDA) functional was introduced as a complementary method to integrate with DFT for resolving Schrödinger equation for multiple electron configurations. While the HF method is capable of predicting the energy of quantum many-body systems and the wave function in a stationary state, it is important to note that the calculated values for the total energy and the energy associated with electron-electron repulsion may be overestimated as a result of the assumption of independent electron migration [5].

Two key mathematical theorems established by Kohn and Hohenberg and a series of equations derived by Kohn and Sham in the mid-1960s provided the foundation for the whole area of DFT [6]. In the first theory, the ground-state energy derived from Schroedinger's equation is a special function of electron density. The ground-state energy E can be stated as $E[n_r)]$, where $n(r)$ presents the electron density. However, the main flaw of this theory is the absence of a functional definition. Therefore, the second theorem of Kohn and Sham proposed the special character of functional which minimizes the energy of the total functional and ensures the correct electron density associated with the entire solution of the Schrödinger equation. This energy functional can be described as follows:

$$E[\{\Psi_i\}] = E_{known}[\{\Psi_i\}] + E_{xc}[\{\Psi_i\}] \qquad (2.4)$$

where $E[\{\Psi_i\}]$ denotes the function that defines the exchange-correlation functional and $E_{xc}[\{\Psi_i\}]$ presents the exchange correlation functional. The $E_{known}[\{\Psi_i\}]$ arises from four contributions which are described as follows:

$$E_{known}[\{\Psi_i\}] = -\frac{h^2}{m}\sum_i \int \Psi_i^* \nabla^2 \Psi_i d^3r + \int V(r)n(r)d^3r$$
$$+ \frac{e^2}{2}\int\int \frac{n(r)\, n(r')}{|r-r'|}d^3r\, d^3r' + E_{ion} \qquad (2.5)$$

The first right term is the electrons kinetic energies, the second is interactions of Coulomb between the electrons and nuclei, the third is the Coulomb interactions among sets of electrons, and the last is the Coulomb interactions among sets of nuclei.

Kohn and Sham demonstrated that in order to determine the proper electron density, a series of equations including a single electron need to be solved as follows:

$$\left[\frac{h^2}{2m} \nabla^2 + V(r) + V_H(r) + V_{XC}(r) \right] \Psi_i(r) = \varepsilon_i \Psi_i(r) \tag{2.6}$$

where $V(r)$ expresses the potential that describes the set of electrons to atomic nuclei interaction. $V_H(r)$ is the Hartree potential defining the Coulomb repulsion among one individual electron accounted for one of the Kohn–Sham equations and the total electron density assumed by all electrons, which is defined by:

$$V_H(r) = e^2 \int \frac{n(r')}{|r - r'|} d^3 r' \tag{2.7}$$

$V_{XC}(r)$ denotes a potential addressing contribution of exchange and correlation to the single-electron equations, which is known as a functional derivative and is defined as follows:

$$V_{XC}(r) = \frac{\delta E_{XC}(r)}{\delta n(r)} \tag{2.8}$$

Equation 2.6 is similar to Equation 2.2, with the exception of the presence of summations in the complete Schroedinger's equation. This is due to the fact that the solution of the Kohn–Sham equations is a single-electron wave function that only depends on three spatial variables of $\Psi_i(r)$.

To determine the Schroedinger equation's ground-state energy, there are several concerns, which make it quite challenging. Therefore, it is necessary to specify the exchange-correlation function, $E_{xc}[\{\Psi_i\}]$, in order to solve the Kohn–Sham equations, which is difficult to define by Equations 2.4 and 2.5. There is just one situation in which exchange correlation functional may be deduced precisely in a uniform electron gas which offers a useful strategy for using the Kohn–Sham equations in practise. In this case, $n(r)$ is constant because the electron density is constant across space. This situation can exist with limited value in every real material due to the fact that changes in electron density establish chemical bonds and allow materials to be attractive. Therefore, the exchange–correlation potential can be applied at each point from the uniform electron gas and can be defined as follows:

$$V_{XC}(r) = V_{XC}^{electron\ gas}[n(r)] \tag{2.9}$$

This approach implements only local density to express the exchange–correlation functional which is known as local density approximation (LDA). The Kohn–Sham equations may be fully defined using the LDA. However, it is important to consider that the data provided by these equations do not perfectly resolve the real Schrödinger equation since the real exchange–correlation functional is not being used.

Besides the LDA functional, there is another functional called generalized gradient approximation (GGA) which incorporates more physical data than the LDA and is thus expected to be more accurate. GGA implements information regarding the local electron density and the local gradient in the electron density and to mitigate the difficulties of the HF. In 1993, Becke employed GGA along with DFT and combined with Lee-Yang-Parr (LYP) gradient-corrected correlation functionals [7]. Additionally, in another study, Xu et al. [8] incorporated the Becke three-parameter LYP (B3LYP) hybrid functional to develop the HF method to determine electron correlation.

2.2.1.1 Emerging Applications of DFT in ALD

Generally, an ALD cycle includes two half cycles which are two precursors incorporated to form one thin layer. Understanding the detailed mechanism of possible reactions during an ALD process is necessary for selecting an appropriate precursor. The DFT can be utilized to investigate the reaction mechanism during ALD process between the precursor molecule and substrate. The compounds' reactions lead to cause different chemical phenomenon including decomposition, transition, and formation of new compounds. Therefore, by virtue of DFT, the nature of atoms interaction, reaction mechanism, and final by-product can be predicted. Geometry equilibrium at absolute zero is one of the usual quantities that can be obtained from ground-state DFT computations. At this state, DFT can determine the electronic properties including partial charge of atoms, electron migration, band gap, density of state, mechanical properties, formation energy, magnetic properties, and optical properties. Moreover, DFT can be used to probe the reaction pathway during initial, transition, and final state of an ALD first and second half cycles. DFT can be used to provide a basis for research, distinguish between potential exploration possibilities, and validate the results of experimental studies. In addition, DFT is helpful in learning about a material's surface characteristics in surface science and catalysis.

Surfaces are significant in many technical sectors, including semiconductor manufacture catalysis, gas separation membranes, and interfaces. It's critical to understand the electrical structure and geometry of surfaces. DFT techniques have been crucial in revealing and better understanding the processes behind the relevant reactions on catalytic metals, oxides, and zeolites which have improved the efficiency and mitigated the cost of catalytic converter design. DFT and surface science experiments have frequently performed together on highly successful projects. DFT has been implemented to analyze the surface structure of nanoparticles, metals, metal oxides, sulfides, and carbides in conjunction with ultra-high vacuum surface science experiments like temperature programmed desorption, scanning tunneling microscopy (STM), X-ray photoelectron spectroscopy, and X-ray diffraction. In the following section, we highlighted the basics of some important analysis which can be accommodated by DFT.

2.2.1.1.1 Electronic Structure

DFT is basically a theory of electronic ground-state structure expressed in terms of the electronic density distribution $n(r)$. First-principles DFT simulations have traditionally been utilized to derive the density of states (DOS). The electronic DOS is one of the basic quantities used to define a material's electronic state. In material science and solid-state physics, the electronic DOS plays a crucial role in determining the characteristics of metals such as density distributions of free electrons in materials. A plane-wave DFT calculation's fundamental concept is to represent the electron density using functions of the type exp *(ik.r)*. This type of plane wave's associated electrons has energy $E = (hk)^2/2m$. The result of DOS can be obtained by integrating the resultant electron density in k space. The DOS for electrons, photons, or phonons can be computed and provided as a function of either energy or the wave vector k, depending on the quantum mechanical system. The system-specific relation between E and k for the energy dispersion must be understood in order to convert between the DOS as a function of energy and the DOS as a function of the wave vector. The number of modes per unit frequency range is expressed by the DOS of a system and described as follows:

$$D(E) = \frac{N(E)}{V} \tag{2.10}$$

where $N(E)\delta E$ denotes the quantity of states in the system with volume V, where the energy range is between E and $E+\delta E$. The dispersion relationships of the system's properties are directly correlated with the DOS. High DOS at a certain energy level indicates that there are several states that are occupiable; for example, in a semiconductor, by examining the DOS of electrons at the boundary between the conduction and valence bands. So, it is obvious that an increase in electron energy expands the range of states that an electron in the conduction band can occupy. Nevertheless, when the DOS is discontinuous at an interval of energy, means that no states are accessible for electrons to occupy inside the material's band gap. Therefore, an electron at the conduction band edge is unable to migrate to a different valence band state without losing a minimum band gap energy of the material. Figure 2.1 demonstrates

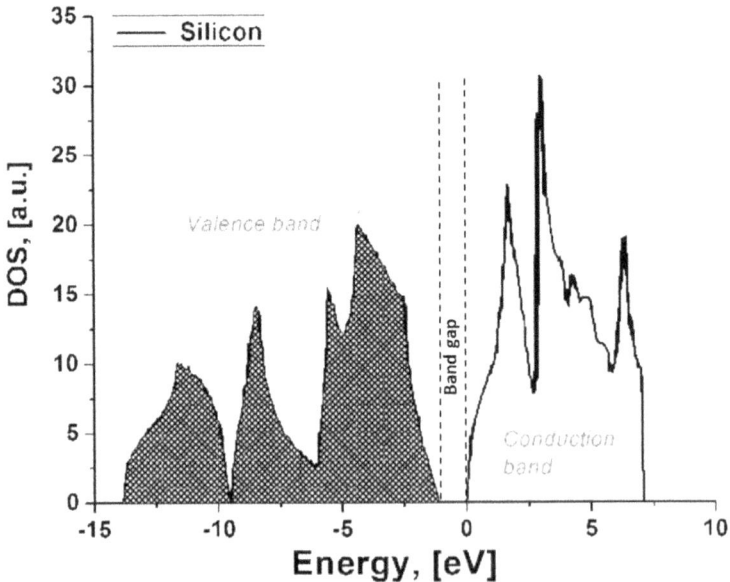

FIGURE 2.1 Density functional theory (DFT) calculated density of states (DOS) for bulk Si [9].

the DFT calculated DOS plot for a bulk Si. As seen in the figure, the valence band and the conduction band are the two distinct zones that establish the DOS. The valence band contains all occupied electronic states, whereas the conduction band contains all unoccupied states (at $T=0$ K). The band gap is the area of energy between the valence and conduction bands that includes no electronic states at all.

2.2.1.1.2 Magnetic Properties

The nonzero spin of electrons has a direct impact on magnetism. Each electronic state in diamagnetic materials has two electrons, one with spin up and the other with spin down. Figure 2.2 depicts this scenario graphically for a periodic material. However, there are electronic states that only have one electron in magnetic materials. Unpaired electrons can be arranged in a wide variety of ways, with each establishing a diverse magnetic state. Figure 2.2b and c displays the two most prevalent magnetic states. On nearby atoms, there is the antiferromagnetic state, in which electron spins alternate, and the ferromagnetic state, in which all electron spins point in the same direction. In Figure 2.2d, a straightforward example shows that more delicate spin

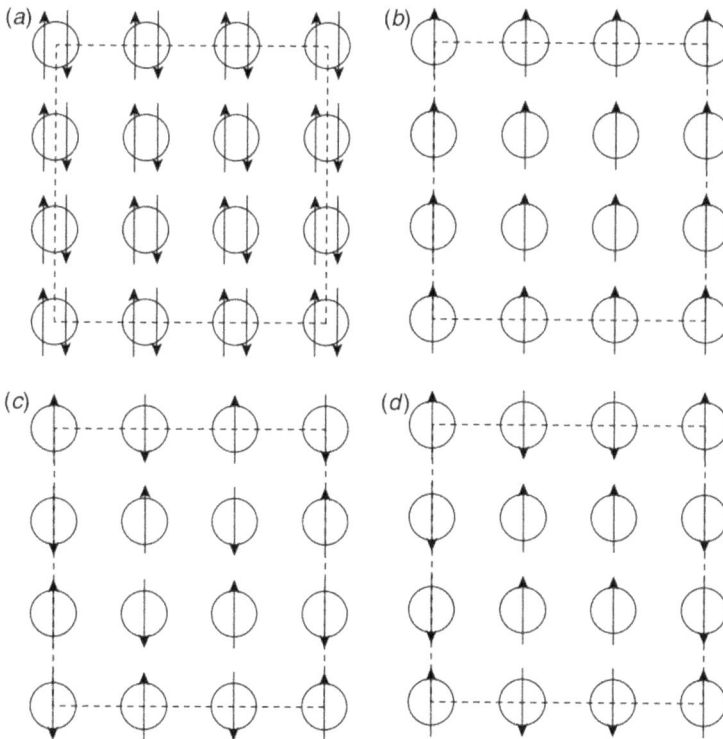

FIGURE 2.2 Schematic diagrams of two-dimensional periodic material's spin states. Individual atoms are depicted by circles, while a single supercell is shown by dotted lines. (a) On each atom, every pair of electrons is present. In the subsequent cases, each atom has a single unpaired electron. (b–d) Depicts examples of ferromagnetic states, antiferromagnetic states, and more complicated magnetic states, respectively [10].

orderings are also conceivable. The magnetic moment is the measure of the typical electron spin per atom. All other cases have magnetic moments of zero, except for the ferromagnetic condition depicted in Figure 2.2b, which has a magnetic value of 1.

Due to the fact that this approximation speeds up computations, electron spin is generally not explicitly taken into account in DFT calculations. Therefore, it is essential that spin is considered in materials where spin effects are significant. For instance, the metal iron is widely recognized for having magnetic characteristics. Figure 2.3 compares computations with and without ferromagnetic spin ordering to illustrate the energy of bulk Fe with bcc crystal structure. Electron spins significantly reduce energy and raise the estimated equilibrium lattice constant by 0.1 A, which shows a significant difference.

Figure 2.3 presents an important effect of applying spin in DFT calculations meaning that there are a variety of conceivable magnetic orderings for electron spins. The optimization of electron spins within a single DFT computation is analogous to the optimization of atom locations inside a crystal structure. The spins are approximated first, and the computation then identifies a local minimum associated with this first approximation. A ferromagnetic state is often employed as a first approximation. This technique can only provide a local minimum on the energy surface defined by all feasible spin orderings, which is a critical finding in this context. However, it is more challenging to determine the spin ordering related to the global minimum on this energy surface and similar to finding the ideal crystal structure for a substance from all potential crystals. Thus, it is advisable to look at more than one starting estimate for the spin ordering while looking at magnetic ordering, similar to when using DFT to study crystal structures. Figure 2.3 illustrates the energy of

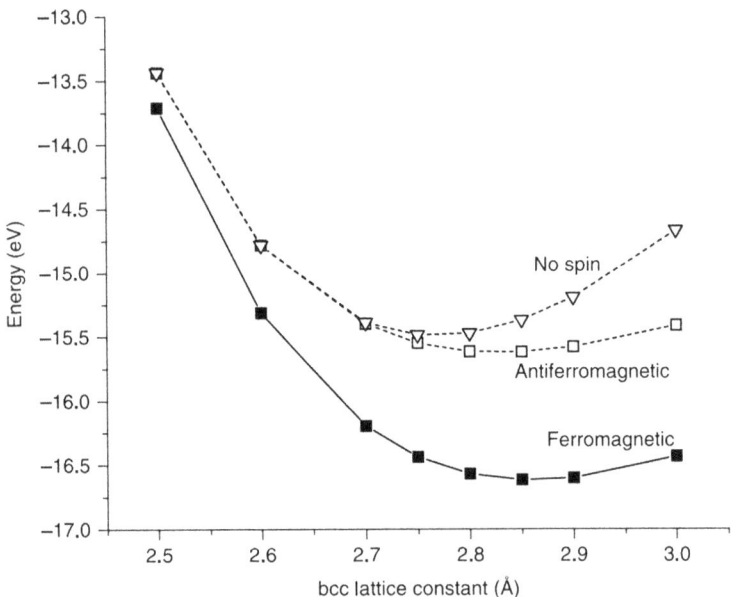

FIGURE 2.3 An analysis of the energy of bcc Fe as a function of the Fe lattice constant using density functional theory (DFT) simulations with various spin states. Plot without spin polarization denoted as "No spin." [10].

fcc-Fe with antiferromagnetic ordering calculated from a series of computations in which this ordering was utilized as a starting estimate for the spin states.

According to experimental evidence, Fe is ferromagnetic, and the antiferromagnetic energies are far greater in energy than the ferromagnetic outcomes. It should be noted that these calculations don't reveal anything regarding the potential existence of ordered spin states with structures that are more sophisticated than basic ferromagnetic or antiferromagnetic ordering.

Notably, this introduction has only raised a small portion of the extremely large topic of magnetic characteristics. Most methods for digital data storing rely on magnetic characteristics. Moreover, intense global studies have been conducted on the different magnetic phenomena for decades.

2.2.1.1.3 Mechanical Properties

Understanding a material's mechanical characteristics is crucial to the great majority of material scientists and engineers. A material's mechanical characteristics provide information about its durability, capacity to withstand damage, and prospective applications including flexible electronics [11], sensors [12], ferroelectrics [13], biomaterials [14], and pharmaceutical nanocrystals [15]. Therefore, for their exploitation, design, and screening, it is essential to understand the mechanical characteristics of crystalline materials. In the hunt for novel materials for hard coating applications, Pugh mechanical analysis expansions have recently been employed to derive hardness descriptors [16].

DFT continues to be one of the most useful computational techniques for quantitatively forecasting and rationalizing the mechanical reaction of the materials. Several experimental methods, including nanoindentation, high-pressure X-ray crystallography, impedance spectroscopy, and spectroscopic ellipsometry, have been proven to quantitatively correspond with DFT predictions. DFT simulations can be used to determine bulk mechanical parameters, and this computational approach enables a thorough comprehension of the elastic anisotropy in complicated crystalline structures. Elastic tensors may also be used to identify materials with particular thermal characteristics since they make it possible to predict trends in thermal conductivities and heat capacity [17]. When paired with mathematical homogenization concepts, the elastic behavior of composite materials may be anticipated by knowing the complete anisotropic elastic tensor. This has enabled the development of materials with controlled stiffness [18,19]. The several DFT strategies that may be used to determine a crystal's mechanical characteristics are covered in this book, along with additional computational chemistry techniques that are currently being employed in this attempt.

Second-order elastic constant calculations may be used to describe the elastic and mechanical characteristics of various materials. Under the scope of Hooke's law, they serve as a gauge of the proportionality between stress and strain. By applying a strain on a crystal and measuring the resultant stress, elastic constants can be generally calculated [20]. However, ultrasonic techniques that depend on elastodynamics can be used to experimentally estimate the elastic constants of molecular crystals [21]. In order to direct experiments and reduce the need for trial-and-error examinations, theoretical studies have recently been employed to determine elastic constants with

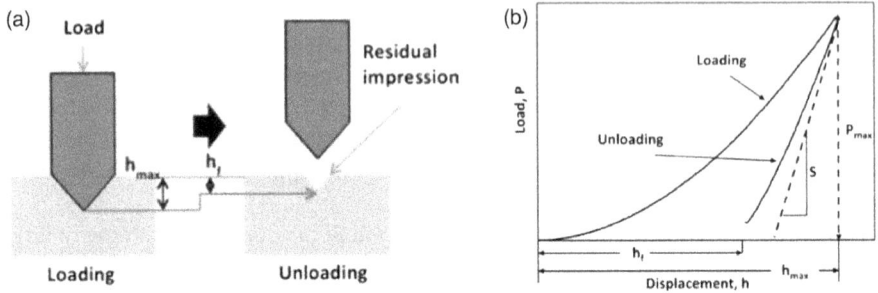

FIGURE 2.4   Method of nanoindentation and the measurement. (a) Diagram illustrating the loading and unloading processes used to imprint the sample. (b) A similar load-displacement curve illustrative of the influence of the loading and unloading procedure. S indicates the contact stiffness of unloading [21].

the use of first-principles approaches. Increased computer capacity has made it possible for DFT-led investigations to find materials with the best qualities before they are synthesized or developed in the lab.

Discussions of their hardness, plasticity, yield strength, and fracture behavior are also essential. For the materials, high-pressure X-ray crystallography may be used to compare anticipated elastic characteristics including compressibility and bulk moduli [22,23]. Recently, nanoindentation data have been utilized to develop machine learning algorithms. Nanoindentation has become a fundamental technique for assessing the mechanical characteristics of crystalline materials as illustrated in Figure 2.4.

Moreover, DFT is capable of database creation and mechanical property screening. Several researches have made efforts to create databases of elastic moduli using first-principles computational techniques [24,25]. These computational methods are useful because they enable clear comparisons across several kinds of materials and allow for the consistent extraction of all data. For example, De Jong et al. [26] have developed on this strategy, creating the biggest library of computed elastic characteristics of crystalline inorganic compounds to date, spanning from metallic complexes to insulators and semiconductors. The computations were carried out as part of a wider high-throughput effort conducted by the Materials Project [27]. Using DFT, it was regularly demonstrated that the predicted elastic constants were within 15% of the experimental values, which, in some circumstances, indicates a lower dispersion than that seen in experimental measurements. As illustrated in Figure 2.5, the calculations conducted in this study result in elastic characteristics that exhibit a strong correlation with experimental values, according to Spearman (ρ) and Pearson (r) and coefficients which is enabling the database to be used for screening materials having elastic tensor-based characteristics.

2.2.1.1.4   Optical Properties

New research has been motivated by the need to find stable materials with improved nonlinear optical characteristics for application in computer information transmission and storage and in telecommunications [28]. Recently, the nonlinear optical characteristics of molecules have attracted the interest of chemists [29,30]. DFT is

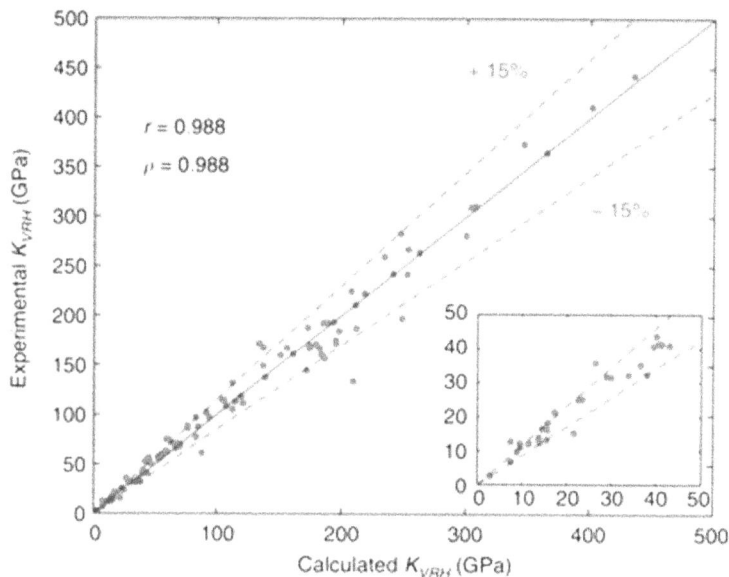

FIGURE 2.5 Comparison of calculated experimental and bulk moduli for a modeled set of systems, with calculated Spearman correlation coefficient ρ and Pearson correlation coefficient r [26].

a potential approach for quantitatively calculating the optical characteristics of molecules in a feasible size range. The approach depends on employing two sets of localized, atom-centered, in situ optimized orbitals to represent the optical response of the system and to accurately capture both the electron and hole wavefunctions of the excitations.

DFT can calculate the optical properties such as optical conductivity, dielectric function, absorption coefficient, refractive index, electron energy loss and reflectivity. DFT has the benefits of an ab initio technique and, while requiring less computing work, often produces results for a wide range of features that are superior to those obtained via the HF approximation. While DFT requires more calculation than semi-empirical approaches, it produces results that are significantly more trustworthy when a wide variety of molecule types and attributes are considered. Hence, DFT is a viable method for treating the optical characteristics of molecules in systems that are sophisticated enough to be of interest to chemists and material scientists.

Experiment optical spectra can be regarded as the outcome of averaging across multiple low-energy configurational structures of the system under consideration and are frequently well replicated by numerical simulations on a singular, averaged low-energy structure of that system.

Baek et al. [31] employed DFT to clarify deposition properties and the mechanism of the first $TiCl_4$ half-cycle reaction for TiO_2 deposition on the surface of Pt nanoparticles. In this work, the selective ALD process of the precursor $TiCl_4$ on Pt catalyst surfaces is explained quantitatively. They demonstrated that the release of HCl from the Pt (111) surface occurs through a favorable pathway, improving the material's reactivity to the surface reaction. In another work, Lee et al. [32] used DFT techniques to describe the silicon oxide ALD processes over the tungsten oxide substrate. They compare the surface reactions of various aminosilane precursors with a range of amino ligands, such as diisopropylaminosilane (DIPAS), bis(diethylamino)silane (BDEAS), and tris(dimethylamino)silane (TDMAS), over the hydroxyl-terminated WO_3 (001) surface to evaluate the ligand influence of precursors.

In another study, Shirazi et al. [33] implemented DFT to study the MoS_2 nanolayer formation on SiO_2 surface by ALD. ALD mechanism of MoS_2 is accommodated by the exposure of metal precursor $Mo(NMe_2)_2(NtBu)_2$ (bis(tert-butylimido) bis (dimethylamido) molybdenum) to the SiO_2 surface presented in Figure 2.6. Moreover, Figure 2.7 shows the chemical adsorption of $Mo(NMe_2)_4$ at S sites. Diethyl disulfide, diethyl trisulfide, dimethyl tetrasulfide, and dimethyl pentasulfide all exhibit exothermic chemical adsorption of $Mo(NMe_2)4$ at S with exothermic energies of 0.69, 0.13, 0.50, and 0.15 eV, respectively.

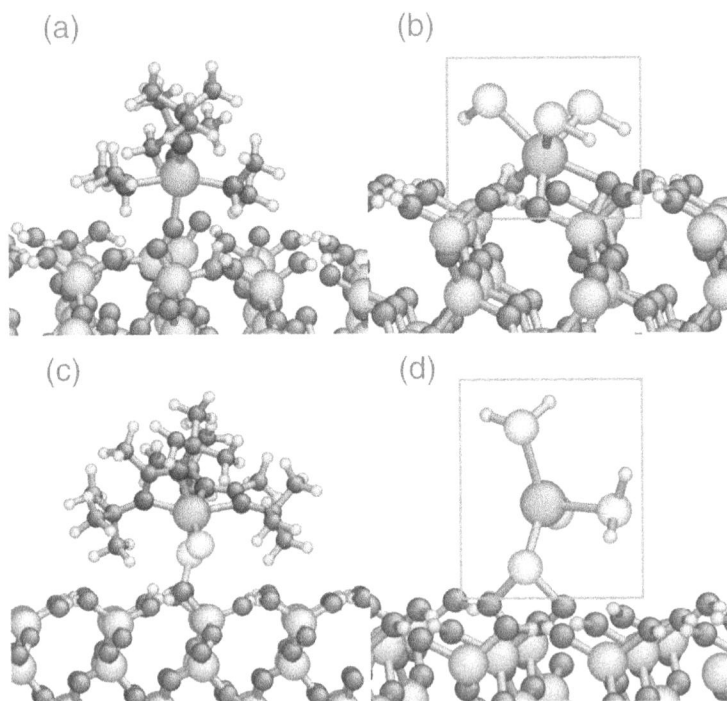

FIGURE 2.6 Demonstration of building block formation and underpinned building block from different metal precursors. (a) Adsorption of $Mo(NMe_2)^2(N^tBu)_2$ precursor at the terminal O at the SiO_2 (001) surface. (b) Indicates the coordinated Mo and S atoms at the SiO_2 surface after the H2S pulse. (c) Adsorption of $MoS_2(tBuAMD)_2$ at the terminal O site at the surface. (d) Presents the coordinated Mo and S atoms at the SiO_2 (001) surface which compose an underpinned building block at the end of H_2S pulse.

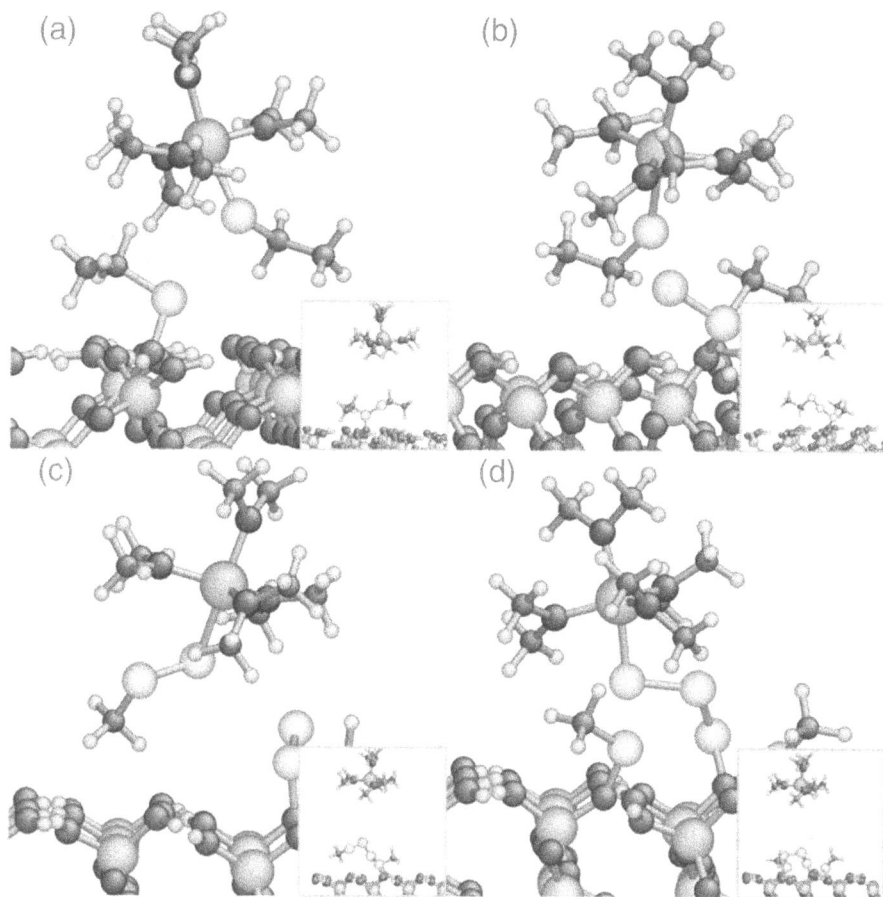

FIGURE 2.7 Chemical adsorption of $Mo(NMe_2)_4$ at the terminal S sites, deposited from sulfur co-reagents pulse. (a) The chemical adsorption at S of diethyl disulfide is exothermic by 0.69 eV. (b) The chemical adsorption at S of diethyl trisulfide is exothermic by 0.13 eV. (c) The chemical adsorption at S of dimethyl tetrasulfide is exothermic by 0.50 eV. (d) The chemical adsorption at S of dimethyl pentasulfide is exothermic by 0.15 eV [33].

2.2.1.2 Challenges of DFT

One of DFT's limitations is its failure to estimate energy gap values for semiconductors and insulators, underestimating by up to 50%. DFT simulations require estimations since the Schrödinger equation for a multi-body system cannot be solved accurately. However, the utility of their DFT simulations for numerous characteristics is dependent on the accuracy of the chosen approximation [34]. Keeping some kind of simplicity as its foundation presents one of DFT's biggest problems. The theory starts to lose one of its key characteristics, namely, its simplicity, if DFT functionals grow as complex as full configuration interaction. Through DFT calculations, in order to provide a more thorough understanding of chemistry, it is required to explain weakly interacting molecules and transition states in chemical processes in addition to molecules at their equilibrium geometry.

2.2.2 Molecular Dynamic Simulation

The fact that the atoms in the materials that surround us are constantly moving is an inevitable fact. A vast number of subjects require knowledge of how atoms in a material move as a function of time to describe some attribute of practical importance. This section focuses on MD techniques, which are a set of computational tools that allow us to track the motion of flowing atoms. In other words, MD simulation determines the forces between the atoms at each time step by using Newton's equation of movements, and it updates the atoms' locations for the subsequent time step [35]. MD explores the concept of atoms, molecules, and crystals in the fluid and condensed phases [36]. MD methods depend on the description of the force field and molecular interaction to identify each atom or molecule's motion and equilibrium conditions. It is utilized in various engineering and science areas, including physics, materials, and chemistry engineering. The MD approach is applicable to simulating different physical processes in both equilibrium and nonequilibrium states.

Kinetic properties are time-dependent transformations of compounds. The kinetic characteristics of molecules can be predicted using the MD approach [37]. MD can predict the thermodynamic properties, melting and boiling point, free binding energy, pressure, heat of vaporization, and free energy perturbation. These estimations are based on intermolecular interactions. MD also can be used for prediction of diffusion coefficients and mean-square displacements in materials. The MD may be used to simulate protein shapes and optimize X-ray structures, and stability of protein–ligand interactions. Another advantage of MD simulations is the ability to calculate vibrational DOS using the velocity auto-correlation function or power spectral density analysis. Multiple simulation softwares were developed for MD calculation including LAMMPS [38], GROMACS [39], NAMD [40], AMBER [41], FORCITE [42], and ReaxFF [43]. In MD simulations, the parameters of Lennard-Jones and columbic potentials are used to determine the nonbonded interactions between atoms [44,45] and are defined as follows:

$$E = E_{vdW} + E_{Columbic} \tag{2.11}$$

$$E = 4\epsilon \left[\left(\frac{\sigma}{r} \right)^{12} - \left(\frac{\sigma}{r} \right)^{6} \right] + \frac{q_i q_j}{\varepsilon_d r} \tag{2.12}$$

where ϵ and σ parameters refer to the potentials and distance between any two atoms, respectively. $q_i q_j$ denote the electronic charges between atoms I and j. ε_d is the dielectric constant and r shows the distances between atoms I and j. MD can incorporate constant Energy for the calculation. The classical MD is a well-established method that is frequently employed in various areas of computational chemistry and materials modeling. For understanding the dynamic behavior of the modeled system, it will account for a case in which there are N atoms inside of a volume V. Therefore, in order to characterize the arrangement of the atoms at any given instant in time, we need to define 3N locations, $(r_1, ..., r_{3N})$ and 3N velocities, $(v_1, ..., v_{3N})$. The total kinetic energy and total potential energy of our system are two metrics that are helpful for defining its overall condition as follows:

$$K = \frac{1}{2} \sum_{i=1}^{3N} m_i v_i^2 \tag{2.13}$$

where m_i is the atom's mass corresponding to the ith coordinate. Moreover, MD can be performed in the Canonical Ensemble. In the many cases, we need to compare the experimental results with theoretical outcome. The atoms of a material can exchange heat with their environment under standard experimental settings. In this case, the atoms form a canonical ensemble in which N, V, and T are constants. There are several methods to modify the microcanonical MD algorithm described above to simulate a canonical ensemble. Nose was the first to use the extended Lagrangian for the microcanonical ensemble defined by Equation 2.14:

$$L = \frac{1}{2} \sum_{i=1}^{3N} m_i s^2 v_i^2 - U\left(r_1, \ldots, r_{3N}\right) + \frac{Q}{2}\left(\frac{ds}{dt}\right)^2 - gk_b T \ln s \tag{2.14}$$

Ab initio MD makes it possible to study chemical processes in condensed phases objectively and accurately, opening up new paradigms for the explanation of microscopic mechanisms, the justification of experimental results, and the development of testable predictions of novel phenomena. The classical MD description above was supposed to highlight how atom dynamics may be explained if the atoms' potential energy is known as $U = U(r_1, \ldots, r_{3N})$, which is the function of the atomic coordination. However, it uses common DFT calculation. In other words, quantum mechanics can be used to determine the potential energy of the system of interest. This is the fundamental idea behind ab initio MD and the Lagrangian can be expressed as follows:

$$L = K - U = \frac{1}{2} \sum_{i=1}^{3N} m_i v_i^2 - E\left[\varphi\left(r_1, \ldots, r_{3N}\right)\right] \tag{2.15}$$

where $\varphi(r_1, \ldots, r_{3N})$ is the full set of Kohn–Sham one-electron wave functions for the electronic ground state of the system. This Lagrangian recommends performing the calculations in the following order: first, compute the ground-state energy; next, advance the locations of the nuclei by one MD step, then the new ground-state energy is computed, and so on. Ab initio MD is the term used to describe any technique that uses forces derived from DFT to move nuclear locations along paths specified by classical mechanics.

2.2.2.1 Application of MD in ALD Research

Due to high expenses and challenging chemical management associated with ALD investigations, most of the researchers prefer to perform numerical modeling of the ALD process to comprehend and investigate the deposition process.

Therefore, more understanding of the ALD process is achieved by modeling, which minimizes precursor usage and waste as well as potential environmental effects in future industrial outputs [46]. MD can predict the overall ALD process in the reaction chamber such as the stability of occurred bonds between the substrate and precursor, their binding energy, and kinetic properties of molecules, structure, and complex. Similarly, in the second half cycle, MD can predict the mechanism of the purging process, precursor reduction mechanism, and by-product formation. However, nature of reaction and stability of formed bonds between molecules can be accommodated by MD. Moreover, based on intermolecular interactions, MD can be utilized to determine the thermodynamic characteristics of the ALD thin-film material.

In 2008, Turner et al. [47] performed MD simulation to investigate the growth dynamic of amorphous Al_2O_3 films at the atomic scale. They showed MD simulation can reveal the growth mechanism of ALD AL_2O_3 thin film and several important properties which can link the atomic scale data with experimental information. Moreover, as demonstrated in Figures 2.8–2.10, the effect of operating condition on film formation showed that according to the models, the growth rate, surface roughness, and growth mode of the deposited films would be strongly influenced by the starting surface composition and process temperature. Furthermore, they showed that the Al_2O_3 growth rate is about 0.06 nm per cycle, which is in agreement with the experimental values (around 0.08 nm per cycle).

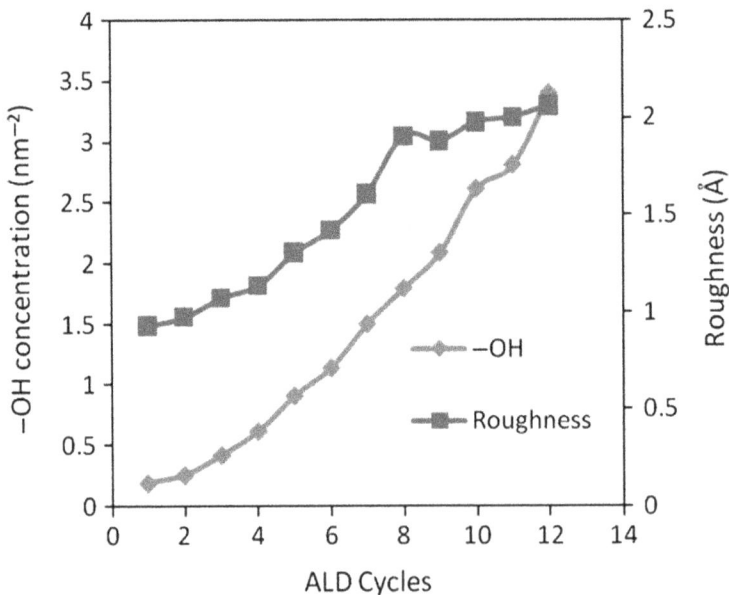

FIGURE 2.8 Surface –OH density and roughness of the Al_2O_3 film at 300°C as a function of ALD cycles [47].

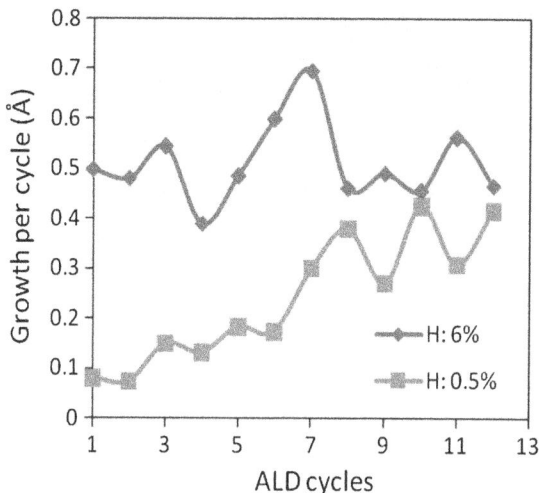

FIGURE 2.9 The evolution of the growth rate during the Al_2O_3 atomic layer deposition (ALD) starting with two different initial –OH densities at 300°C [47].

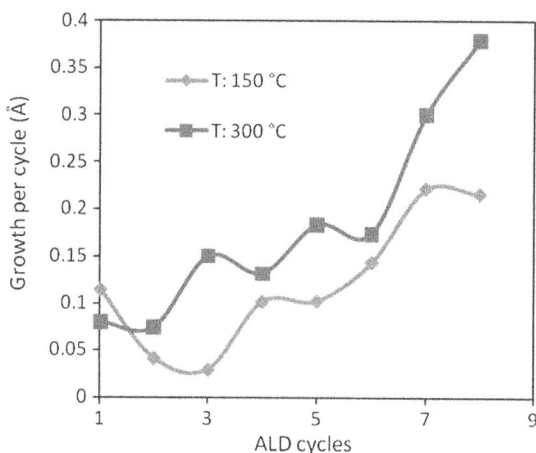

FIGURE 2.10 The evolution of the surface roughness during the atomic layer deposition (ALD) process at two different temperatures [47].

2.2.3 MONTE CARLO

The use of Monte Carlo (MC) techniques to solve molecular issues is known as MC molecular modeling. The MD approach may also be used to model these issues. The distinction is that, as opposed to MD, this method is based on equilibrium statistical mechanics. The MC method has the capacity to deal with far wider temporal and space scales. MC creates states in accordance with the proper Boltzmann distribution rather than attempting to replicate the dynamics of a system. As a result, it is a subset of the broader MC approach in statistical physics. It uses a Markov chain approach to derive a new state for a system from an old one. The acceptance of this new state is stochastically determined by its stochastic character. Accordingly, each trial counts as one move. In comparison to continuum-type models like the reaction-diffusion models, MC also provides a far more in-depth understanding of the surface processes.

In MC/kMC, the system is transferred stochastically through the phase-space while the particles (atoms, molecules, and beads) move in accordance with predetermined rules (events/processes), roughly approaching the mean values of different attributes. Unlike other molecular techniques like MD, the system in MC may be transferred to different states by including complex events, thus it is less likely to become stuck in local energy minima. In addition, kMC runs across considerably greater spatial and temporal scales than MD because it excludes vibrational motions. Images from the MC/kMC simulation may be readily compared to those from STM, particularly in the context of film growth. Several software programs have been developed especially for the Metropolis MC method's use to molecular simulations including: BOSS, MCPro, Faunus, Sire, and CP2K.

Most deposition processes include physicochemical surface phenomena, which are studied using MC and kinetic Monte Carlo (kMC) models. This includes atomic layer, chemical vapor, and physical deposition, as well as electrochemical deposition. MC/kMC has significant applications in the research of film deposition processes, such as in the production of semiconductor devices. Deposition processes have been widely investigated due to their relevance using MC/kMC models either alone or in combination with other methods in the framework of multiscale modeling [48]. Sequential situations are processed randomly in the MC and kMC techniques. The steady-state Master Equation is solved by MC, whereas the transient one is solved by kMC. The temporary ME is defined as follows:

$$\frac{\partial_{p_j}(t)}{\partial t} = \sum_{i \neq j} p_i(t) T_{ij} - \sum_{i \neq j} p_{j(t)} T_{ji} \tag{2.16}$$

Where $\rho_i(t)$ denote the probability of the system to be found in state i at time t. T_{ij} and T_{ji} are the transition probabilities or transition rates from state i to j and vice versa. Lattice theory is crucial when considering the deposition processes. The lattice symbolizes the deposition surface and is made up of locations where all events take place, simplifying the rate collection generation. Depending on the physical/chemical phenomena of interest, different numbers and types of events at a lattice site appear in MC/kMC. The mechanisms of adsorption, surface reaction, desorption, and surface diffusion are engaged in the deposition of materials. This sequence of actions as presented in Figure 2.11 has been specifically designed to record the physical and chemical mechanisms particular to each deposition phase. The main objective of a MC/kMC model is to specify the growth rate, predict the profile of the surface-growing film, and describe how particles interact with the surface.

Shirazi et al. [50] utilized the MC method incorporated with DFT to investigate the atomistic kinetics of ALD. They examined the ALD precursor's early stages of

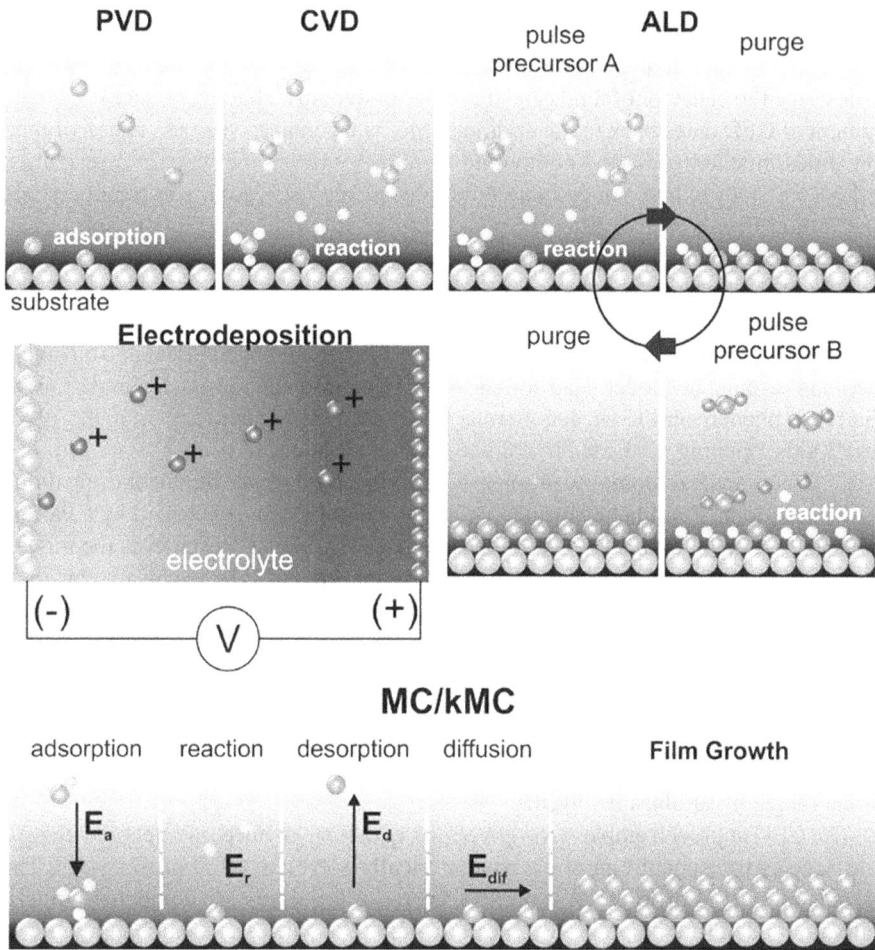

FIGURE 2.11 Schematic illustration of the fundamental concepts underlying deposition processes at the molecular level of Physical vapor deposition (PVD), Chemical vapor deposition (CVD), and atomic layer deposition (ALD) methods [49].

adsorption, surface proton kinetics, steric effects, the impact of leftover fragments on adsorption sites, the precursor's compression, migration, and cooperative behavior of the precursors. They came to the conclusion that the local environment at the surface affects the fundamental chemistry of the ALD processes. In another work, Knoops et al. [51] employed the MC approach to analyze the conformality of plasma enhanced ALD (PEALD) utilizing the recombination possibility, reaction occurrence, and particle diffusion rate to monitor conformal deposition in high aspect ratio structures. They found three deposition states: a reaction restricted regime, a diffusion limited regime, and a recombination limited regime. According to their results, conformal deposition may be obtained in high aspect ratio structures with a minimal recombination chance.

2.2.4 Computational Fluid Dynamic (CFD)

CFD is the method that studies the multiphysics systems by using numerical models to describe the behavior of fluids and their thermodynamic characteristics. The development of CFD dates back to the middle of the 20th century, when the fusion of fluid dynamics, mathematics, and computing technology paved the way for a revolution in how we analyze and comprehend fluid movement. Then, John von Neumann and R. Courant began investigating the possibility of utilizing computers to solve challenging fluid dynamics issues in the 1940s and 1950s. The 1960s witnessed the development of the finite volume method (FVM) and the finite difference method (FDM), two potent methods for discretizing fluid flow equations. In 1970–1980, faster computers, along with the development of the finite element method (FEM) and advancements in turbulence modeling, allowed for more precise and thorough simulations of fluid flow phenomena. Then, commercial CFD software began to appear in the 1990s and 2000s, opening up the technology to a larger audience of users.

CFD can analyze complex problems involving fluid–fluid, fluid–solid, or fluid–gas interactions. Fluid behavior may be highly complicated, influenced by a variety of variables including pressure, temperature, and viscosity. CFD enables the modeling of complicated fluid flow problems that might otherwise be unsolvable by splitting the fluid domain into a finite number of cells or components. CFD gives helpful details on the fundamental mass, momentum, or heat transfer phenomena in chemical and biological processes. Scientists and engineers employ mathematical formulas, such as the Navier–Stokes equations using computers and advanced algorithms, to define fluid movement to comprehend these intricacies. However, solving these equations analytically is often unfeasible due to their inherent complication and the broad range of variables involved.

The CFD approach employs the governing equations of mass, momentum, energy, and species transport in an effort to numerically solve the ALD process [52]. The associated partial differential equations for these processes are computed numerically on predetermined nodes in a mesh domain. Nonetheless, research in the literature have favored and focused only on the mechanical or chemical aspects because of the intricacy of this production process [53–55]. Despite this, researchers have used this approach in recent years to merge mechanical and chemical features, where the required chemical data are provided by DFT chemical simulation tools or through chemical reaction experiments [56,57].

The DFT method is essential to the CFD modeling approach. The DFT approach contains the necessary information to model the heterogeneous and homogeneous processes involved in the ALD process, including differentiating between the adsorption/desorption kinetics aspects of various chemical recipes [58]. The study using combined DFT/CFD is now scarce but is becoming more significant as ALD advances into the industrial sphere. By integrating and extracting the source terms of the chemical processes from the earlier governing equations, these features are then applied in the following equations:

$$\frac{\partial \rho}{\partial t} + \nabla \rho \vec{v} = 0.17 \qquad (2.17)$$

$$\frac{\partial \rho \vec{v}}{\partial t} + \nabla(\rho \vec{v} \vec{v}) = -\nabla P + \nabla_\tau + \rho \vec{g} + \vec{F} \qquad (2.18)$$

$$\frac{\partial \rho E}{\partial t} + \nabla(\vec{v}(\rho E + \rho)) = \nabla\left(k_{eff}\nabla T - \sum J_i^{h_i}\right) + R_r \qquad (2.19)$$

$$\frac{\partial(\rho Y_i)}{\partial t} + \nabla \rho \vec{v} Y_i = -\nabla \vec{J}_i + R_i \qquad (2.20)$$

where R, J, and k_{eff} represent the reaction source term, diffusion flux term, and effect of conductivity, respectively.

Accordingly, the rth irreversible surface reaction can be defined as follows:

$$\sum_{i=1}^{N_g} g_{i,r}' G_i + \sum_{i=1}^{N_b} b_{i,r}' B_i + \sum_{i=1}^{N_s} s_{i,r}' S_i = \sum_{i=1}^{N_g} g_{i,r}'' G_i + \sum_{i=1}^{N_b} b_{i,r}'' B_i + \sum_{i=1}^{N_s} s_{i,r}'' S_i \qquad (2.21)$$

where B, S, and G represent the bulk, site, and gaseous species, respectively. The formula below can be used to calculate the molar reaction rate for the irreversible surface reaction:

$$R_r = k_{f,r}(\Pi_{i=1}^{N_g}[C_i]_{wall}^{n_{i,gr}})(\Pi_{i=1}^{N_s}[S_i]_{wall}^{n_{i,sr}}) - k_{b,r}(\Pi_{i=1}^{N_g}[C_i]_{wall}^{n_{i,gr}})(\Pi_{i=1}^{N_s}[S_i]_{wall}^{n_{i,sr}}) \qquad (2.22)$$

With an emphasis on macroscopic factors, these equations may be used to determine desirable characteristics of the ALD process, such as the mass deposition rate ($\dot{m}_{dep}$) that can be measured at the substrate surface as follows:

$$\dot{m}_{dep} = \sum_{i=1}^{N_b} M_{w,i} \hat{R}_{i,bulk} \qquad (2.23)$$

In a research, Gakis et al. [59] investigated the special flow characteristics of the Ultratech Fiji F200 reactor as represented in Figure 2.12. Their report showed how the reactor's configuration and operating circumstances affected temperature fields, gas flow, and the spreading of species on the heated substrate surface. However, they ignored the substrate's chemical processes. Moreover, their result showed that non-uniform flow during pulse and purge process injected to the reactor had an impact on the reactants' content and temperature on the surface of substrate. As seen in Figure 2.13a, the gas enters the reactor at a temperature that is similar to that of the walls (270°C), with the exception of the cold entrance zones and the substrate, which is hotter (300°C), respectively. Due to the cooling supplied by the gas entering from the nearby areas of the reactor walls, the substrate is not isothermal (Figure 2.13b).

The design concept of the reactor is also very important to comprehend the ALD process. These investigations may disclose the ideal parameter settings, an error in the ALD recipe, chemical dose and distribution parameters, the heat flow impact, etc. Coetzee et al. [60] explored the internal behavior of an ALD reactor (Gemstar 6 square model). They examined the influence of flow in these reactors and the buffer

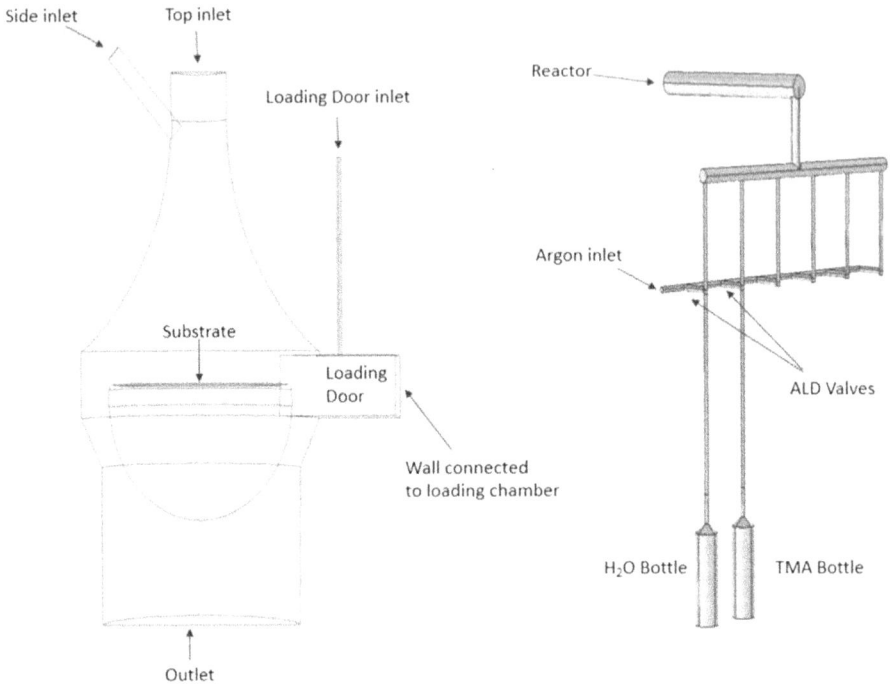

FIGURE 2.12 Schematics of the atomic layer deposition (ALD) system (Ultratech® Fiji F200). (a) The ALD reactor chamber. (b) The reactant feeding system [59].

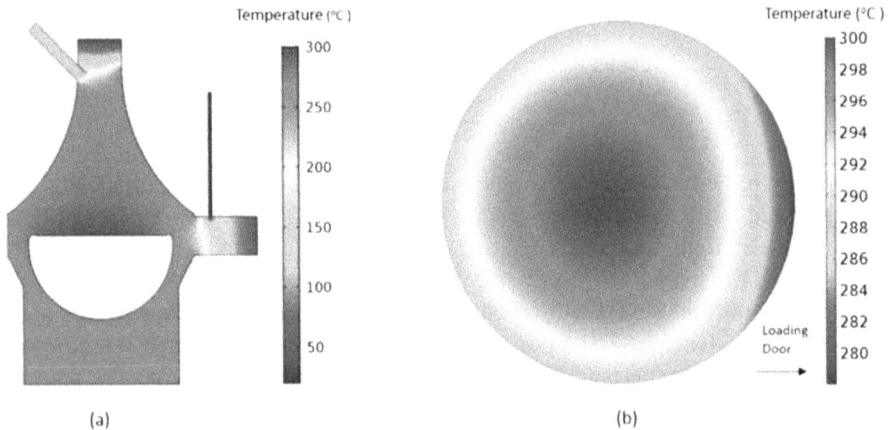

FIGURE 2.13 (a) Temperature field inside the reactor chamber. (b) Temperature profile on the substrate surface for the substrate center at 300°C [59].

layers that were generated on the substrate surface as a result of the injection sequence used to form the Al_2O_3 thin film. The previously described investigations by Pan et al. [61] and Shaeri [62] were carried out using a standard Cambridge Nanotech Savannah S100 reactor. More recently, CFD simulations were used to analyze a few different reactor designs. Moreover, Shaeri [63] used the CFD approach to study the impacts of high substrate temperatures, multiple inlets, and reactor inlet position. These investigations demonstrated the impacts of the growth rate, deposition rate, coverage, and mass fraction which were comparable to experimental results.

In an effort to optimize the ALD process, Deng et al. [64] quantitatively analyzed the temperature, pressure, mass flow, and precursor mass fractions. They came to the conclusion that greater temperatures enhanced the processes of surface coating and improved growth rates. In another work, Holmqvist et al. [53] studied the factors that influenced the deposition of a thin layer of ZnO in a continuous cross-flow ALD reactor. Their study was focused on the optimization and control of the film thickness profile by analyzing the effects of altering operational factors including the process cycles, local coordinate variable, and temperature.

The researchers are presently examining these studies as well as those that include adsorption/desorption kinetics, alternative ALD film recipes, system optimization, reactor and substrate mass/fluid flow behaviors, and the development of enhanced ALD processes. It is worth mentioning that the process analysis approach is rare in most other ALD reactor designs and manufacturers. Because of the various designs of the reactors, future ideas in the refinement of the ALD process will have to be explored similarly in order to produce optimal thin film manufacturing processes for the different kinds of thin films ALD recipes.

2.3 CHALLENGES AND FUTURE DIRECTIONS IN ALD SIMULATION AND MODELING

Simulations and modeling have improved our understanding of ALD processes, allowing researchers to accurately forecast and optimize thin film properties. The accuracy and dependability of ALD simulations and modeling various hurdles remain. The process of ALD process complexity is a major issue. ALD involves many chemical reactions involving substrates, precursor molecules, and reaction intermediates. Temperature, pressure, and other environmental conditions make these processes complex and hard to model. These models and simulations for ALD need experimental data to validate. Experimental data can be scarce. This makes realistic models and simulations of ALD film characteristics and deposition mechanisms difficult to build. The simulations and modeling in ALD demand enormous computational resources. Simulations must account for thousands of atom interactions in ALD processes. This is computationally expensive and time-consuming. Different research groups use different ALD modeling and simulation methods. This makes it difficult to compare study results and create ALD modeling and simulation best practices. New ALD modeling and simulation methods could progress nanotechnology to solve these difficulties. Multi-scale modeling can connect atomic-scale simulations to macroscopic ALD process models. This method can simulate atomic-scale and macroscopic ALD processes. Data-driven modeling is another trend. Artificial neural networks can predict ALD film qualities from little experimental data. Predicting ALD film characteristics with data-driven modeling is faster. Experimental and theoretical methods can validate and improve ALD simulations and modeling. This method helps researchers understand ALD processes and test their models and simulations. Finally, standardizing ALD modeling and simulation may increase accuracy and dependability. A uniform strategy can assist compare study results and develop ALD modeling and simulation best practices. Simulations and models have substantially improved our understanding of ALD processes, but several difficulties remain to increase their accuracy and dependability. However, multi-scale modeling, data-driven modeling, integration of experimental and theoretical techniques, and standardization offer enormous potential for ALD simulations and modeling.

2.4 CONCLUSION

Simulations and modeling have transformed our understanding of ALD processes, allowing researchers to precisely forecast and optimize thin film features. ALD simulations and models face problems such as the complexity of ALD processes, the scarcity of experimental data, and the requirement for considerable computational resources, yet new modeling and simulation paths seem promising. Multi-scale modeling, data-driven modeling, integration of experimental and theoretical methodologies, and standardization of modeling and simulation techniques can solve these problems and increase ALD simulations and modeling accuracy and dependability. Simulations and modeling must improve as ALD continues to develop better materials and devices to harness its full potential. Modern ALD simulations and modeling are promising and will continue to impact materials science and engineering. The combination of simulation and experimental methodologies, the development of new simulation techniques, and the standardization of modeling and simulation approaches will increase ALD simulation and modeling accuracy and dependability. Thus, ALD will continue to generate improved materials and devices with desirable features, impacting electronics, energy, and catalysis.

REFERENCES

[1] A. D. Kulkarni, and D. G. Truhlar, "Performance of density functional theory and Møller-Plesset second-order perturbation theory for structural parameters in complexes of Ru," *Journal of Chemical Theory and Computation*, vol. 7, pp. 2325–2332, 2011. https://doi.org/10.1021/CT200188N/SUPPL_FILE/CT200188N_SI_001.PDF.

[2] D. Sholl, *Density Functional Theory: A Practical Introduction*, 2022. https://www.wiley.com/en-us/Density+Functional+Theory%3A+A+Practical+Introduction%2C+2nd+Edition-p-9781119840862 (accessed April 14, 2023).

[3] W. Kohn, and L. J. Sham, "Self-consistent equations including exchange and correlation effects," *Physical Review*, vol. 140, pp. A1133, 1965. https://doi.org/10.1103/PHYSREV.140.A1133/FIGURE/1/THUMB.

[4] D. R. Douglas, and R. Hartree, *The Calculation of Atomic Structures*, New York: J. Wiley, 1957.

[5] R. A. van Rutger, A. Santen, and P. Sautet, *Computational Methods in Catalysis and Materials Science*, p. 455, 2009. https://www.wiley.com/en-us/Computational+Methods+in+Catalysis+and+Materials+Science%3A+An+Introduction+for+Scientists+and+Engineers-p-9783527320325 (accessed April 14, 2023).

[6] M. Penz, and R. van Leeuwen, "Density-functional theory on graphs," *Journal of Chemical Physics*, vol. 155, p. 244111, 2021. https://doi.org/10.1063/5.0074249.

[7] M. D. Halls, J. Velkovski, H. B., and Schlegel, "Harmonic frequency scaling factors for Hartree-Fock, S-VWN, B-LYP, B3-LYP, B3-PW91 and MP2 with the Sadlej pVTZ electric property basis set," *Theoretical Chemistry Accounts*, vol. 105, pp. 413–421, 2001. https://doi.org/10.1007/S002140000204/METRICS.

[8] X. Xu, and W. A. Goddard, "The X3LYP extended density functional for accurate descriptions of nonbond interactions, spin states, and thermochemical properties," *Proceedings of the National Academy of Sciences*, vol. 101, pp. 2673–2677, 2004. https://doi.org/10.1073/PNAS.0308730100.

[9] N. Medvedev, and B. Rethfeld, "Dynamics of electronic excitation of solids with ultra-short laser pulse," *AIP Conference Proceedings*, vol. 1278, pp. 250–261, 2010. https://doi.org/10.1063/1.3507110.

[10] D. S. Sholl, and J. A. Steckel, *Density Functional Theory: A Practical Introduction*, pp. 1–238. 2009. https://doi.org/10.1002/9780470447710.

[11] P. S. Maydannik, T. O. Kääriäinen, K. Lahtinen, D. C. Cameron, M. Söderlund, P. Soininen, P. Johansson, J. Kuusipalo, L. Moro, and X. Zeng, "Roll-to-roll atomic layer deposition process for flexible electronics encapsulation applications," *Journal of Vacuum Science & Technology A: Vacuum, Surfaces, and Films*, vol. 32, p. 051603, 2014. https://doi.org/10.1116/1.4893428.

[12] X. Du, and S. M. George, "Thickness dependence of sensor response for CO gas sensing by tin oxide films grown using atomic layer deposition," *Sensors Actuators B: Chemical*, vol. 135, pp. 152–160, 2008. https://doi.org/10.1016/J.SNB.2008.08.015.

[13] R. Alcala, C. Richter, M. Materano, P. D. Lomenzo, C. Zhou, J. L. Jones, T. Mikolajick, and U. Schroeder, "Influence of oxygen source on the ferroelectric properties of ALD grown Hf1-xZrxO2 films," *Journal of Physics D: Applied Physics*, vol. 54, p. 035102, 2020. https://doi.org/10.1088/1361-6463/ABBC98.

[14] S. M. Lee, E. Pippel, and M. Knez, "Metal infiltration into biomaterials by ALD and CVD: A comparative study," *ChemPhysChem*, vol. 12, pp. 791–798, 2011. https://doi.org/10.1002/CPHC.201000923.

[15] J. Wang, N. Muhammad, T. Li, H. Wang, Y. Liu, B. Liu, and H. Zhan, "Hyaluronic acid-coated camptothecin nanocrystals for targeted drug delivery to enhance anticancer efficacy," *Molecular Pharmaceutics*, vol. 17, pp. 2411–2425, 2020. https://doi.org/10.1021/ACS.MOLPHARMACEUT.0C00161/SUPPL_FILE/MP0C00161_SI_001.PDF.

[16] M. De Jong, W. Chen, T. Angsten, A. Jain, R. Notestine, A. Gamst, M. Sluiter, C. K. Ande, S. Van Der Zwaag, J. J. Plata, C. Toher, S. Curtarolo, G. Ceder, K. A. Persson, and M. Asta, "Charting the complete elastic properties of inorganic crystalline compounds," *Scientific Data*, vol. 21, no. 2, pp. 1–13, 2015. https://doi.org/10.1038/sdata.2015.9.

[17] W. Bao, D. Liu, P. Li, and Y. Duan, "Elastic anisotropies and thermal properties of cubic TMIr (TM=Sc, Y, Lu, Ti, Zr and Hf): A DFT calculation," *Materials Research Express*, vol. 6, 2019. https://doi.org/10.1088/2053-1591/AB1F01.

[18] S. Ryu, S. Lee, J. Jung, J. Lee, and Y. Kim, "Micromechanics-based homogenization of the effective physical properties of composites with an anisotropic matrix and interfacial imperfections," *Frontiers in Materials*, vol. 6, p. 21, 2019. https://doi.org/10.3389/FMATS.2019.00021/BIBTEX.

[19] R. Penta, and A. Gerisch, "The asymptotic homogenization elasticity tensor properties for composites with material discontinuities," *Continuum Mechanics and Thermodynamics*, vol. 29, pp. 187–206, 2017. https://doi.org/10.1007/S00161-016-0526-X/METRICS.

[20] D. A. Papaconstantopoulos, and M. J. Mehl, "Tight-binding method in electronic structure," *Encyclopedia of Condensed Matter Physics*, pp. 194–206, 2005. https://doi.org/10.1016/B0-12-369401-9/00452-6.

[21] E. Kiely, R. Zwane, R. Fox, A. M. Reilly, and S. Guerin, "Density functional theory predictions of the mechanical properties of crystalline materials," *CrystEngComm*, vol. 23, pp. 5697–5710, 2021. https://doi.org/10.1039/D1CE00453K.

[22] X. Guo, X. Lü, J. T. White, C. J. Benmore, A. T. Nelson, R. C. Roback, and H. Xu, "Bulk moduli and high pressure crystal structure of U3Si2," *Journal of Nuclear Materials*, vol. 523, pp. 135–142, 2019. https://doi.org/10.1016/J.JNUCMAT.2019.06.006.

[23] J. Sánchez-Martín, R. Turnbull, A. Liang, D. Díaz-Anichtchenko, S. Rahman, H. Saqib, M. Ikram, C. Popescu, P. Rodríguez-Hernández, A. Muñoz, J. Pellicer-Porres, and D. Errandonea, "High-pressure x-ray diffraction and DFT studies on spinel FeV_2O_4, *Crystals*, vol. 13, p. 53, 2022. https://doi.org/10.3390/CRYST13010053.

[24] P. R. C. da Silveira, C. R. S. da Silva, and R. M. Wentzcovitch, "Metadata management for distributed first principles calculations in VLab-A collaborative cyberinfrastructure for materials computation," *Computer Physics Communications*, vol. 178, pp. 186–198, 2008. https://doi.org/10.1016/J.CPC.2007.09.001.

[25] C. R. S. da Silva, P. R. C. da Silveira, B. Karki, R. M. Wentzcovitch, P. A. Jensen, E. F. Bollig, M. Pierce, G. Erlebacher, and D. A. Yuen, "Virtual laboratory for planetary materials: System service architecture overview," *Physics of the Earth and Planetary Interiors*, vol. 163, pp. 321–332, 2007. https://doi.org/10.1016/J.PEPI.2007.04.018.

[26] M. De Jong, W. Chen, T. Angsten, A. Jain, R. Notestine, A. Gamst, M. Sluiter, C. K. Ande, S. Van Der Zwaag, J. J. Plata, C. Toher, S. Curtarolo, G. Ceder, K. A. Persson, and M. Asta, "Charting the complete elastic properties of inorganic crystalline compounds," *Scientific Data*, vol. 21, no. 2, pp. 1–13, 2015. https://doi.org/10.1038/sdata.2015.9.

[27] A. Jain, S. P. Ong, G. Hautier, W. Chen, W. D. Richards, S. Dacek, S. Cholia, D. Gunter, D. Skinner, G. Ceder, and K. A. Persson, "Commentary: The materials project: A materials genome approach to accelerating materials innovation," *APL Materials*, vol. 1, p. 011002, 2013. https://doi.org/10.1063/1.4812323.

[28] C.J. Bottcher, "Theory of electric polarization: dielectrics in static fields," Elsevier Science, Department of Physical Chemistry, University of Leiden, The Netherlands, p. 398, 1973.

[29] D. M. Burland, "Optical nonlinearities in chemistry: Introduction," *Chemical Reviews*, vol. 94, pp. 1–2, 1994. https://doi.org/10.1021/CR00025A600/ASSET/CR00025A600. FP.PNG_V03.

[30] H. Sekino, and R. J. Bartlett, "New algorithm for high-order time-dependent hartree-fock theory for nonlinear optical properties," *International Journal of Quantum Chemistry*, vol. 43, pp. 119–134, 1992. https://doi.org/10.1002/QUA.560430111.

[31] J. Baek, K. Nam, J. yeon Park, and J. H. Cha, "Adsorption selectivity of TiCl4 precursor on Pt surfaces for atomic layer deposition via density functional theory," *Applied Surface Science*, vol. 606, p. 154695, 2022. https://doi.org/10.1016/J.APSUSC.2022.154695.

[32] K. Lee, and Y. Shim, "First-principles study of the surface reactions of aminosilane precursors over WO3(001) during atomic layer deposition of SiO 2," *RSC Advances*, vol. 10, pp. 16584–16592, 2020. https://doi.org/10.1039/D0RA01635G.

[33] M. Shirazi, W. M. M. Kessels, and A. A. Bol, "Strategies to facilitate the formation of free standing MoS2 nanolayers on SiO2 surface by atomic layer deposition: A DFT study," *APL Materials*, vol. 6, p. 111107, 2018. https://doi.org/10.1063/1.5056213.

[34] P. Verma, and D. G. Truhlar, "Status and challenges of density functional theory," *Trends in Chemistry*, vol. 2, pp. 302–318, 2020. https://doi.org/10.1016/J.TRECHM.2020.02.005.

[35] D. Cohen-Tanugi, *Nanoporous Graphene as a Water Desalination Membrane*, *Technology*, 2015. https://dspace.mit.edu/handle/1721.1/98743 (accessed April 13, 2023).

[36] P. Brault, "Multiscale molecular dynamics simulation of plasma processing: Application to plasma sputtering," *Frontiers in Physics*, vol. 6, p. 59, 2018. https://doi.org/10.3389/FPHY.2018.00059/BIBTEX.

[37] A. Hospital, J. R. Goñi, M. Orozco, and J. L. Gelpí, "Molecular dynamics simulations: Aadvances and applications," *Advances and Applications in Bioinformatics and Chemistry*, vol. 8, pp. 37–47, 2015. https://doi.org/10.2147/AABC.S70333.

[38] S. Karimzadeh, B. Safaei, and T. C. Jen, "Predicting phonon scattering and tunable thermal conductivity of 3D pillared graphene and boron nitride heterostructure," *International Journal of Heat and Mass Transfer*, vol. 172, p. 121145, 2021. https://doi.org/10.1016/j.ijheatmasstransfer.2021.121145.

[39] S. Karimzadeh, B. Safaei, and T.C. Jen, "Prediction effect of ethanol molecules on doxorubicin drug delivery using single-walled carbon nanotube carrier through POPC cell membrane," *Journal of Molecular Liquids*, vol. 330, p. 115698, 2021. https://doi.org/10.1016/j.molliq.2021.115698.

[40] J. C. Phillips, R. Braun, W. Wang, J. Gumbart, E. Tajkhorshid, E. Villa, C. Chipot, R. D. Skeel, L. Kalé, and K. Schulten, "Scalable molecular dynamics with NAMD," *Journal of Computational Chemistry*, vol. 26, pp. 1781–1802, 2005. https://doi.org/10.1002/JCC.20289.

[41] B. Leimkuhler, and C. Matthews, *Molecular Dynamics*, vol. 39, 2015. https://doi.org/10.1007/978-3-319-16375-8.

[42] K. Ledwaba, S. Karimzadeh, and T. C. Jen, "Enhancement in the hydrogen storage capability of borophene through yttrium doping: A theoretical study," *Journal of Energy Storage*, vol. 55, p. 105500, 2022. https://doi.org/10.1016/J.EST.2022.105500.

[43] K. Chenoweth, A. C. T. Van Duin, and W. A. Goddard, "ReaxFF reactive force field for molecular dynamics simulations of hydrocarbon oxidation," *The Journal of Physical Chemistry A*, vol. 112, pp. 1040–1053, 2008. https://doi.org/10.1021/JP709896W/SUPPL_FILE/JP709896W-FILE005.PDF.

[44] A. Hospital, J. R. Goñi, M. Orozco, and J. L. Gelpí, "Molecular dynamics simulations: Advances and applications," *Advances and Applications in Bioinformatics and Chemistry*, vol. 8, pp. 37–47, 2015. https://doi.org/10.2147/AABC.S70333.

[45] M. Heiranian, Y. Wu, and N. R. Aluru, "Molybdenum disulfide and water interaction parameters," *The Journal of Chemical Physics*, vol. 147, p. 104706, 2017. https://doi.org/10.1063/1.5001264.

[46] D. Pan, T. Li, T. C. Jen, and C. Yuan, "Numerical modeling of carrier gas flow in atomic layer deposition vacuum reactor: A comparative study of lattice Boltzmann models," *Journal of Vacuum Science & Technology*, vol. 32, p. 01A110, 2013. https://doi.org/10.1116/1.4833561.

[47] Z. Hu, J. Shi, and C. Heath Turner, "Molecular dynamics simulation of the Al_2O_3 film structure during atomic layer deposition," *Molecular Simulation*, vol. 35, pp. 270–279, 2009. https://doi.org/10.1080/08927020802468372.

[48] N. Cheimarios, G. Kokkoris, and A. G. Boudouvis, "Multiscale modeling in chemical vapor deposition processes: Models and methodologies," *Chemical Engineering Science*, vol. 282, no. 28, pp. 637–672, 2020. https://doi.org/10.1007/S11831-019-09398-W.

[49] N. Cheimarios, D. To, G. Kokkoris, G. Memos, and A. G. Boudouvis, "Monte Carlo and Kinetic Monte Carlo models for deposition processes: A review of recent works," *Frontiers in Physics*, vol. 9, p. 165, 2021. https://doi.org/10.3389/FPHY.2021.631918/BIBTEX.

[50] J. I. Hochstein, and A. L. Gerhart, *Young, Munson and Okiishi's A Brief Introduction to Fluid Mechanics*, 2021. https://books.google.com/books?hl=en&lr=&id=RNoPEAAAQBAJ&oi=fnd&pg=PA1&ots=FcTJZOWCqa&sig=ticiptic0DhJ3VECnlXs2fLQKzk (accessed April 13, 2023).

[51] H.C.M. Knoops, E. Langereis, M.C.M. van de Sanden, W.M.M. Kessels, "Conformality of plasma-assisted ALD: physical processes and modeling," *Journal of the Electrochemical Society*, vol. 157, G241, 2010. https://doi.org/10.1149/1.3491381

[52] R. A. M. Coetzee, T. C. Jen, M. Bhamjee, and J. Lu, "The mechanistic effect over the substrate in a square type atomic layer deposition reactor," *International Journal of Modern Physics*, vol. 33, 2019. https://doi.org/10.1142/S0217979219400186.

[53] A. Holmqvist, T. Törndahl, and S. Stenström, "A model-based methodology for the analysis and design of atomic layer deposition processes-Part I: Mechanistic modelling of continuous flow reactors," *Chemical Engineering Science*, vol. 81, pp. 260–272, 2012. https://doi.org/10.1016/J.CES.2012.07.015.

[54] P. Peltonen, V. Vuorinen, G. Marin, A. J. Karttunen, and M. Karppinen, "Numerical study on the fluid dynamical aspects of atomic layer deposition process," *Journal of Vacuum Science & Technology A*, vol. 36, p. 021516, 2018. https://doi.org/10.1116/1.5018475.

[55] R. G. Gordon, D. Hausmann, E. Kim, and J. Shepard, "A kinetic model for step coverage by atomic layer deposition in narrow holes or trenches," *Chemical Vapor Deposition*, vol. 9, pp. 73–78, 2003. https://doi.org/10.1002/CVDE.200390005.

[56] J. Lu, J.W. Elam, and P. C. Stair, "Atomic layer deposition-sequential self-limiting surface reactions for advanced catalyst "bottom-up" synthesis," *Surface Science Reports*, vol. 71, pp. 410–472, 2016. https://doi.org/10.1016/J.SURFREP.2016.03.003.

[57] S. Suh, S. Park, H. Lim, Y.-J. Choi, C. S. Hwang, H. J. Kim, and S.-J. Won, "Investigation on spatially separated atomic layer deposition by gas flow simulation and depositing Al_2O_3 films," *Journal of Vacuum Science & Technology A*, vol. 30, p. 051504, 2012. https://doi.org/10.1116/1.4737123.

[58] D. Pan, D. Guan, T. C. Jen, and C. Yuan, "Atomic layer deposition process modeling and experimental investigation for sustainable manufacturing of nano thin films," *Journal of Manufacturing Science and Engineering*, vol. 138, 2016. https://doi.org/10.1115/1.4034475/375126.

[59] G. P. Gakis, H. Vergnes, E. Scheid, C. Vahlas, B. Caussat, and A. G. Boudouvis, "Computational fluid dynamics simulation of the ALD of alumina TMA and H_2O in a commercial reactor," *Chemical Engineering Research and Design*, vol. 132, pp. 795–811, 2018. https://doi.org/10.1016/J.CHERD.2018.02.031.

[60] R. M. Kumar, K. Rajesh, S. Halder, P. Gupta, K. Murali, P. Roy, and D. Lahiri, "Surface modification of CNT reinforced UHMWPE composite for sustained drug delivery," *Journal of Drug Delivery Science and Technology*, vol. 52, pp. 748–759, 2019. https://jglobal.jst.go.jp/en/detail?JGLOBAL_ID=201902211438034076 (accessed February 23, 2021).

[61] D. Pan, "Numerical and experimental studies of atomic layer deposition for sustainability improvement," Thesis and Dissertation, The University of Wisconsin, 2016.

[62] M. R. Shaeri, "Reactor scale simulation of atomic layer deposition," Thesis and Dissertation, The University of Wisconsin, 2014.

[63] M. R. Shaeri, T. C. Jen, C. Y. Yuan, and M. Behnia, "Investigating atomic layer deposition characteristics in multi-outlet viscous flow reactors through reactor scale simulations," *International Journal of Heat and Mass Transfer*, vol. 89, pp. 468–481, 2015. https://doi.org/10.1016/J.IJHEATMASSTRANSFER.2015.05.079.

[64] Z. Deng, W. He, C. Duan, B. Shan, and R. Chen, "Atomic layer deposition process optimization by computational fluid dynamics," *Vacuum*, vol. 123, pp. 103–110, 2016. https://doi.org/10.1016/J.VACUUM.2015.10.023.

3 Characterization Methods in ALD

3.1 INTRODUCTION

The requirements and methods for the characterization of thin films have grown greatly along with the number of technologies available for producing a wide range of thin films, especially atomic layer deposition (ALD) thin films, due to the atomic scale of film formation. The proper way to characterize a generated thin film relies on the fabrication process, required endurance, and the intended or possible applications. Moreover, techniques with high spatial resolution are becoming more essential due to the growing significance of nanostructures and microstructures by ALD method. In most of the cases, accurate characterization necessitates the employment of multiple tools. Accordingly, in situ analytical techniques for ALD process are imperative which can enable direct inspection of ALD chemistry and a number of film-growing process features. They are the correct option to precisely track the reaction processes and minimize the change in film properties by avoiding air exposure to the sample. In addition, for in situ analysis to yield reliable data on reaction processes, there must be a certain setup and interpretation criteria. In this respect, this chapter provides more insight into in situ analytical methods to explore the thin film growth during ALD process and also presents an overview of some of the most frequently implemented analytical methods for chemical and physical characterization of ALD films and surfaces as well as highlighting their fundamental techniques and benefits.

3.2 SIGNIFICANCE OF RIGHT SELECTION OF ALD CHARACTERIZATION TECHNIQUES

The quality and usefulness of ALD films for certain applications depend on the characterization method chosen. Thickness, composition, and electrical, optical, mechanical, and tribological properties may vary for ALD applications. Thus, accurate characterization methods are needed to measure these features. ALD layer thickness and composition affect device performance in electronics and optoelectronics. Ellipsometry, quartz crystal microbalance, and spectroscopic ellipsometry (SE) are used to measure film thickness and composition nondestructively. For accurate film property control in electrical and optoelectronic devices, these methods can give sub-nanometer measurements while ALD film performance in electronic, photovoltaic, and display applications depends on electrical and optical characterization. Hall effect, four-point probe resistivity, and capacitance-voltage measurements can reveal film electrical properties like conductivity, carrier concentration, and mobility. Photoluminescence spectroscopy may also characterize film optical properties such

DOI: 10.1201/9781003346234-4

as bandgap, defect density, and luminescence efficiency. The durability and wear resistance of ALD films in protective coatings and Micro-electromechanical systems (MEMS) depend on mechanical and tribological evaluation. Nanoindentation and scratch testing can reveal the film's hardness, modulus, and adhesion while tribometer testing can reveal the film's friction coefficient and wear rate. Likewise, the quality and functioning of ALD film for specific applications depend on characterization methodologies. Accurate film property measurements can optimize the ALD process and improve product performance. Thus, application-specific characterization approaches must be carefully evaluated and chosen.

An improper characterization technique can give misleading or partial information about the ALD film, resulting in poor device performance or failure. A semiconductor device's conductivity and carrier concentration can be affected by an ALD film's thickness. A protective covering with poor adherence might prematurely wear and fail the substrate. The procedure and materials of ALD may require distinct characterization methods. Ellipsometry or quartz crystal microbalance may struggle to estimate layer thickness in various ALD processes because of high roughness or porosity. Rutherford backscattering spectroscopy or X-ray reflectivity may work better. X-ray diffraction is challenging to use on amorphous ALD films. Transmission electron microscopy can then reveal the film's microstructure. The correct characterization technique can optimize the ALD process and decrease manufacturing costs. Ellipsometry, a nondestructive method for measuring film thickness, can eliminate the requirement for transmission electron microscopy, which is expensive and time-consuming. Scratch testing can reliably determine the adhesion strength of a protective coating, eliminating the need for costly and time-consuming environmental testing. Finally, selecting the right characterization methods for ALD applications is crucial. Using appropriate methods to measure the film's properties helps optimize the ALD process and improve the product's performance. Thus, application-specific characterization approaches must be carefully evaluated and selected.

3.3   ALD CHEMISTRY

ALD includes alternate precursors reacting with a substrate's surface to generate a monolayer of the desired substance. The film's qualities depend on these precursors' chemistry. ALD chemistry depends on the precursors—metalorganic chemicals, metal halides, or metal alkyls. These precursors react with the substrate surface self-limitingly, depositing a homogeneous layer each cycle. ALD precursors affect film composition, crystallinity, and density. The chemicals processed involved in ALD are characterized using numerous methods. These methods investigate deposited films' properties and guarantee their composition and structure, thereby meeting standards. ALD is characterized by methods like X-ray photoelectron spectroscopy (XPS), Fourier transform infrared spectroscopy (FTIR), and ellipsometry, amongst others. The XPS surface examination reveals material chemical composition. XPS can identify the film's elements and chemical states by studying its surface X-ray emissions. However, FTIR uses spectroscopy to identify functional groups in a substance. To study the chemistry, it can reveal the chemical bonds in the deposited film. ALD also uses ellipsometry to quantify light polarization after reflection from a sample surface. ALD requires measuring thin film thickness and refractive index, which this method does well.

In conclusion, ALD chemistry determines film properties, hence understanding it is crucial for achieving desired film qualities. XPS, FTIR, and ellipsometry help examine ALD's chemistry and ensure film qualities satisfy specifications. In order to comprehend the connection between ALD process and material characteristics, a more thorough understanding of the reaction mechanism is required. The process and the formed film's properties may then be adjusted using this knowledge. The reaction processes are strongly affected by the selected precursors and co-reactants, as well as other factors like temperature, pressure of deposition, and pulse and purge times. However, understanding the surface reaction products at each stage of the process is crucial to understanding the ALD chemistry of the process.

3.4   CHARACTERIZATION METHODS IN ALD

ALD characterization procedures ensure that deposited films have the correct thickness, composition, and crystallinity. This section discusses some of the most common ALD characterization approaches and how they help us understand and optimize the process.

3.4.1   QUADRUPOLE MASS SPECTROMETRY (QMS)

QMS is a method which is commonly used for the tracking and monitoring of gaseous species reaction in the reactor during the ALD process. The volatile by-products produced during the ALD process undergo ionization and decomposition through the utilization of electrons generated by a filament source. These ionized by-products are subsequently detected and analyzed by the quadrupole analyzer. In other words, it can estimate which reactions have occurred in the reactor. Therefore, it makes it possible to change the parameters of the process. High vacuum conditions are necessary for the QMS, and these conditions are attained by differentially pumping a turbomolecular pump via an aperture. The signal is obtained as a function of the mass to charge ratio (m/z). The typical visual representation of a mass spectrum acquired by QMS is in the form of a bar graph, where each bar indicates an ion with a certain mass-to-charge ratio (m/z). Since most ions have a charge of +1, the m/z ratio is often equal to the mass. The major drawback of this method is the challenge of distinguishing a single segment from an m/z signal. For example, multiple by-products generation from various sources, such as hydrogenated ligands and ozone combustion, can produce different compositions at the same m/z, so the signal detected by the QMS may consist of several participations. These signals can be distinguished by comparing their occurrence with the signal of one component. The interpretation is still challenging, particularly when combustion processes are taking place. Another limitation of QMS is tracking few numbers of masses simultaneously through ALD cycle that can be mitigated by employing a different type of analyzer, such as a Time-of-Flight. Figure 3.1 shows a schematic representation of the mass spectrometer working mechanism. At first a little amount of gas species is captured by the

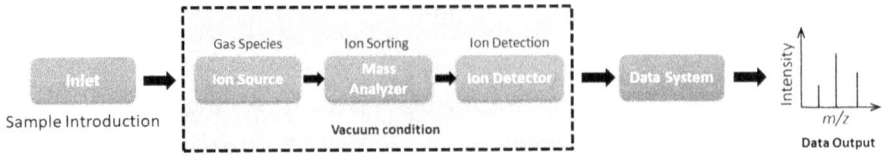

FIGURE 3.1 Schematic diagram of a mass spectrometer. The ionized gas species are classified and segregated based on their mass and charge. The processed ions are then analyzed by the detector. The obtained data are reflected on a graph.v

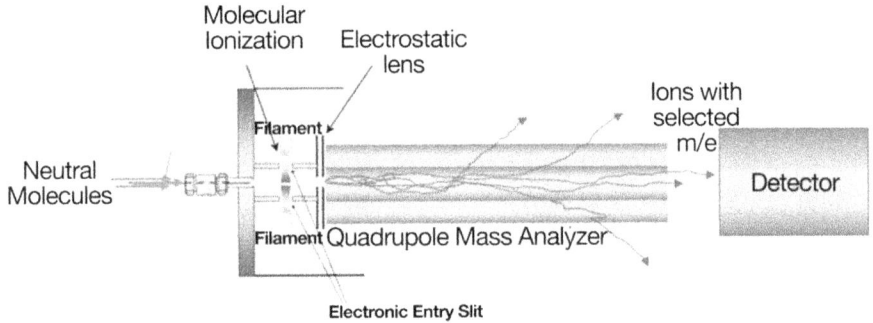

FIGURE 3.2 Functional components of a quadrupole mass spectrometer (Residual Gas Analyzer, RGA) [1].

orifice in QMS system. Then, electrons with an energy of around 70 eV impact with the gas molecules to ionize them as described in Equation 3.1:

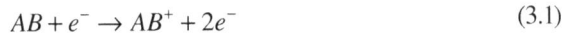

$$AB + e^- \rightarrow AB^+ + 2e^- \tag{3.1}$$

However, a large portion of the molecules will be split into smaller segments of the original molecule as indicated in Equation 3.2:

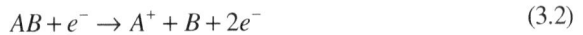

$$AB + e^- \rightarrow A^+ + B + 2e^- \tag{3.2}$$

Since ions are unstable and highly reactive, and a high vacuum condition is required for their creation and processed. The Residual Gas Analyzer (RGA) is an instrument that has a similar working mechanism as that of a quadrupole mass analyzer and consists of four parallel cylindrical rods, a mass spectrometer, an ion source, and a measurement part as illustrated in Figure 3.2. Contaminants can be found in vacuum chambers owing to the hydrocarbons or system leakages that backstream into the process chamber from the vacuum system. RGA is capable of analyzing small amounts of residual gas that remain in a vacuum chamber after pump down. When the thermo-electrons discharged from the high-temperature filament collide with the remaining gas, the generated ions which are separated by mass to charge ratio (m/z) accelerated and then converted as a mass spectrometer.

3.4.2 QUARTZ CRYSTAL MICROBALANCE (QCM)

QCM method has evolved as a powerful instrument for monitoring and measuring real mass change of particles vs time on the reaction surface throughout the ALD process. It can accurately detect deposition of less than a monolayer and provide information such as reaction occurrence and molecular interactions. QCM is a sensitive method that determines changes in a quartz crystal's resonance frequency in terms of mass variation per unit area with accuracy of nanogram to microgram. It is therefore appropriate for gas detection applications at low concentration levels. A quartz crystal's resonance frequency fluctuates in accordance with the amount of mass deposited on it and this variation in frequency depends on the different factors such as deposition region and process temperature. QCM system consists of quartz disk which is constructed by piezoelectric materials that can oscillate at a certain frequency when the proper voltage is applied through metal electrodes. QCM has been employed in vacuum and gas phase for 60 years, and around 40 years ago it was demonstrated that this approach was also suitable in liquid media.

Figure 3.3 depicts the fundamental working mechanism of QCM, which is based on the piezoelectric effect of quartz crystal. The oscillation or mechanical vibration is generated by the quartz crystal of the QCM when an alternating voltage is introduced to its two poles. The thickness of the crystal on the electrode system and the acoustic frequency have a significant impact on this resonance. The resonance frequency is influenced by mass variation on quartz crystal and then the generated output electric signal is transferred into the computer system for data interpretation.

Figure 3.4 represents an example of the generated QCM data during the ALD process of TMA and H_2O^*. In the pulsing of the metal precursor, a sharp mass

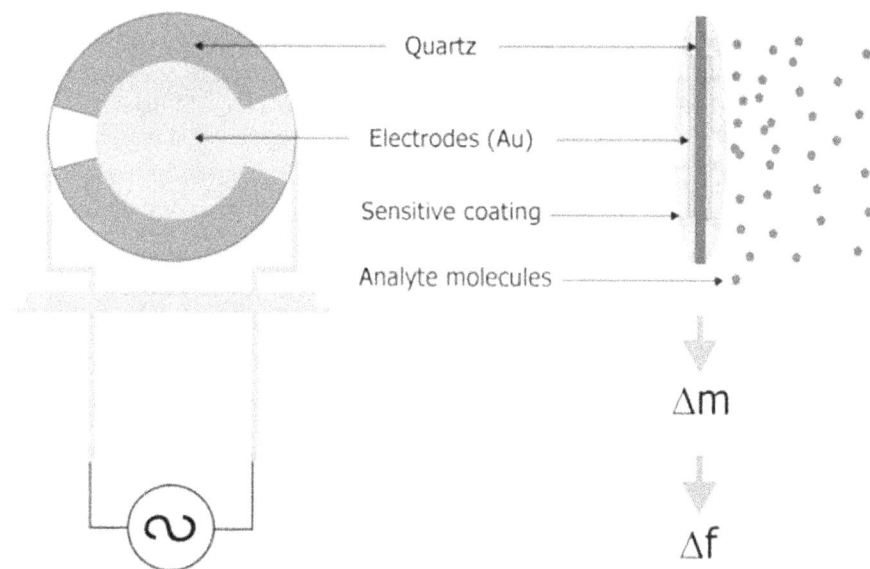

FIGURE 3.3 Schematic working mechanism of a quartz crystal microbalance (Quartz Crystal Microbalance, QCM) sensor [2].

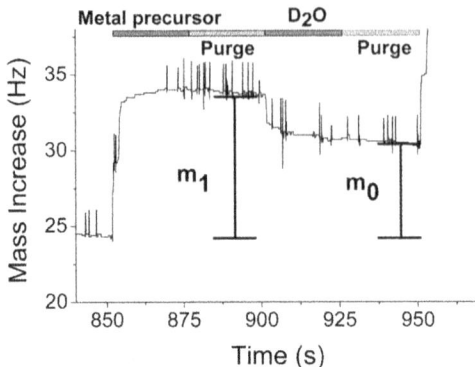

FIGURE 3.4   The example Quartz Crystal Microbalance (QCM) data result of one atomic layer deposition (ALD) cycle. m_1 represents the mass increase during pulsing of metal precursor. m_0 shows the overall mass change following the full cycle [3].

increase at initial times has been observed. The mass stabilization in plateau during the metal precursor pulse illustrates the self-limiting characteristic of the performed ALD process. The slight reduction in mass during the pulse process occurred in terms of dissociation of ligands or desorption of residual molecules from the surface. The subsequent oscillation in mass change is due to the introduction of the oxygen co-reactant (D_2O) pulse during the process of surface oxidation, along with the subsequent decomposition of ligands. An increment in mass change depends on the mass ratio between the ligand and the reactive surface species such as active oxygen and hydroxyl groups. As seen from Figure 3.4, typically in the majority of operations, the elimination of heavy ligands causes mass reduction during the oxygen precursor pulse. Furthermore, QCM data can be affected by some other factors such as abrupt temperature or pressure change and the nature of the precursor. For example, ozone pulsing induces a prominent error to the QCM data; therefore, further stabilization time is required for the purging of residual that is formed after ozone injection.

3.4.3   Spectroscopic Ellipsometry (SE)

SE is a powerful and most accurate tool for investigation of coated films which can measure thickness from sub Å to tens of microns. SE technique is nondestructive, contactless, surface sensitive, and noninvasive optical which relies on the alteration in the polarization state of light upon diagonally reflecting from a thin film surface. For a variety of application requirements, SE can be carried out in situ or ex situ, kinetic or in static scheme. SE is a critical instrument for the ALD thin film analytical research and can be employed for the characterization of film thickness, surface and interface roughness, composition, electrical conductivity, anisotropy, optical properties, doping concentration, and crystalline nature by area and depth. Figure 3.5a demonstrates the principle of SE method. As seen, the light source emits electromagnetic radiation, then it is linearly polarized by a polarizer, the compensator as a quarter wave plate or retarder can be supplied before exposure on sample

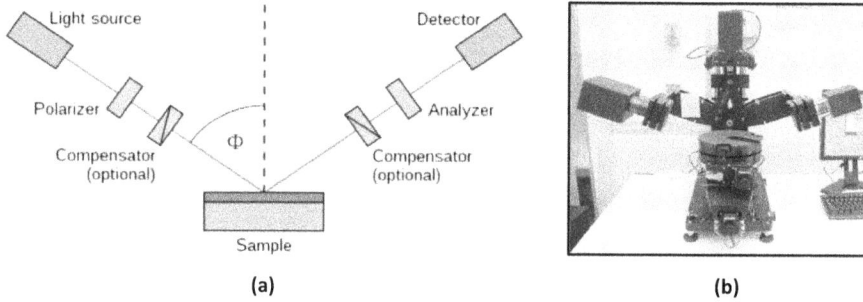

FIGURE 3.5 (a) Schematic principle of spectroscopic ellipsometry (SE) system. (b) SE instrument [4,5].

surface. Then, the reflection passes through the second compensator and analyzer and finally reaches the detector. The polarization-change is defined as an amplitude ratio (Ψ) and a phase difference (Δ). Phase modulator also can be used instead of compensator in the path of an incident light beam.

3.4.4   X-Ray Photoelectron Spectroscopy (XPS)

XPS is known as a nondestructive and surface-sensitive technique, which is an extensively employed method to determine the surface chemistry including elemental composition except hydrogen and helium, within the depths of the substrate when integrated with ion-beam etching or on the surface outermost at ~10 nm. In addition, the chemical state of the material is determined by measuring the bonding strength and the electronic state of elements with surrounding atoms. The value of atomic percentage for each element is determined in the base of relative sensitivity coefficient. Figure 3.6 demonstrates the basic setup of an XPS instrument. The fundamental working of XPS is in the base of photoelectric detection which is firstly reported by Hertz in 1887 [6] and further introduced by Einstein in 1905 and was later developed to surface characterization by K. Siegbahn's research team in 1960 at Uppsala university [7]. When electromagnetic radiation colloids with an atom and causes electrons to be emitted from the atom, which is known as an electron emission. The energy of impacting photons exceeds the binding energy of electrons in that substance, photoelectrons would be generated from that material. Energy and frequency are inversely related. Since atoms contain several orbitals at various energy levels, the binding energy of each electron influences the kinetic energy of an emitted electron. The obtained result as an XPS spectrum shows a variety of released electrons with various kinetic and binding energies. Equation 3.3 represents these relationships:

$$E_{binding} = E_{photon} - E_{kinetic} - \phi \qquad (3.3)$$

where $E_{binding}$ is the binding energy of a given electron, E_{photon} is the incident photon energy of the used X-ray, $E_{kinetic}$ is the kinetic energy of the photoelectron determined

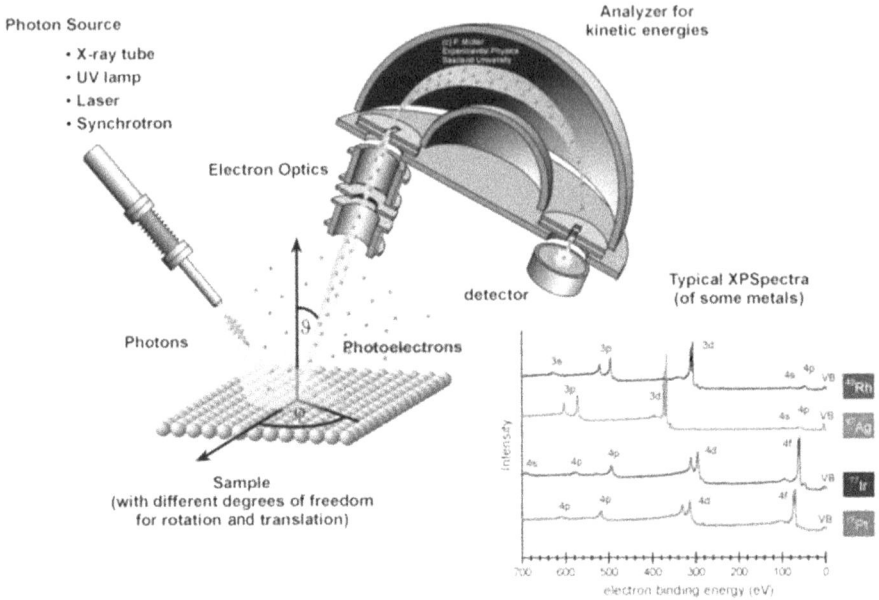

FIGURE 3.6   Setup of an X-ray photoelectron spectroscopy (X-ray photoelectron spectroscopy, XPS) instrument. Four different metals' XPS (intensity vs. binding energy) are displayed together with the distribution of core-level photoemission unique to each element [9].

by the detector of XPS tool, and φ is the work function which refers to the minimum energy at which an electron must be emitted from an atom—the energy difference between the vacuum energy (E_v) level and the Fermi (E_f) level of a solid.

XPS is operated in the ultra-high vacuum $(p < 10^{-7}$ Pa) condition, which occasionally overestimates the real state of elements. XPS is utilized to analyze the surfaces of a wide range of materials, including coated thin films, semiconductors, metal alloys, catalysts, organic compounds, and biomaterials. In addition, surface-mediated processes such as adsorption, dissolution/precipitation, catalysis, corrosion, redox, and evaporation/deposition type reactions are studied using XPS. Since the ALD deposition under vacuum condition may disrupt the XPS setup, XPS analyses do not even perform directly in situ. To overcome this issue between each cycle or half cycle, the sample is transferred from the reaction chamber to the XPS chamber for characterization without disturbing the vacuum. In addition to mitigating substrate contamination through the avoidance of air exposure, it is essential to maintain ultra-high vacuum (UHV) conditions within the XPS chamber to facilitate accurate data acquisition. An alteration in the appearance of an XPS oscillation demonstrates a change in the atom's surroundings state which provides accurate information regarding the quality of initial growth rate on substrate and creation of interface layer [8].

In order to probe the specific element transfer such as dopant through in-depth profile of the bulk or thick films, the ion beam etching analysis (IBA) can be applied for quantitative composition measurement in ex situ XPS by determining the successive XPS spectra versus etching time. Thus, ion beam analysis integrated with

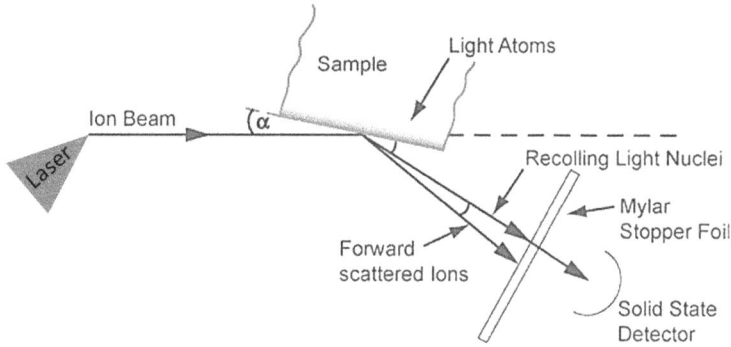

FIGURE 3.7 Schematic basic principle of ion beam etching analysis (IBA).

XPS to further supply the thin film characterization. IBA technique is very sensitive and is capable of scanning few nanometers to few micrometers in depth of material. However, for determining the mass density and hydrogen atoms' presence in composition of thin films, the use of other integrated methods, such as elastic backscattering (EBS), elastic recoil detection (ERD), and Rutherford backscattering spectrometry (RBS), are required to verify the XPS results. Figure 3.7 elucidates the basic principle of IBA in a high-energy beam light (~1 MeV/amu). The emitted heavy ion which is typically (He^+) ions should have a larger mass than the target atom located at the sample. This hard sphere collision should be elastic enough to prevent energy migration between the incident atoms. The energy distribution and efficiency of the reflected He+ ions are measured at a certain angle. RBS can be applied for the films up to 1 μm thickness. This reflection is appeared for the atoms which are heavier than the emitted ions. So, the RBS method cannot detect the hydrogen atom. However, for probing hydrogen atoms and its isotopes, the ERD method can be performed. In addition, the EBS method can be used to measure the elements more accurately in layer. Unlike RBS, this method uses high-energy protons as projectiles rather than He^+ ion, which has the benefit of enhancing the cross sections for light elements. The obtained result data of IBA are presented by the areal density (at·cm^{-2}) of the atoms in the deposited film and can be verified by simulation data.

3.4.5 SCANNING ELECTRON MICROSCOPY (SEM)

SEM is broadly used and is a versatile method for characterizing the morphology and composition of solid surfaces' texture. SEM generates high-resolution images of surface by using the focused beam of electrons with high kinetic energy which are emitted to atoms of surface in different depths and lead to generating multiple signals. This signal can be found in various forms such as backscattered, second scattered electrons, characteristic X-ray photons, lights, and current of sample which are utilized to learn about the physical properties, topography, and composition of the surface. Each of these electrons or photons can be detected using a different kind of detector. SEM is capable of achieving resolutions greater than 1 nm. SEM by imaging the cross section of thin film can determine its thickness. High vacuum

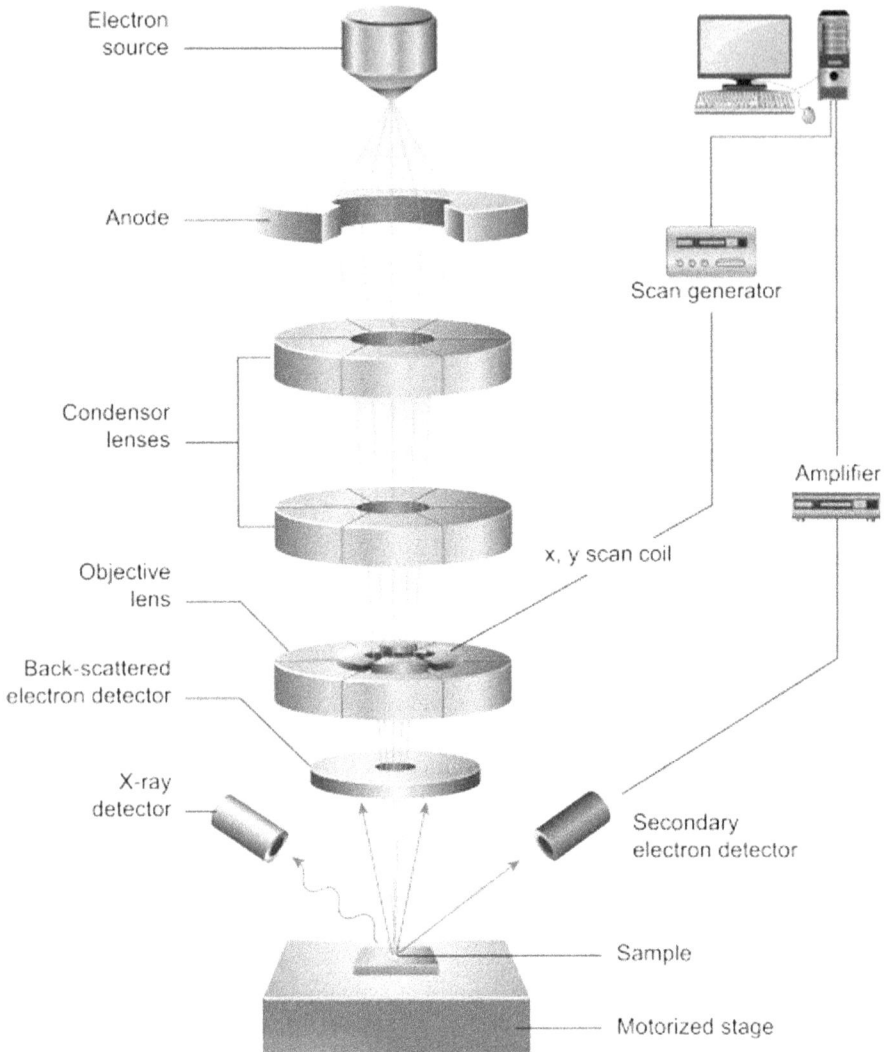

FIGURE 3.8   Major components of a scanning electron microscopy (SEM) instrument [10].

condition is essential for SEM operation to guaranty high-quality imaging, protecting electron source from noise and vibration, and avoid interaction between electron beams and atoms of surface. Schematic component of SEM tool, including electron source, anode, condenser lens, scanning coils, and objective lens, is demonstrated in Figure 3.8. The emission of high-energy electrons from an electron source is followed by their absorption by the positively charged anode plate, resulting in the formation of a beam. The condenser lens or apertures regulate the beam's size and counts the electrons in the beam. The image resolution is determined by the beam's size. The scanning coil inspects the surface of the specimen in raster pattern by reflecting the beam in the X and Y directions. Finally, the electromagnetic objective lens focuses the beam to a very tiny spot in the sample.

3.4.6 ENERGY-DISPERSIVE X-RAY SPECTROSCOPY (EDS)

EDS is a qualitative analytical method referred to as energy dispersive X-ray (EDX) spectroscopy implemented for the elemental determination and chemical analysis of the materials. The EDS is typically integrated with the SEM instrument. When a high-intensity electron beam interacts with the electron structure of a material, it induces excitation of electrons and subsequently leads to the emission of x-rays from core-shell electrons. These x-rays possess energies that are determined by the orbital configuration of each element present in the sample. Although the EDS is a powerful method, it has some drawbacks that hinder its utility. The first, EDS is interaction volume, or the quantity of material that is activated by the incident electrons, and is a downside of employing EDS on thin films. Second, EDS is not accurate enough to identify too small proportions of trace elements with concentration lower than 0.01 wt. %. Third, EDS cannot measure the elements with low atomic number such as Hydrogen and Helium elements since both only have a $n = 1$ shell, and no core electrons need to be removed for generating X-ray emission. Meanwhile, the typical outcome result of an EDS is in the form of appearance of the peaks in the spectrum, which indicates the elements, whilst the signal's strength reflects the element's concentration. Moreover, the image of elemental mapping can be generated by EDS.

3.4.7 TRANSMISSION ELECTRON MICROSCOPY (TEM)

TEM is a powerful microscopy tool for imaging of materials through a beam of electrons as an ejection source. In the fields of physical, chemical, and biological research, TEM is a crucial analytical technique. Typically, the material is an ultra-thin sample with the thickness of less than 100 nm; thin films and many other types of specimens can be examined using electron microscopy which makes it multi-technique. TEM provides high-resolution image by means of interaction between electrons emission and specimen as beam passes through the sample and then the image magnified. TEM is a quantitative technique to verify the growth of films, shape, morphology, composition, size, and diffusion of the nanoparticles. TEM provides much higher quality image compared to light microscope tools and SEM , as the wavelength of electrons is substantially shorter than that of light. The major difference between TEM and SEM is in the method of image generation. TEM provides the image by transmitted electrons and SEM through detection of reflected electrons. However, better spatial resolution of (0.0001 μm) which is two times higher than SEM and the great capacity of analytical measurements are TEM's benefits over SEM. Although TEM has advantages, there are also some limitations including sample preparation, which is difficult and time-consuming, and the proper preparation process affects the quality of generated images. The samples containing lithium could not be subjected to TEM examination and destructive analysis is probable. However, to overcome this issue using higher acceleration

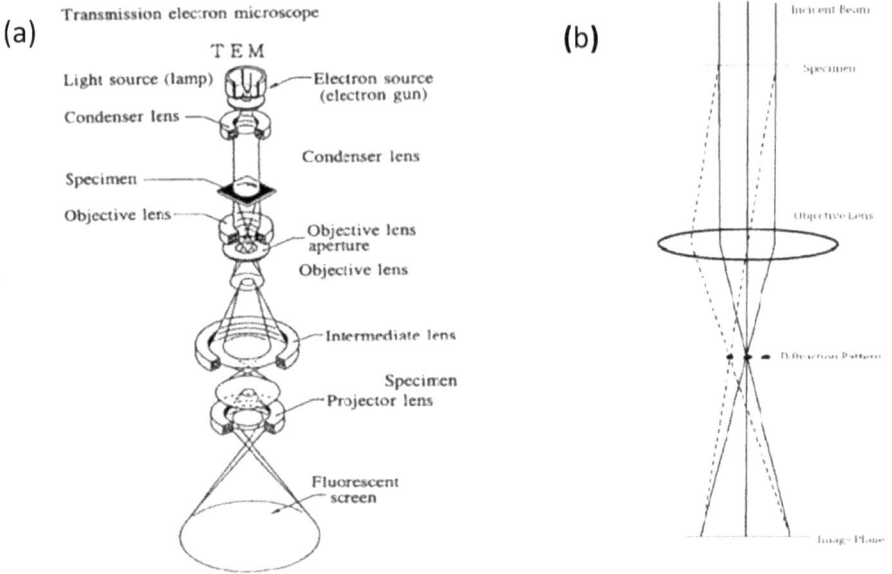

FIGURE 3.9 (a) Typical layout of a Transmission Electron Microscopy (TEM) describing the path of an electron beam in a TEM. (b) A ray diagram for the diffraction mechanism in TEM [11].

voltage could be helpful. The general layout of a TEM and Ray diagram is illustrated in Figure 3.9.

3.4.8   X-Ray Diffraction (XRD)

XRD is a broadly used technique for analyzing the crystallinity and periodical structures of thin films and materials such as metals, polycrystal ceramics, minerals, semiconductors, and powders, with their component molecules, atoms, and ions aligned in a regular pattern. XRD is a useful technique for determining the thickness and atomic structures, and internal stress and strain of deposited thin films. When using XRD, a material is exposed to emitted X-rays, as waves of electromagnetic radiation in nanometer and the intensity and scattering angles of the X-rays that backscattered from the atoms are then measured. This is schematically indicated in Figure 3.10. The components of a crystalline substance are represented by dots. The incident X-ray beam scatters at various planes of the substrate. The interferences phenomena occurred between the different diffracted wavelengths, which are commonly destructive. However, a few constructive interferences occurred in specific directions, which are defined by the famous Bragg's law [12]:

$$n\lambda = 2d\sin\theta \tag{3.4}$$

where λ is the beam wavelength, n is any integer, d is the spacing between diffracting planes, and θ is the incident angle. By this law, the atomic spacing, density of

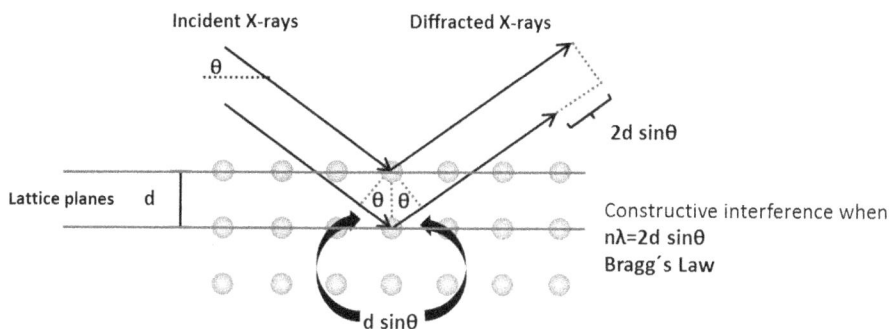

FIGURE 3.10   Schematic representation of Bragg's Law reflection.

electrons, and crystal structure can be measured. It is possible to infer several details from the electron density, including the average location of the atoms inside the crystal, the type and length of chemical bonds, crystallographic disorder, etc. The measurement's output of XRD is generated as a diffractogram. This is a graph that shows the relationship between the X-ray intensity on the y-axis and the angle 2θ (where 2θ is the angle between the incident and diffracted beams) on the x-axis. The obtained diffractogram provides specific qualitative crystal information, like a fingerprint, that can be used to identify any type of material. In the case of impurity of the sample, the relative quantities of the various components can also be determined. In addition, XRD can determine the orientation and unit cell lattice dimension of crystal films also in a nonambient condition which can verify DFT calculation results.

3.4.9   X-Ray Fluorescence Spectroscopy (XRF)

XRF is in situ and a nondestructive analytical technique, which can quantitatively and qualitatively determine the elemental composition of ALD thin film and bulk material. During an ALD cycle, XRF can accurately monitor the crystallinity and the number of deposited atoms by means of high intensity synchrotron, which illuminates the secondary X-rays on thin film. In addition, this method is critical for studying the effect of reaction temperature on crystalline film formation and nucleation process. Figure 3.11 represents the fundamentals of the XRF process. XRF works based on the atoms behavior which are excited and interacted with radiation. The substrate is ionized and excited by high intensity, and short wavelength emission of X-rays. If the radiation's energy is high enough to shift an inner electron with its stable position, an outer electron alters the missing inner electron. Then, due to the inner electron orbital's lower binding energy compared to the outer one, the energy as fluorescent radiation is generated. This obtained radiation consists of data about a feature of an element's transition between a certain set of electron orbitals. In addition, each element of the substrate generates the specific fluorescence X-ray (fingerprint) which makes XRF an unique method for determining of material composition. Therefore, XRF can be a very useful method in ALD for investigating the initial growth or nucleation phenomenon on the nanoporous and planar materials. However, XRF has few limitations including identification

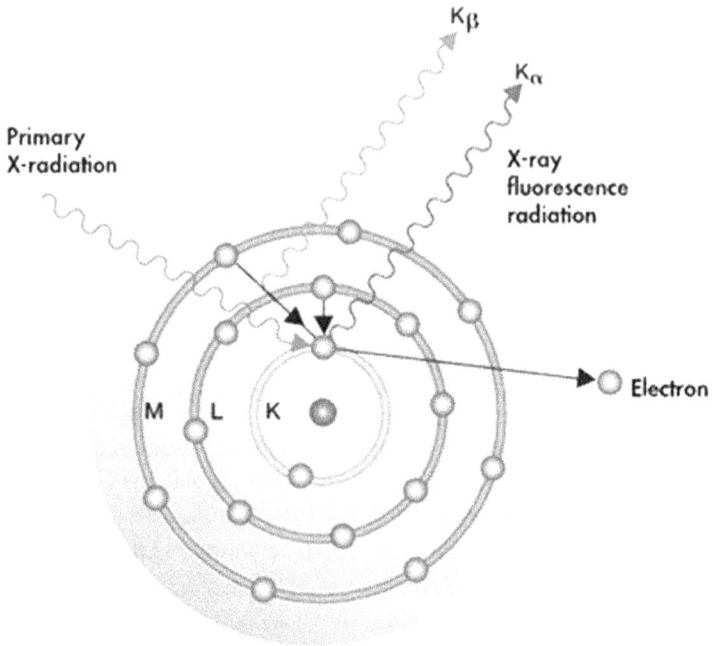

FIGURE 3.11   Principle of X-ray fluorescence spectroscopy (XRF) [13].

of light atoms such as Hydrogen, Nitrogen, Carbon, and Oxygen. Second, there is limited access to XRF facilities and they require very expensive instruments that limit further use of this technique.

3.4.10 ATOMIC FORCE MICROSCOPY (AFM)

AFM is a powerful analytical tool and a form of scanning probe microscope (SPM) technique. It provides atomic resolution images of nano-/microstructured thin films' coating and many other inorganic and organic materials. AFM is typically applied in chemistry and physics to probe the formation of atoms over surface. AFM is essential for the fabrication, improvement, and inspection of thin film growth processes as well as for simplifying design routes to obtain required functional attributes including mechanical, magnetic, and electric (conductivity) properties. However, AFM can be used for three main applications including topographic imaging, force calculation, and manipulation. The mechanical properties such as Young's modulus, stiffness, strength, scratch resistance, friction coefficient, adhesion, elastic, and inelastic displacement of thin films can be measured by AFM and these properties can be very much influenced by interatomic potentials, grain topography, surface energy, and crystallography of thin films. It provides qualitative and quantitative data on various physical properties of thin film such as roughness, morphology, thickness, spatial distribution, and surface area. In addition, AFM can measure many types of forces including chemical bonding,

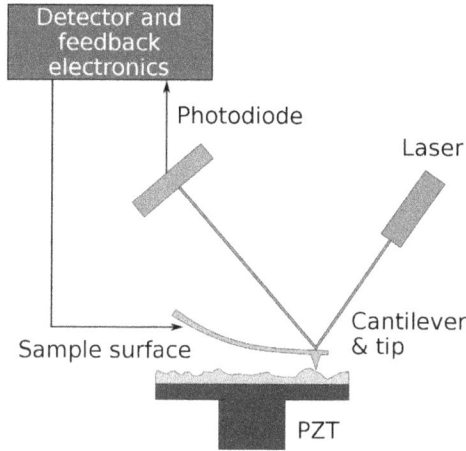

FIGURE 3.12   Simplified schematic fundamental of atomic force microscopy (AFM) [15].

electrostatic, Van der Waals, mechanical, capillary, magnetic, etc. AFM has also been shown to be an effective method for analyzing the nanomechanical characteristics of thin coatings and employed in nanotribological research for the fabrication of microelectromechanical systems (MEMS). AFM can also integrate with infrared spectroscope (AFM-IR) to provide super resolution characterization of nanocomposites and polymeric surfaces. AFM is a developed conventional scanning tunneling microscope (STM) technique and can be applied to a variety of substrates and does not have the limitations of the STM, which is only applicable to conductive or semiconducting materials.

Figure 3.12 demonstrates the simplified basic principles of AFM. The tool scans the sample surface in raster pattern using a cantilever that contains a sharp tip with a nanometer scale at the end. The cantilever typically is silicone and bends as the tip comes into touch with the sample, and the cantilever is deflected by the generated force between the tip and sample (Hook's law) [14], leading to regulating the amount of laser light which is reflected into the detector to convert to electrical signal. The height of the cantilever is then modified in accordance with the response signal, allowing the topography of the sample to be determined.

3.4.11   THERMOGRAVIMETRIC ANALYSIS (TGA)

TGA or thermography analysis is a versatile and useful experimental technique for investigating the precursor of deposition systems such as ALD and Chemical vapor deposition (CVD) methods, with their kinetically controlled process strongly depending on the volatility, reactivity, and thermal stability of the precursors that verify the desired film quality. TGA can help in designing successful precursors of ALD by determining the precursor stability and decomposition during the heating process from room temperature to over 1,000°C. TGA works in the base of continuous recording of mass fluctuation of precursor molecules as a function of temperature change or time corresponds to the chemical reaction in the inert atmosphere,

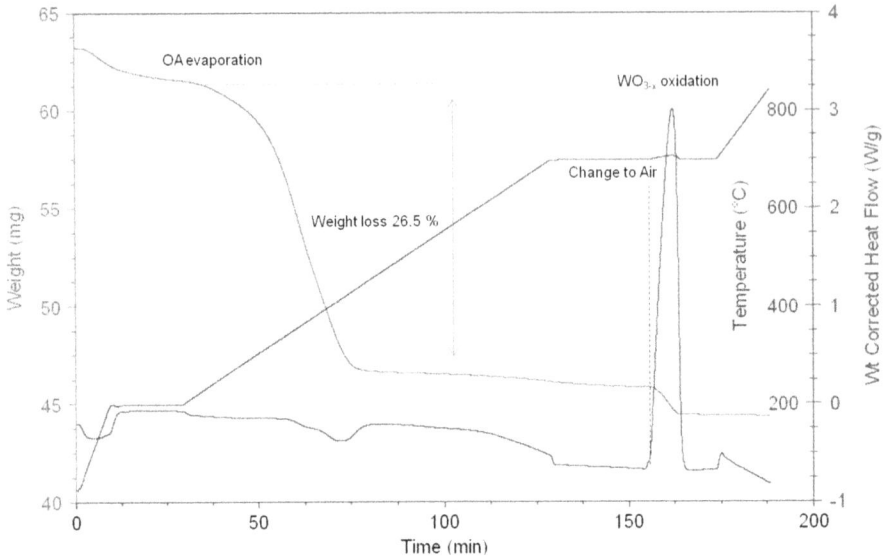

FIGURE 3.13   The example of data obtained by the thermogravimetric analysis (TGA) technique [16].

as most of ALD precursors are not stable in an ambitious condition. However, other different atmospheric conditions such as oxidizing and reactivity can be accommodated along with the measurement. Furthermore, the heat capacity and phase change can be measured by TGA. The heat subjection commonly can be applied in three variations:

1. **Dynamic TGA** records the mass change during continuous heat treatment and reveals the decomposition temperature and amount of gas reduction.
2. **Static TGA** records the mass change at the fixed temperature which determines more information of precursor decomposition at specific temperature and explores the precursor durability at a certain temperature versus time.
3. **Quasistatic TGA** is employed to examine the behavior of a precursor substance that can undergo decomposition under varying conditions. This analysis includes subjecting the precursor to heat treatment at multiple temperature intervals, maintaining each interval for a specific duration until a mass stabilization is attained.

The choice of the heat treatment method depends on the type of the demanded information about the precursor. Nevertheless, TGA cannot be a sufficient method for characterization of ALD precursors; therefore, chemical testing or differential calorimetry must be used in conjunction with TGA to confirm the identification of residuals. The obtained result data of a TGA (Figure 3.13) is plotted as a TGA curve, which is the displayed amount of mass or percentage mass change versus temperature or time. In addition, another complementary curve known as the differential TGA (DTG) is plotted. It represents the rate during mass change and facilitates easy identification of the point at which various mass changes happen. This curve is typically significant for multiple decompositions that occur simultaneously.

3.5 CONCLUSION

This chapter has provided an overview of the most utilized characterization techniques for ALD thin film samples and surface technology and is intended to guide the readers for implementing appropriate analytical methods by understanding their principles for various applications. Moreover, their capabilities in investigating chemical, structural, physical, and electronic properties of ALD thin films and coating in situ and ex situ relying on electron, ion, and photon approaches were highlighted. The fundamental working mechanism of the characterization techniques was discussed. Their advantages and disadvantages and their capability to integrate with other methods as a multi-technique to enhance the sensitivity, resolution, and application were discussed. Instruments with improved sensitivity and the capability to function in realistic conditions (in situ) and combination of tools are being developed by researchers and industries to meet the evolving and current research demands. These multidisciplinary engagements are anticipated to have a big influence on material characterization and support the ongoing dynamic progress of technologies that heavily depend on thin films and coating technology.

REFERENCES

[1] Residual Gas Analysis, n.d. https://www.mks.com/n/residual-gas-analysis (accessed April 15, 2023).
[2] A. Sabdo Yuwono, and P. Schulze Lammers, "Odor pollution in the environment and the detection instrumentation," *CIGR E-Journal*, 2004. https://ecommons.cornell.edu/handle/1813/10399 (accessed April 15, 2023).
[3] A. Rahtu, T. Alaranta, and M. Ritala, "In situ quartz crystal microbalance and quadrupole mass spectrometry studies of atomic layer deposition of aluminum oxide from trimethylaluminum and water," *Langmuir*, vol. 17, pp. 6506–6509, 2001. doi: https://doi.org/10.1021/LA010103A.
[4] Ellipsometry - Wikipedia, n.d. https://en.wikipedia.org/wiki/Ellipsometry (accessed April 25, 2023).
[5] NanoFab Tool: J. A. Woollam M-2000 DI Spectroscopic Ellipsometer. | NIST, n.d. https://www.nist.gov/laboratories/tools-instruments/nanofab-tool-j-woollam-m-2000-di-spectroscopic-ellipsometer (accessed April 25, 2023).
[6] D. R. Baer, and S. Thevuthasan, "Characterization of thin films and coatings," in *Handbook of Deposition Technologies for Films and Coatings*, pp. 749–864, 2010. doi: https://doi.org/10.1016/B978-0-8155-2031-3.00016-8.
[7] A. Taylor, "Practical surface analysis," in *Auger and X-Ray Photoelectron Spectroscopy*, 2nd ed., vol. I, D. Briggs, and M. P. Seah, Eds., New York: John Wiley, p. 657, 1990. doi: https://doi.org/10.1002/JCTB.280530219.
[8] D. N. G. Krishna, and J. Philip, "Review on surface-characterization applications of X-ray photoelectron spectroscopy (XPS): Recent developments and challenges," *Applied Surface Science Advances*, vol. 12, p. 100332, 2022. doi: https://doi.org/10.1016/J.APSADV.2022.100332.
[9] A. G. Jacobs, n.d. https://jacobs.physik.uni-saarland.de/home/index.php?page=steinbeiss/home_cms_steinbeissdet3-1&navi=service (accessed April 15, 2023).

[10] A Brief Introduction to SEM (Scanning Electron Microscopy). | SciMed, n.d. https:// www.scimed.co.uk/education/sem-scanning-electron-microscopy/ (accessed April 25, 2023).

[11] Transmission Electron Microscopy (TEM), n.d. https://warwick.ac.uk/fac/sci/physics/ current/postgraduate/regs/mpagswarwick/ex5/techniques/structural/tem/ (accessed April 25, 2023).

[12] D. D. Le Pevelen, "Small molecule x-ray crystallography, theory and workflow," *Encyclopedia of Spectroscopy and Spectrometry*, pp. 2559–2576, 2010. doi: https://doi. org/10.1016/B978-0-12-374413-5.00359-6.

[13] XRF: X-Ray Fluorescence Spectroscopy | Hi Rel Parts | Alter Technology, n.d. https:// wpo-altertechnology.com/xrf-x-ray-fluorescence-spectroscopy-hi-rel-parts/ (accessed April 25, 2023).

[14] F. Dell''Isola, G. Sciarra, and S. Vidoli, "Generalized Hooke''s law for isotropic second gradient materials," *Proceedings of the Royal Society A: Mathematical, Physical and Engineering Sciences*, vol. 465, pp. 2177–2196, 2009. doi: https://doi.org/10.1098/ RSPA.2008.0530.

[15] Atomic Force Microscopy - Wikipedia, n.d. https://en.wikipedia.org/wiki/Atomic_ force_microscopy (accessed April 25, 2023).

[16] 2.8: Thermal Analysis - Chemistry LibreTexts, n.d. https://chem.libretexts.org/ Bookshelves/Analytical_Chemistry/Physical_Methods_in_Chemistry_and_Nano_ Science_(Barron)/02%3A_Physical_and_Thermal_Analysis/2.08%3A_Thermal_ Analysis#Thermogravimetric_Analysis_(TGA).2FDifferential_Scanning_ Calorimetry_(DSC) (accessed April 15, 2023).

4 Industry 4.0, Manufacturing Sector and Thin Film Technology

4.1 INTRODUCTION

Artificial intelligence (AI) and other Industry 4.0 technologies have been a game changer in upgrading the critical metrics of manufacturing industries. The interconnected dynamics of Industry 4.0, manufacturing and thin film deposition is critical to the future trajectories of companies that are currently faced with new manufacturing hurdles as a result of social, economic, and technical advances. Industry 4.0 utilization provides physical and virtual frameworks to businesses, enabling collaboration and swift adaption along the entire value chain [1]. Its integration has assisted business areas that generate value, and the entire value chain by adopting the idea of digital production, or "smart factory" [2]. The sector has been revolutionized with potential benefits of the AI such as improved operation efficiency, enhanced quality, minimization of downtimes, and cost effectiveness. In this chapter, we shall examine the key enabling technologies of the Industry 4.0, and their potential impacts on thin film deposition. Further to this, an overview of the pros and cons of Industry 4.0 in thin film are presented while issues in its adoption in thin film and manufacturing sector are addressed. According to Schwab [3], unlike previous industrial revolutions, Industry 4.0 is increasing exponentially as opposed to linearly, changing not only "what" and "how" we do things but also "who" we are. Significant transformation is ongoing in today's industries encompassing novel business models and shifts in production and consumption [4].

4.2 A BRIEF OVERVIEW OF INDUSTRY 4.0

A swift shift from the old nomadic lifestyle was significantly influenced by agricultural and industrial revolution [5]. To better comprehend the trend and emergence of the present version Industry 4.0, it is essential to obtain a grasp of these earlier versions vis-à-vis "1.0", "2.0" and "3.0". The generation of mechanical energy in steam engines was the most significant development in the industry around the 18th century. This heralded the advent of Industry 1.0, which resulted in significant improvements in human living conditions and the economy. The increased demand for raw materials was a prominent feature of this expansion. The Industry 2.0 was associated with the second revolution in the industrial growth in the 19th century which involves the generation of electricity and massive production instead of individual hand-made tools and products. The birth of steam engine in the first revolution

DOI: 10.1201/9781003346234-5

73

enhanced the railway transportation development which facilitated easy movement of raw materials from place to place, factory to factory, thus boosting production speed and capacity.

The Industry 3.0 emerged in the early 20th century, and characterized by electricity generation. The later part of the century was delineated by automations and expansion in information and communication technology. A new improved computerized production line emerged with automated and programmable devices, equipment, and machines. In some developed regions of the world, their factories had begun the use of automated processes with robots [6]. In the era of Industry 4.0 in the 21st century, the production sector now looks very different thanks to ICT advancements like AI, the internet, and big data. All facets of life now employ digital technologies. Without the involvement of humans, production in the industries has become editable [5].

According to its high-tech strategy 2020, the German Federal Government first introduced "Industry 4.0" at its Hannover Fair in 2011 with a goal to work at a higher level of automation, achieving a higher level of operational productivity and efficiency by connecting the "real" and "virtual" worlds [7]. The trajectories of the manufacturing sector were re-invented through a considerable use of these technologies with regards to this project [8]. Industry 4.0 serves as the framework for the integration of tangible items, machinery, systems, and processes over a networked environment [1,9]. Industry 4.0 is transforming the global industrial environment by integrating digital technology into how businesses create, enhance, and distribute their goods. Industry 4.0 introduces nine cutting-edge technologies, which enables and drives the automation and digitalization trend in Industry 4.0 [10]. As shown in Figure 4.1, these technologies include:

- Internet of things (IoT)
- Cloud computing
- Big data and analytics
- Additive manufacturing
- Autonomous robotics
- Cybersecurity

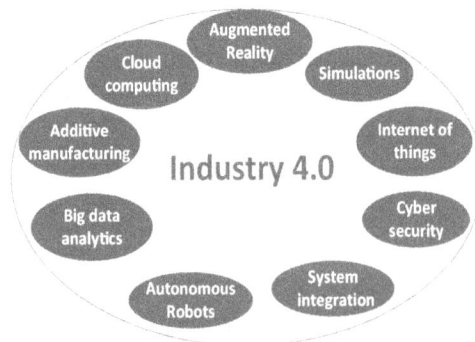

FIGURE 4.1   Enabling technologies of Industry 4.0.

- Augmented reality
- Simulation technologies
- System integration

4.3 FEATURES OF INDUSTRY 4.0

4.3.1 INTERCONNECTIVITY AND INTEROPERABILITY

Interconnectivity is one of Industry 4.0's core characteristics. Interconnectivity entails the real-time ability to share information amongst several equipment, sensors, and machine components. Better visibility and control throughout the whole manufacturing process, from the supply chain to production and logistics, are made possible by this interconnectedness. This movement links computers not just to distant users but also to one another [11].

4.3.2 ADVANCED AUTOMATION AND EXPONENTIAL TECHNOLOGY

Another important aspect of Industry 4.0 is intelligent automation. A manufacturing system which integrates these modern AI techniques would be able to achieve intelligent process automation. Smart automation paves the way for the creation of autonomous equipment and systems that can improve their performance over time. This could boost manufacturing operations' effectiveness, precision, and productivity. The businesses that benefit the most from digital transformation spend up to 60% more on machine learning than their rivals. Organizations can adopt a mindset of continuous improvement when they concentrate on technologies that offer long-term optimization advantages. As a result, surroundings for Industry 4.0 are constantly changing.

4.3.3 DATA-DRIVEN DECISION AND POLICY MAKING

Big data and analytics are also key components of Industry 4.0. These businesses leverage historical and real-time information on daily activities, sales, and surrounding areas to make better decisions. These insights assist businesses in better anticipating the future, but they are not possible without cutting-edge technologies like the IOT and AI [11]. One of the most typical applications for this data-driven decision-making is sales forecasting. This technique builds virtual replicas of factory floors using data from the floors to simulate what-if scenarios that show how various adjustments could enhance operations

4.3.4 CUSTOMIZATION

Exciting benefits of 3D printing and digital manufacturing have been explored in customization of goods and services for addressing specific needs of different customers towards a personalized and flexible production process. This can result in significant cost reductions and improved manufacturing efficiency. Manufacturers may develop virtual versions of their products and test them under various scenarios thanks to digital simulation, allowing them to spot and fix potential problems before the product is ever built. This lowers the possibility of defects and raises the standard of the product.

4.3.5 TRANSPARENCY OF INFORMATION

Information transparency is one of Industry 4.0's other major components. This speaks to the capacity to gather, examine, and communicate data among various systems and procedures. As a result, manufacturers can gain a deeper understanding of their business operations, including their production capacity, inventory holdings, and resource usage. Manufacturers can improve their operations for greater effectiveness and productivity by using this information to make better-informed decisions. Real-time data-sharing options through cloud computing are necessary to achieve this level of transparency. In addition, it highlights the necessity for trustworthy cybersecurity in Industry 4.0 operations as the risk of data breaches increases with increased information sharing. But, with adequate security, this transparency produces value chains that are much more effective and adaptable [11].

4.4 INDUSTRY 4.0 IN MANUFACTURING AND SERVICE SECTOR

Upon the introduction of the high-tech strategy by the German government, other nations have adopted the paradigm by launching the USA's "Smart Manufacturing," "Made in China 2025," "Future of Manufacturing" in the United Kingdom, "Smart Advanced Manufacturing and Rapid Transformation Hub (SAMARTH)—Udyog Bharat 4.0" in India, among others [12,13]. Through contemporary technologies like cloud computing, the Internet of Things (IoT), and cyber-physical systems, Industry 4.0 aims to create intelligent organizations [14]. The advantages of implementing Industry 4.0 should outweigh the efforts in both nonfinancial and financial elements in order to persuade traditional firms of the value of such a large scale [15].

When manufacturing businesses prioritize the adoption of the industrial 4.0 technologies, they can concentrate on various needs they may have. However, recent findings have demonstrated that the benefits anticipated by those technologies for industrial performance vary by industry, and businesses should consider systemically using such technologies to reach a higher Industry 4.0 maturity level [16]. One of the key advantages of Industry 4.0 in production is the ability to customize products to particularly address the specific demands of the different customers using 3D printing, intelligent simulations, and modeling. By doing so, products and service are created, as well as virtual models and prototypes. Supply chain management has improved as a result of Industry 4.0 technology in the service industry. For instance, real-time tracking and monitoring of product location and condition with IoT sensors enables more effective logistics and inventory management.

Industry 4.0 has empowered the development of novel models and services for businesses in the service industry by utilizing the Industry 4.0 enabling technologies. For example, using machine learning algorithms and AI has made it less difficult to create specialized services, such as customer-centered specifications and recommendations. Consequently, businesses are now able to provide their clients with experiences that are more distinctive and exciting. In order to process data from intelligent and dispersed system interaction from the perspectives of production and service management, Industry 4.0 (4IR) aims at creating a system that is able to communicate intelligently between machine and humans [16]. 4IR promotes independent interoperability, responsiveness, adaptability, decision-making, effectiveness, or reductions in expenses in addition to other features [17]. The adoption of these technologies which are the building blocks of 4IR in manufacturing and service sector are briefly defined in Table 4.1.

A new type of production that combines sensors, computer platforms, communication technology, control, simulation, data-intensive modeling, and predictive engineering with today's and tomorrow's manufacturing assets is termed "smart manufacturing" [13]. The major building block technologies of the 4IR are driving principles in cyber-physical systems that are used in this system. Once put into practice, these ideas and innovations would make smart manufacturing the defining feature of the upcoming industrial revolution. The ideas that were mostly established in the field of computing have served as inspiration for smart manufacturing. Figure 4.2 depicts a broad outline of a smart manufacturing. The manufacturing equipment layer and the cyber layer are the two fundamental layers of the idea in Figure 4.2.

They are connected through an interface. Manufacturing equipment has its own intelligence, but the cyber layer provides system-wide intelligence. Smart manufacturing will frequently use dynamic predictive models, and virtual and augmented reality. The expanding volume of data in smart manufacturing will inevitably provide opportunities for the distribution of value derived from the data. Data-driven modeling techniques will become more popular as they make it possible to incorporate parameters from other domains (such as products, processes, and logistics) into models that would be challenging to create using the conventional technique.

Smart manufacturing has its own identity, which is defined in six pillars that are described below. The ultimate pillars could be explicitly established in a variety of ways, such as by using text and data mining algorithms to cluster academic articles, industry reports, and information about emerging technologies. According to Kusiak et al. [13], some of the pillars of smart manufacturing are as follows:

- Data
- Materials
- Predictive engineering
- Sustainability
- Resource sharing and networking
- Manufacturing technology and processes

4.5 ADVANTAGES OF INDUSTRY 4.0 IN MANUFACTURING INDUSTRIES

Some of the benefits of the implementation of Industry 4.0 in the manufacturing and service sector are highlighted as follows.

TABLE 4.1
Description of Industry 4.0 Enabling Technologies in Manufacturing [7,16]

S/N	4IR Technology	Description
1	Big data and analytics	It details the acquisition, storage, distribution, management, and analysis of enormous volumes of high velocity, complex, and variable data requiring advanced procedures and techniques. Manufacturing businesses can now make data-driven decisions that enhance operational effectiveness, product quality, and customer pleasure thanks to big data and analytics, which are revolutionizing the sector. Predictive maintenance, quality control, supply chain, and production optimization are other benefits of this technology to this manufacturing industry.
2	System integration	Industry 4.0 is driven to function at its best by the fusion of diverse computing systems and software packages in order to form a larger system. System integration adds value to a system by combining subsystems and software applications to provide new functionality. Information technology supports systems integration in product development and manufacturing for information sharing. System integration has many advantages in the context of manufacturing sector, such as enhanced flexibility, better decision, and reduced cost.
3	Cloud computing/ service	Cloud computing describes a broad network access, on demand self-service, and resource pooling with quick elasticity and measured service. Utilizing cloud computing in products to increase their functionality and provide associated services. The manufacturing sector is changing because of cloud-based services, which allow businesses to optimize operations, store and back-up data, enhance collaboration and communication, and get real-time business performance data.
4	Additive manufacturing	This is a flexible manufacturing system that uses versatile manufacturing equipment to turn digital 3D models into real products. It includes rapid prototyping, solid freeform manufacturing, layer manufacturing, digital manufacturing, and 3D printing. With the use of additive manufacturing, manufacturers may build goods and parts just when they are needed, as opposed to keeping vast stockpiles. Rapid prototyping is made possible by additive manufacturing, allowing companies to test and improve their product designs rapidly and affordably. Additive manufacturing can reduce waste by using only the material needed to produce a part.
5	Augmented reality	This technology is still in the development stage; however, there has been a rise in the rate of its use and adoption in recent times. Augmented reality technology, which uses reality operators, enhances the awareness of reality based on unreal knowledge regarding the environment. Augment reality is compatible with different kinds of technology provided. It involves human senses. In the manufacturing industry, augmented reality has the potential to alter how products are designed, produced, and maintained.
6	Autonomous robots	The needed reconfigurable automation technologies as the manufacturing paradigm rapidly shifts output from mass to customized production is the autonomous robots. Adaptive robots are highly helpful in manufacturing systems for operations like product development, manufacture, and assembly. In the manufacturing sector, autonomous robots are being used to carry out a range of duties, from material handling to assembly and inspection.

(Continued)

TABLE 4.1 (*Continued*)
Description of Industry 4.0 Enabling Technologies in Manufacturing [7,16]

S/N	4IR Technology	Description
7	Simulations	The computer simulation such as finite element, computational fluid dynamics, etc. are crucial and effective instruments for the adoption of digital manufacturing for engineering projects and model-based system design, where synthesized models imitate the characteristics of the implemented model.
8	Internet of things (IoT)	IoT refers to physical things (or groups of such things) equipped with sensors, computing power, software, and other technologies that communicate with one another and exchange data over the Internet or other communication networks. Different things or items interact and work together for a shared goal, digitalizing all physical systems. IoT is being utilized in the manufacturing sector to connect machines, sensors, and devices to enhance operational effectiveness, cut costs, and raise product quality.
9	Cyber security	As fundamental components of the supply chain, the IoT must be constructed on the foundation of safety communications at every stage of the manufacturing and safety interoperability between facilities. Technology to identify phishing scams and how to create strong passwords has helped industries to prevent attacks and improve the overall security of the manufacturing environment.
10	Computer-aided design and manufacturing (CAD/CAM)	Creation of projects and work schedules for manufacturing and product development based on system. CAD wouldn't exist without CAM. The design of a product or part is the main emphasis of CAD. How it seems and works. CAM emphasizes how to create it.
11	Flexible manufacturing lines	This Reconfigurable Manufacturing Systems (RMS) are encouraged by digital automation with sensor technology in manufacturing processes (such as radio frequency identification, or RFID, in product components and raw materials), allowing for cost-effective product integration and rearrangement with the industrial environment.
12	Data automation with sensors (sensoring)	This is an automation system for monitoring via data collection with embedded sensor technologies. The output of the sensor could be used to direct a process, serve as input to another system, or give information to a final user. Almost any physical element may be found with sensors.

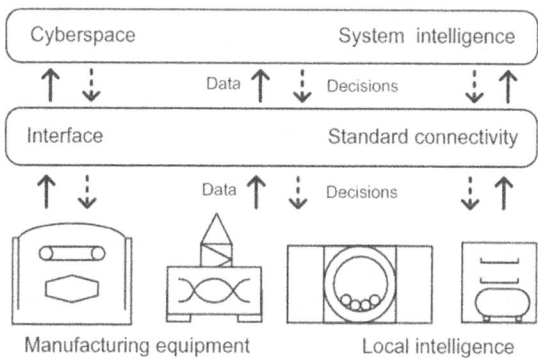

FIGURE 4.2 General concept of smart manufacturing [13].

4.5.1 IMPROVED PRODUCTION EFFICIENCY, FLEXIBILITY, AND AGILITY

The use of technologies associated with Industry 4.0 will increase the efficiency of many parts of the production line such as the ability to produce more products more quickly and experience less machine downtime [15]. Faster batch changeovers, automatic track & trace procedures, and automated reporting are additional instances of increased efficiency. NPIs (New Product Introductions) and business decision-making both improve in efficiency. Industry 4.0 ensures manufacturing process configuration flexibility to produce customized goods. Industry 4.0 improves organizational effectiveness by vertically, horizontally, and end-to-end integrating industrial processes. Additional advantages of Industry 4.0 include increased adaptability and agility. For instance, scaling production up or down is simpler in a smart factory. In addition, it is simpler to add new goods to the production line, opening up possibilities for high-mix manufacturing, one-off production runs, and other processes.

4.5.2 CUSTOMER SATISFACTION

Industry 4.0 ensures that product failures are decreased by the use of digital technologies in manufacturing and procedures. There are less faults and problems in the items thanks to real-time monitoring employing intelligent sensors, software, IoTs, etc. [15]. The digital supply chain is a vital component of every business that can enable organizations to fully realize their Industry 4.0 objectives. The digital supply chain, which spans design, manufacturing, asset management, and shipping, can be used to play a crucial part in guaranteeing great customer satisfaction. Leading businesses, however, are integrating the digital supply chain across formerly distinct business sectors, production, supply chain planning, logistics, and after-sales support and maintenance, in order to get there

4.5.3 REAL-TIME DATA-BASED DECISIONS

The integration of real-time data is crucial to align the supply chain with consumer expectations. The speed at which data is captured and transmitted in critical locations is directly influenced by the increased value that clients obtain as they progress through the supply chain. Critical and intelligent data-driven decisions in the manufacturing sector are essential metrics of transformation in improved efficiency and service delivery. The value generation chain may now be organized and managed in entirely new ways across the entire product life cycle and beyond. Its core is a never-ending flow of pertinent, real-time data. As soon as the parties in the horizontal and vertical value creation chains are networked, the streams of real-time data ensure that they have immediate access to the knowledge they need to carry out their tasks successfully.

4.5.4 CUSTOMIZATION OF SMART PRODUCTS

Industry 4.0 makes manufacturing processes more adaptable so that they can create items with a degree of customization that is similar to the period of crafts manufacturing. Personalization is a differentiation tactic used by manufacturers to stay unique in the face of increased worldwide competition. Small- and medium-sized businesses (SMEs) can meet the need for customized products where individualization adds value by adding personalized products alongside their core offerings. Yet, because of limited production runs or even products that are uniquely unique, personalization is accompanied by substantial variety [17,18] Smart and highly personalized products are increasingly in demand. Industry 4.0–related technologies like as automation, simulations, collaborative robots (COBOTS), and others make it possible to build such items with the flexibility of a made process [15].

4.5.5 Enhanced Productivity and Less Downtime

Industry 4.0 technologies' optimization and automation result in higher production rates, less waste production, fewer manufacturing defects, and higher profitability. Industry 4.0 technologies let you accomplish more with less resources increasing throughput while more effectively and economically deploying your resources. Deeper levels of integration, improved machine monitoring, and automated/semi-automated decision-making will all reduce downtime on your manufacturing lines. As your facility gets closer to becoming an Industry 4.0 Smart Factory, overall equipment effectiveness increases.

4.6 DISADVANTAGES OF INDUSTRY 4.0 IN THE MANUFACTURING INDUSTRY

There are certain drawbacks of automations in Industry 4.0 to the manufacturing industry. Some of these demerits of Industry 4.0 to the sector are highlighted as follows.

4.6.1 Inequality Challenge

The beneficiaries of these technologies and the significant results they facilitate in the manufacturing process are of notable consequences. In actuality, innovators and investors are typically the largest beneficiaries when they invest in people and physical capital. The decline in income can be attributed to technology, particularly for the high-income region of the world. Eventually, it won't be able to bridge the demand discrepancy between highly qualified and educated employees and lesser qualified and skilled employees. Therefore, this can also lead to future job losses. It is imperative to address the difficulties that emerging countries face in terms of technology, infrastructure, and talent.

4.6.2 Cyber Threat and Insecurity

There is a higher likelihood that data will be hacked, altered, or used maliciously when everything is interconnected. There is less containment as it once was. We are hearing the dreadful news of a new data security breach more and more frequently. In fact, it happens so frequently that it no longer surprises us. In addition, it poses fundamental concerns about privacy and identification, particularly in the light of the expanding use of data analytics and machine learning.

4.6.3   Job Insecurity and Threat

While it is simple to tout the advantages of 4IR, there is a disconnect between company leaders' optimism and employees' concerns about the potential effects on their jobs. In a UK survey by the Social Market Foundation (SMF), 45% of workers voiced worries about their job security. The same proportion of people (49%) expressed concern over "machines and software making decisions that humans once made" [19]. In contrast, 4IR technologies were seen as having the potential for employment generation by 69% of business leaders. However, there was a general agreement that traditional career trajectories were altering due to the automation of middle management positions in addition to entry-level and low-skilled occupations. Employers can optimize resourcing by using data analytics to reduce employment during slow periods and approach human capital as a "just-in-time" good. Although switching workers to zero-hour contracts may be economically advantageous, it leads to social unrest and income volatility [19].

4.6.4   Moral and Ethical Challenges

There are new ethical issues and moral issues that are coming up with the emergence and adoption of improved AI, genetic engineering, and increasing automation, which varies from individual to individual, and industry to industry. The possibility of exploring more information on a person or a group of people for one's own benefit or to manipulate others increases with increased access to such information. For instance, the Cambridge Analytica data crisis from the beginning of 2018, when it was exposed that Cambridge Analytica had improperly acquired the personal information from millions of Facebook profiles and utilized it for political advertising. In addition, this is just one of the data misuses that we are aware of [20].

4.7   INDUSTRY 4.0 AND THIN FILM TECHNOLOGY

According to Sinha [21], nanostructured materials have been highlighted by the World Economic Forum (WEF) as a technology that will enable Industry 4.0. Cyberphysical infrastructure supported by a cluster of computing and communication technologies is one example of an Industry 4.0 technology. That moment when industry acquires cutting-edge nanomanufacturing methods in order to deposit the best nanothin films on nano-gadgets has arrived, thanks to recent advancements in higher computation. This effort resulted in thin-film fabrication methods that produce a thin film of extremely high uniformity, conformity, and pin-hole-free quality.

There exists an indispensable relationship between thin film technology and critical elements of Industry 4.0. The recent acceleration of research and application in thin film technology has been identified as a key driver for Industry 4.0. Atomic layer deposition (ALD) technology, which creates a multilayer thin film with a hybrid inorganic-organic structure, has undergone vigorous research and development for 10 years, preparing for the fourth industrial revolution. According to Jae-hac [22], the South Korean engineering firm creates ALD machinery for coating objects with an incredibly thin layer, a crucial step in the mass manufacture of semiconductors, solar cell, display, and biotech devices. It is believed that these devices will be a key component of the future with a wide range of nanotechnology applications.

Some of the components of Industry 4.0 which relies on the thin films technology developments are as follows [23]:

- Sensor
- Real-time monitoring
- Smart materials
- Efficient energy systems
- Smart devices/gadgets
- Self-healing gadgets
- Nano-devices

The advancement in digital technology has caused a paradigm shift from the application of traditional sensors to high precision nano-sensors. Different industries employ sensors for a variety of everyday and commercial applications. Sensor system use in industrial demonstrations has recently increased, showcasing their extraordinary capabilities. Ordinary sensors have become intelligent sensors thanks to the Internet of Things (IoT), which allows complicated calculations to be made locally in a sensor module from observed data [24]. High accuracy sensors can be made using thin film technology, which is well known for doing so. For example, thin film methods like ALD, ion beam, and sputtering help to produce biological sensors for a stable health monitoring of various states. High-performance digital devices offered by sensors are needed for remote surgical procedures. A crucial function is played by thin film coating.

The potential benefits, impacts, and roles of Industry 4.0 and the thin film technology are intertwined. The Industry 4.0 components like AI, big data analytics, and machine learning would greatly benefit from thin film technology. Through other 4IR digital technology, enormous amount of data are generated through sensors. The complex thin film technology process can be modeled utilizing data-driven machine learning and process optimization for maximum output. Utilizing these models will enable material, cost, and process output optimization. In the chapters in the next section, these are covered in more detail. By adopting the digital technologies made available by Industry 4.0, miniaturized devices are vital in these complex systems with little input power supply. With great success, the potential of thin film solar and fuel cells has been assessed. In order to create efficient nano- and micro-power sources for these devices, thin film technology is being developed and will continue to be developed [23]. Several other Industry 4.0–driven technologies are contingent on the application of thin film technology. The desire for smart phones, smart watches, and other handmade smart devices is rising. The use of smart technologies in building for services like power supply, security, control, and consumption surveillance, among others, has gained traction, while the demand for solar-powered jackets and backpacks for remote harvesting and powering portable devices has been growing. These technologies all utilize thin film technology [23,24].

4.8   CHALLENGES OF ADOPTING INDUSTRY 4.0 AND THIN FILM TECHNOLOGY IN MANUFACTURING

Some of the key difficulties that businesses encounter while using Industry 4.0 and thin film technologies in the industrial sector are highlighted as follows.

4.8.1   High Initial Investment Cost

The large initial investment needed to integrate Industry 4.0 and thin film technology is one of the key obstacles to its adoption. This high cost involves the cost of new equipment, software, and workers developmental programs. This cost limitation might pose a significant barrier to small and medium-scale enterprises since they could lack the resources to make the necessary start-up investments.

4.8.2   Paucity of Competent and Skillful Labor

Another vital challenge that can be encountered while adopting Industry 4.0 and thin film technologies in the manufacturing sector is the paucity of skilled labor. The successful adoption of these technologies is contingent on employees with sophisticated skills in cutting-edge 4IR technologies like robotics, AI, and data analytics. However, the inability of businesses to employ individuals with these skills limits the businesses' ability to use these individuals in their enterprises.

4.8.3   Interoperability with Legacy Systems

Integration of new Industry 4.0 technologies with legacy systems is another challenge that many businesses must deal with. Although new hardware and software are frequently needed for these technologies, the procedure can be complicated and incompatible with current systems. The shift to Industry 4.0 and thin film technology may therefore need businesses to make investments in new hardware or software.

4.8.4   Data Privacy and Security

Using Industry 4.0 and thin film technology presents significant hurdles in terms of data security and privacy. Huge amounts of data, some of which may be confidential and discrete, must be gathered and processed in order for these technologies to function. Businesses must guarantee the safety of these data and safeguard the confidentiality of both customers and employees. The creation of new policies and procedures as well as large investments in cybersecurity measures may be necessary.

4.8.5   Compliance to Regulations

Businesses need to take steps to ensure they are abiding by a number of regulations, such as those that deal with confidentiality and safety of information, environmental sustainability, and the welfare and security of employees. It might be required to develop novel regulations and processes and make substantial investments to implement the measures.

4.9 CONCLUSION

The paradigm shift in the manufacturing sector heralded by the industrial revolutions and the emergence of cutting-edge technologies in the Industry 4.0 era has been a game changer in transforming the trajectories of service delivery, production quality and efficiency. This chapter presented an overview of the far-reaching effects of the industry 4.0 on the thin film deposition and manufacturing processes. However, these technologies grapple with some hurdles such as data security and privacy difficulties, high initial expenditures, integration with existing systems, a lack of qualified staff, and regulatory compliance. To fully harness these inherent benefits in the adoption of these technologies in the manufacturing, the dynamics of these critical challenges must be given reasonable considerations. Moreover, companies must invest in these technologies, learn new skills, and overcome implementation challenges. This will position them for potential achievement while contributing to the ongoing development of the industrial sector.

REFERENCES

[1] V. Alcácer, J. Rodrigues, H. Carvalho, and V. Cruz-Machado, "Industry 4.0 maturity follow-up inside an internal value chain: A case study, "Industry 4.0 maturity follow-up inside an internal value chain: a case study," *International Journal of Advanced Manufacturing Technology*, vol. 119, no. 7–8, pp. 5035–5046, 2022, doi: 10.1007/s00170-021-08476-3.

[2] B. Rodič, "Industry 4.0 and the new simulation modelling paradigm," *Organizacija*, vol. 50, no. 3, pp. 193–207, 2017, doi: 10.1515/orga-2017-0017.

[3] K. Schwab, *The Fourth Industrial Revolution*, Geneva: World Economic Forum, 2016.

[4] G. L. D. M. M. K. B. and A. P. Marjeta Marolt, "Business model innovation: Insights from a multiple case study of Slovenian SMEs," *Organizacija*, vol. 49, no. 3, pp. 161–171, 2016.

[5] S. Savas, "The effects of artificial intelligence on industry: industry 4.0," in *Current Studies in Basic Sciences, Engineering and Technology*, M. Ozaslan, and Y. Junejo, Eds., ISRES Publishing, Turkey, pp. 95–106, 2021.

[6] Siemens, Endüstri 4.0 yolunda. Siemens," https://siemens.e-dergi.com/pubs/Endustri40/Endustri40/Default.html#p=7, 2021.

[7] V. Alcácer and V. Cruz-Machado, "Scanning the industry 4.0: A literature review on technologies for manufacturing systems," *Engineering Science and Technology, an International Journal*, vol. 22, no. 3, pp. 899–919, 2019. doi: 10.1016/j.jestch.2019.01.006.

[8] A. G. Frank, L. S. Dalenogare, and N. F. Ayala, "Industry 4.0 technologies: Implementation patterns in manufacturing companies," *International Journal of Production Economics*, vol. 210, pp. 15–26, 2019, doi: 10.1016/j.ijpe.2019.01.004.

[9] P. K. Hajoary, "Industry 4.0 maturity and readiness- A case of a steel manufacturing organization," *Procedia Computer Science*, vol. 217, pp. 614–619, 2023, doi: 10.1016/j.procs.2022.12.257.

[10] S. Mittal, M. A. Khan, D. Romero, and T. Wuest, "A critical review of smart manufacturing & industry 4.0 maturity models: Implications for small and medium-sized enterprises (SMEs)," *Journal of Manufacturing Systems*, vol. 49, pp. 194–214, 2018. doi: 10.1016/j.jmsy.2018.10.005.

[11] E. Gabel, "What are the defining characteristics of industry 4.0?" https://revolutionized.com/defining-characteristics-of-industry-4-0/, 2022.

[12] Y. Liao, F. Deschamps, E. D. F. R. Loures, and L. F. P. Ramos, "Past, present and future of Industry 4.0- a systematic literature review and research agenda proposal," *International Journal of Production Research*, vol. 55, no. 12, pp. 3609–3629, 2017. doi: 10.1080/00207543.2017.1308576.

[13] A. Kusiak, "Smart manufacturing," *International Journal of Production Research*, vol. 56, no. 1–2, pp. 508–517, 2018, doi: 10.1080/00207543.2017.1351644.

[14] R. Y. Zhong, X. Xu, E. Klotz, and S. T. Newman, "Intelligent manufacturing in the context of industry 4.0: A review," *Engineering*, vol. 3, no. 5, pp. 616–630, 2017, doi: 10.1016/J.ENG.2017.05.015.

[15] M. Sony, J. Antony, O. Mc Dermott, and J. A. Garza-Reyes, "An empirical examination of benefits, challenges, and critical success factors of industry 4.0 in manufacturing and service sector," *Technology in Society*, vol. 67, 2021, doi: 10.1016/j.techsoc.2021.101754.

[16] L. S. Dalenogare, G. B. Benitez, N. F. Ayala, and A. G. Frank, "The expected contribution of Industry 4.0 technologies for industrial performance," *International Journal of Production Economics*, vol. 204, pp. 383–394, 2018, doi: 10.1016/j.ijpe.2018.08.019.

[17] I. A. R. Torn and T. H. J. Vaneker, "Mass personalization with industry 4.0 by SMEs: A concept for collaborative networks," *Procedia Manufacturing*, vol. 28, pp. 135–141, 2019. doi: 10.1016/j.promfg.2018.12.022.

[18] R. Schmidt, M. Möhring, R. C. Härting, C. Reichstein, P. Neumaier, and P. Jozinović, "Industry 4.0- Potentials for creating smart products: Empirical research results," *Lecture Notes in Business Information Processing*, vol. 208, pp. 16–27, 2015. doi: 10.1007/978-3-319-19027-3_2.

[19] Hellen Phusela, "Job security in industry 4.0," https://www.rosstone.co.za/2020/01/06/job-security-in-industry-4-0/, 2020.

[20] George Firican, "The pros and cons of the 4th industrial revoluation," https://www.lightsondata.com/pros-cons-4th-industrial-revolution/, 2020.

[21] Saurabh Sinha, "Industry 4.0 and nanostructured materials," *EngineerIT*, pp. 5–8, 2018.

[22] Jeong Jae-hack, "CN1 readies for industry 4.0 with competitive ALD technology. Country reports, South Chine monitoring post," https://www.scmp.com/country-reports/country-reports/topics/south-korea-business-report-2021/article/3130575/cn1-readies?module=perpetual_scroll_0&pgtype=article&campaign=3130575, 2021.

[23] T.-C. J. L. Z. Fredrick Madaraka Mwema, "Thin film technology and industry 4.0," in *Thin Film Coatings: Properties, Deposition and Applications*, First edition, Boris I. Kharissov, Ed., London, New York: CRC Press, Taylor and Francis Group, LCC, pp. 271–273, 2022.

[24] M. Javaid, A. Haleem, R. P. Singh, S. Rab, and R. Suman, "Significance of sensors for industry 4.0: Roles, capabilities, and applications," *Sensors International*, vol. 2, 2021. doi: 10.1016/j.sintl.2021.100110.

Part II

Machine Learning Techniques

5 Fundamentals of Machine Learning

5.1 INTRODUCTION

In this chapter, we provide a brief survey of machine learning (ML) basics and introduce its core ideas, algorithms, and applications in diverse fields as a background information for understanding its application in atomic layer deposition and thin film deposition. A swift transition from mere theoretical framework to real-life and practical applications of artificial intelligence (AI) and ML has been observed in recent years. Thus, there is an increasing interest by different businesses, organization, and public sectors in deploying AI techniques in solving most of their complex problems. Despite the fact that it would be nearly impossible to give a comprehensive review of this extremely dynamic topic, this chapter will attempt to give a brief overview of salient information needed to build the foundation for its deployment in atomic layer deposition applications. Many books have been written on ML theories and processes. For a more thorough examination of the subject, we suggest the reader consult one of the more specialized ML textbooks.

5.2 ARTIFICIAL INTELLIGENCE, MACHINE LEARNING, AND DEEP LEARNING

There is an upsurge in the adoption of AI, ML, and deep learning (DL) in contemporary times by different businesses and sectors to construct intelligent devices and products, making them the most talked-about technologies in business. Although these terminologies dominate business conversations worldwide, many people have trouble distinguishing them. Figure 5.1 depicts their interconnectivity. Today, the phrase "artificial intelligence" refers to a wide area of research in computer science that aims to create AI systems that can think, see visually, and solve problems, all of which are generally attributed to human intelligence.

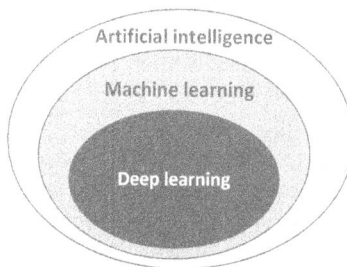

FIGURE 5.1 Interconnection of artificial intelligence, machine learning, and deep learning.

DOI: 10.1201/9781003346234-7

Based on learning capabilities, the following are the categories of AI which also represents the three stages through which AI has evolved:

1. **Artificial narrow intelligence:** The only kind of AI we have so far effectively generated is artificial narrow intelligence (ANI). ANI is goal-oriented, created to carry out a single task, such as driving a car, and is exceptionally intelligent at carrying out the particular task it is taught to do. It is a high-functioning system that duplicates and possibly exceeds human intelligence for making a specific goal possible [1]. All modern types of AI such as pattern, face and image recognition, and autonomous vehicle amongst others are examples of ANI. ANI often performs better than humans; however, a major demerit of this AI type is that it can only perform the tasks for which they were created and can only decide based on the training data.

2. **Artificial general intelligence:** The artificial general intelligence (AGI) otherwise referred to as strong AI enables machines to use knowledge and skills in various contexts. AGI distinguishes itself from the ANI by thinking and carrying an assignment independently on human programming while its process can change in response to various circumstances [2]. AGI aims to build machines that can reason and think like humans, as opposed to ANI applications, which can perform singular, automated, and repetitive activities [3]. Many of the issues with ANI are resolved with AGI. For instance, ANI concentrates on a single task, while the performance of algorithms can be negatively impacted by minor alterations since it is only trained to accomplish its objective without taking any unintentional actions.

3. **Artificial super intelligence:** The stage at which computer intelligence surpasses that of humans is known as artificial superintelligence. Human-impossible ideas and interpretations will be understandable to super intelligent computers [4]. This is attributed to the fact that just a little fraction of the billions of neurons in the human brain are capable of thought. The idea of artificial superintelligence is centered on the capacity to comprehend human feelings and experiences such as to excite its own beliefs and desires through the comprehension of its own functionality. This is in addition to replicating the multifaceted human behavioral intelligence.

Based on functionality, AI can be categorized as follows [5]:

1. **Reactive AI:** Reactive AI is the most fundamental form of AI; it is designed to give a predictable result based on the information it receives. Reactive machines never change how they react to the same circumstances and cannot learn behavior or imagine the past or future. Reactive AI represented a significant advance in AI research, although these AIs are limited to doing the tasks for which they were initially created. They are hence intrinsically constrained and open to improvement. From this basis, other forms of AI were developed. Common examples are spam mails filters in our mail inboxes, recommendation engines for Netflix, and super-computer IBM machine for chess playing amongst others.

2. **Limited memory AI:** Limited memory AI alludes to the capacity of AI to retain past information and forecasts and use it to inform future predictions. The complexity of ML design increases slightly when memory is limited. To create predictions and carry out challenging classification assignment, limited memory AI combines pre-programmed information with historical, observational data. Every ML model is contingent on limited memory to be built, but the model can be used as a reactive machine type, being the fundamental and simplest kind of AI. Machine learning algorithms which deploy the limited memory AI are the reinforcement learning, long-short-term memory (LSTM), and Evolutionary Generative adversarial network (E-GAN).

3. **Theory of mind AI:** Theory of mind (ToM) AI is based on the awareness of distinctiveness of one's intent, views, and emotions from others' and how it influences their actions and behaviors. Sequel to this, machines would be able to develop actual, human-like decision-making abilities with the help of this kind of AI. When interacting with humans, machines with ToM AI will be able to recognize, comprehend, and remember emotions in order to modify their behavior. Because human communication involves a process of behavior adjustment based on rapidly changing emotions, there are still a lot of challenges to the ToM AI. ToM AI is the distinction between present AI machines and those to be built in the future. Although ToM robots are not yet a reality, science has been making steady progress toward this goal.

4. **Self-aware AI:** Self-aware AI is the most sophisticated kind. Machines will resemble humans in consciousness and intellect when they can recognize their own emotions as well as the emotions of those around them. This kind of AI will also have needs, wants, and emotions. This kind of AI will enable machines to be aware of their own feelings and mental states. They will be able to draw conclusions that other forms of AI cannot. More attention should be on efforts to comprehend memory, learning, and the capacity to draw conclusions from the past even though there is a long way to go before building computers that have self-awareness.

ML is simply one subfield of AI that deals with techniques for enabling machines to "learn" or to perform tasks more effectively depending on prior knowledge or supplied data [6]. It is also referred to as a set of techniques within AI that are based on "learning" to model patterns in data using mathematical functions. Despite the enormous impact that other fields of AI like symbolic reasoning, heuristics, and evolutionary algorithms have had on science and technology, ML is undoubtedly the most intriguing and promising branch of AI across all fields of application. Figure 5.2 shows the branches of AI including ML.

ML creates a set of rules automatically using answers and data, as opposed to symbolic AI, which uses rules and data to make responses. In order to facilitate a response or action, these rules can then be applied to fresh, unobserved data [6]. ML is frequently referred to as "statistical learning" since its mathematical foundation is largely based on the ideas of conventional statistics [7,8]. It works on the principle of learning new patterns from data structures instead of having these characteristics

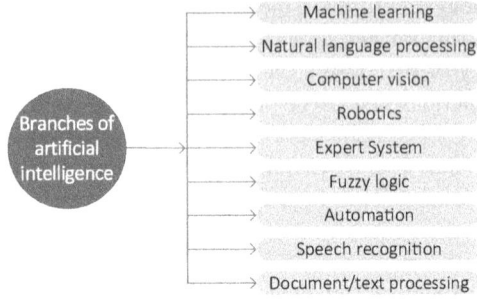

FIGURE 5.2   Branches of artificial intelligence including machine learning.

explicitly programmed. DL method involves the creation of a deep network of artificial "neurons" whose function is meant to be analogous to that of real neurons found in biological systems such as the human brain [1]. The use of features to characterize or extract information from the data is one of the primary contrasts that can be made between DL and more conventional ML techniques. A feature is a quality about the people and event under study and collecting data from. A feature could be simple and straightforward, or it can be more complex and represent transformations of combinations of many variables [9].

According to King et al. [9], it's customary to distinguish between hand-crafted and learned features as a way to distinguish the ordinary ML and DL. Hand-crafted characteristics are those that the algorithm developer has chosen to emphasize utilizing some domain expertise from the application under study. The ML model itself will frequently recognize learned properties as relevant for a particular job. Sequel to this, ordinary ML models employ hand-crafted features, whereas DL models use learned features. Figure 5.3 illustrates this significant difference in the methodologies of the DL and ML. The DL strategy offers a number of significant benefits. First, compared to conventional ML, DL is able to create models from vast volumes of data that are more accurate. In addition, the adoption of DL can drastically shorten training timeframes while removing the need for expert subject knowledge [10].

The common distinguishing features of the DL from the traditional ML are the kind of data it uses and the ways in which it learns. ML develops a predictive model using structured, labeled data, which means that the model's input data are used to extract certain patterns that are then arranged in tables [11]. This doesn't necessary imply that it doesn't employ unstructured data; it only implies that, if it does, it usually goes through some pre-processing to put it in a structured manner. However, some of the data pre-processing steps common with ML are abolished with DL. These algorithms systematize feature extraction and can absorb and interpret unstructured data, including text and images, thus reducing the reliance on human specialists.

Most DL approaches use neural network topologies; hence, DL models are also referred to as deep neural networks (DNNs). The number of hidden layers in the neural network is typically indicated by the term "deep." While deep networks can have up to 150 hidden layers, traditional neural networks only have two to three hidden layers [12]. Convolutional neural networks (CNN or ConvNet) are among

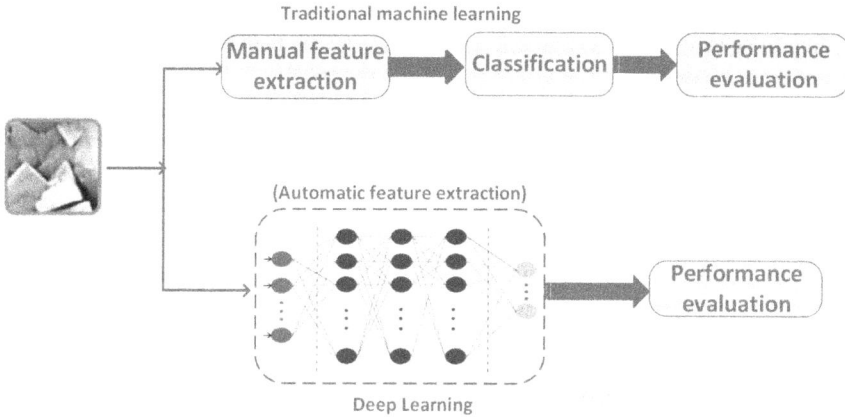

FIGURE 5.3 Traditional machine learning and deep learning compared for a classification task.

FIGURE 5.4 Comparative approaches of classifying a vehicle by machine learning and deep learning [12].

the most commonly used varieties of DNNs. A CNN uses 2D convolutional layers and combines learned features with input data, making it an excellent architecture for processing 2D data, including images [12]. You no longer need to recognize the features that are utilized to categorize photos since CNNs eliminate the necessity for manual feature extraction [12]. A deep ML model is one that uses numerous layers of neural networks, and this is known as DL-based architecture. DL has the features of hierarchical and distributed abstraction since it translates the input data from low-level to high-level to a new feature space during the learning process. This enables the processing of high-dimensional nonlinear input data and the accurate fitting of complex nonlinear functions [13]. MathWorks [12] demonstrated the distinction in the operations of the ML and DL as illustrated by their approaches in categorizing a vehicle as shown in Figure 5.4. In ML, features and a classifier are manually selected to sort images. The phases of feature extraction and modeling are automatic with DL.

5.3 UNDERSTANDING MACHINE LEARNING

The application of techniques based on ML has been shown to be an effective method for extracting hidden features and patterns in a given set of information from data in different approaches [14]. The effectiveness and success of ML utilization is contingent on data accessibility and availability for the purpose of training the computer to carry out the special task and make intelligent decisions. This has therefore positioned it to be especially well-suited to scenarios in which an output is projected from a set of inputs based on some properties of the input [6].

The study of statistical models and techniques that enable computers to "learn" from data without being explicitly programmed to carry out particular tasks is known as ML. Large datasets are used to train ML algorithms to find patterns and relationships that can be utilized to forecast or decide. The model's performance can then be automatically enhanced over time by applying these judgments and predictions to fresh data. The advance in technology, sensors, and IoT has caused an upsurge in the volume of data generated, and its storage. ML has the ability to process these huge data. Thus, it has transformed many different industries, including the health care industry, the financial industry, and the retail sector by developing data-driven-based decisions. It has thus become imperative to ensure that their use is objective and complies with ethical requirement owing to their societal impact.

You may think of the ML paradigm as "programming by example." It learns from experience E about some set of tasks T, and performance metrics P, if its performance at tasks in T, as evaluated using P, gets better with experience E. For instance, we have a certain assignment such as email spam detection and filtering in mind. ML then seeks approaches by which the computer will create its own program based on examples that we offer, rather than programming it to do the problem directly. Therefore, ML generally entails learning to perform better in the future based on prior experiences.

Nowadays, computers may outperform skilled human operators at tasks like picture classification, object identification (such as face detection and recognition), and landmark localization. Numerous viewpoints have been used to discuss the capabilities of machines, including their capacity to gather knowledge [15], think [16], feel [17], be creative [18], and make intelligence-based decisions [19]. ML addresses how machines learn. ML can tackle many real-world problems by using computer systems that can learn to execute a task without being explicitly programmed [20]. ML can transform human history, so everyone expecting to compete in today's fast-paced digital environment must study it.

The great majority of ML models also include operations that are either directly or indirectly connected to Equation 5.1 [21].

$$y = f(x,\beta) + c \tag{5.1}$$

In this equation, y is the response/output we wish to predict, using the function $f(x,\beta)$, while x is a vector of the input variables and β is the model's parameters vector. Since the input data are imperfect in real life, there will always be some noise and/or unobserved data. This is typically denoted by the error component ϵ, which is a random variable, thus making y a random variable by itself. Equation 5.2 depicts a linear form of the function $f(x,\beta)$:

$$f(x,\beta) = \beta_o + \beta_1 x_1 + \beta_2 x_2 + \ldots + \beta_n x_n \tag{5.2}$$

The best combination of the values of β which best approximate the linear relationship with a dataset with pairs of (x, y) is achieved through an optimization process called training, which minimizes the discrepancy between the true values of y and the model's predictions (i.e. the error function).

Some of the critical goals of the ML process are to produce general-purpose, efficient, practical, and useful algorithms, a prediction rule that makes predictions as accurately as feasible, versatile as possible, simple to use with a wide range of learning challenges, interpretability of the prediction rules produced by learning. When faced with a complicated assignment or problem that involves big data and a lot of variables but no formula or equation, think about employing ML.

5.4 BRIEF HISTORY OF MACHINE LEARNING

The creation of general-purpose computers, pioneered during World War II and made accessible for non-military usage in the 1950s, represents the earliest steps toward AI [6]. Fergus and Chalmers [10] reported that the work of Alan Turing [22] which proposed the Turing test in 1950 as a way to gauge AI was the most recognized out of all other references pointing to the emergence of AI. The introduction of Lisp in 1958 made it one of the primary programming languages used to build AI, and by 1964, the program became the game changer for computer programmers to build algorithms that can understand natural language [3]. Because of the increasingly accessible processing power of computing resources, symbolic intelligent programs that use a set of rules to simulate reasoning and make decisions have been made possible [23]. The first-generation chatbots could simulate a natural language dialogue to a large extent, while initially developed checkers and chess algorithms had a favorable performance as early as the 1970s [24–26].

In 1959, a significant contribution was made to the body of knowledge by Arthur Samuel. In his novel publications, he firstly proposed the core idea of "machine learning" [27]. An article was published in 1986 by a prominent researcher, Geoffrey and his research team. The article introduced the concept of artificial neural network (ANN). The study was novel and presented some significant contributions to the knowledge in computer studies and served as a viable alternative to the rule-based systems which existed earlier. This, therefore, formed the foundation for the evolution of ML [28]. As processing and computational capacity improved and a further rise in data availability and accessibility was observed in the 1990s and 2000s, more sophisticated ML algorithms were developed. A range of problems were solved using a variety of ML techniques including neural networks, linear, logistics, and support vector regression (SVR) which demonstrated astonishing accuracy in prediction and decision-making.

The development of ML in its current form is typically credited to Cornel University psychologist Frank Rosenblatt, who built an alphabet character recognition machine using theories about how the human nervous system functions. The machine is named perceptron, being a prototype of the modern day ANN [29]. The previously unknown limits of basic neural network architectures were disclosed in a seminal paper published in 1969 titled Perceptron [30]. The perceptron, which is still widely used today, was one of the first significant developments in ML [31]. Further improvement in the development and evolution of neural network machines was noted after the creation of Spatial-Numerical Association of Response Codes (SNARC) in 1950 [28]. Around this period, the IBM programmers began to create ML algorithms to take on difficult games while the field games continued to progress throughout this time, especially after reinforcement learning was introduced [26]. CNNs were first developed as a result of research on Neocognitron in 1979, while the adoption of the Hopfield network, a branch of the recurrent neural networks (RNN), as content-addressable system firstly introduced the algorithm [32,33]. The emergence of the expert systems, notably the game logic, remained the dominant area of the growth of AI [10].

The support vector machine (SVM) and random forest (RF) algorithms were first introduced in 1995 [33]. In addition, the LSTM RNN was created in 1997 to increase the usefulness and efficacy of RNNs in time series prediction [34]. An interesting aspect of ML, computer vision, gained traction after the Modified National Institute of Standards and Technology database (MNIST) dataset, which comprises a huge quantity of handwritten figures, was conceptualized [35]. The simultaneous improvement in ML over time birthed the development of DL through the adoption of the concept of integrated perceptron with Public Relation (PR)-campaign [29]. A concept and technique known as DNNs were developed after many years of research on multilayer neural networks. Though the exact timeline of its emergence is unclear, it is thought that Dechter [34] first suggested the word "deep learning". DL and large data have fueled ML's rapid growth. Researchers and engineers have pushed the frontiers of AI and ML, resulting in continual innovation and development.

5.5   DESCRIPTION OF MACHINE LEARNING TYPES AND ALGORITHMS

A wide variety of ML algorithms have been developed to assist in resolving challenging situations in the real world in these highly dynamic times. The automatic and self-adaptive ML algorithms get better over time. A comprehensive understanding of algorithms for various purposes is a vital prerequisite to the successful application of ML. There are no one-size-fits-for-all rules for selecting the most suitable ML algorithms for a particular task. Each ML method has advantages and disadvantages for a certain application. The following must be considered when selecting an appropriate ML algorithm;

1. **The type, size, and format of the data:** This is an important consideration in the choice of ML algorithms as the accuracy of training such models greatly relies on the type, quantity, and quality of the data, which is often not practicable to gather [35]. Some algorithms only work best with a large volume of data while other algorithms such as those with low variance and high bias can perform satisfactorily with little data. Sometimes you only need 200 training data samples to produce a solid solution while in other times you need 200,000. Data could be discrete values or continuous, straightforward, or complex, small or medium or large [36]. According to Vidiyala [37], any particular dataset may be entirely of categorical data, entirely of numerical data, or both. If your data is categorical or otherwise nonnumerical in format, you will need to think of a way to turn it into numerical data because algorithms can only function with numerical data.

2. **Model's explainability and interpretability:** A ML model's explainability is its ability to translate its behavior into human terms. It is very crucial to explain a model's outcome and establish its transparency [38]. Sadly, most complex algorithms operate like black boxes, making it difficult to

understand the findings, despite their effectiveness. A trade-off between a model's interpretability or explainability and its performance has been noted. This factor significantly influences the choice of a particular algorithm.

3. **Model's performance:** A critical factor to consider when selecting a model is the quality of the model's outcome. Algorithms that improve the performance should be given top priority. Different performance metrics are used for different tasks [39]. For instance, regression problems use common metrics like mean square error (MSE) and root mean square error (RMSE), while in classification problems, metrics like precision and recall, accuracy, confusion matrix, and F1-score are often used. While considering algorithms selection based on performance, it is important to choose a reliable statistical metrics to assess your model's performance.

4. **Computational speed, time, and cost:** Machine algorithms need more computational time to train on big training data. The longer the training period, the higher the accuracy. The preference of an algorithm is thus contingent on the computational time, speed, and complexity [40]. The choice between a model that is 99% accurate and training cost is $50,000 or the one that is 95% accurate and the training cost is $5,000 is dependent upon your specific situation [35].

As presented in Figure 5.5, ML algorithm can be classified into three major categories based on the task to be carried out and/or the data. These categories are as follows:

1. Supervised learning
2. Unsupervised learning
3. Reinforcement learning

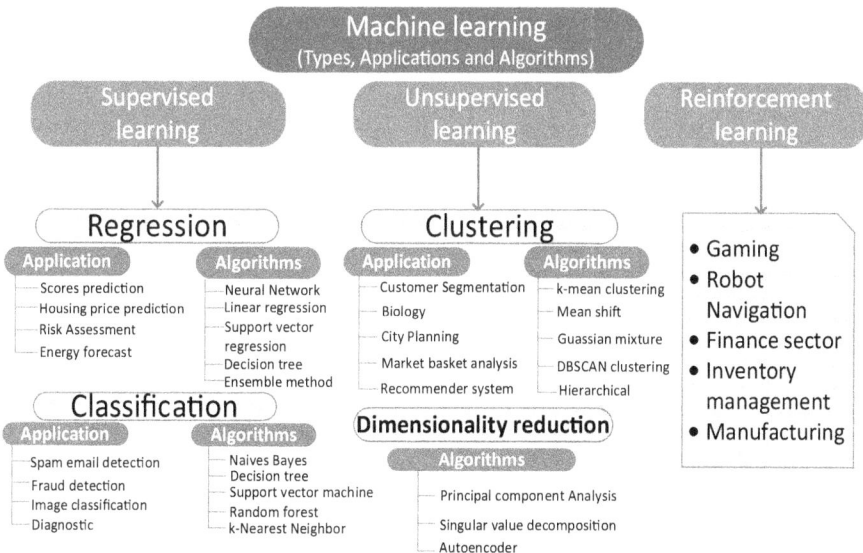

FIGURE 5.5 Diagrammatic representations of the types, tasks, algorithms, and applications of machine learning.

5.5.1 SUPERVISED LEARNING

In supervised learning, models are developed using data in which each observation in the dataset has a label which could be a categorical variable or a continuous variable. These techniques require supervision in the form of labeled examples just like its name suggests, with the objective of predicting the value of one or more outcomes using a variety of input features [14]. In supervised learning, family of functions is modified, or we say that the model is trained, such that the function provides the right output for each input of the dataset. There are inputs $x \in X$ also called the predictors, or covariates and output $y \in Y$ also called response or target. A supervised learning learns from experience E given a set of task T with a training data D. The performance P of the model is estimated with a training dataset, D in Equation 5.4 using empirical risk $L(\theta)$ expressed in Equation 5.3 [41]:

$$L(\theta) = \frac{1}{N} \sum_{n=1}^{N} l\big(y_n, f(x_n; \theta)\big) \tag{5.3}$$

$$\text{Data } (D) = \{(x_n; y_n)\}_{n=1}^{N} \tag{5.4}$$

The difference between the predicted and observed values is denoted as $l\big(y_n, f(x_n; \theta)\big)$, while f is the learning function and θ represents the parameters that define the function f. The proposed processes for creating supervised ML models differ slightly in how they define the phases; however, they all generally entail three phases: the beginning, estimating, and deployment of the model [42]. At the initiation phase, an assignment is defined, the data are prepared and processed, and an appropriate algorithm is selected. At the performance evaluation phase, the validation of several parameter permutations describing the method is carried out while appropriate configuration is selected depending on its performance in carrying out a task. Finally, at the deployment stage, the model is implemented and used to complete a task involving an entirely new dataset.

Supervised learning is often used for regression and classification problems with prominent algorithms such as neural networks, decision tree (DT), ensemble learning, SVM, and k-nearest neighbor (kNN), amongst others. The neural network has gained traction as a viable supervised learning approach owing to its ability to adaptively approximate complex function [43,44]. There are basically three categories of neural network that are prominently applied in ML-based solutions, namely, ANN, CNN, and RNN [45]. Backpropagation optimizes nonlinear problems by computing gradients in neural network algorithms using the chain rule of derivative. Backpropagation trains multilayer perceptron (MLP) or backpropagation neural networks (BPNN) and feed-forward neural networks [46]. Figure 5.6 illustrates supervised ML mode of operation by labeling the training set. The figure shows that the trained model tests the test sample directly.

The DNN learning technique is an approach used in DL with a capacity to automatically extract more features from the input data and having a sophisticated architecture with more layers and neurons. Common examples of DL models which use

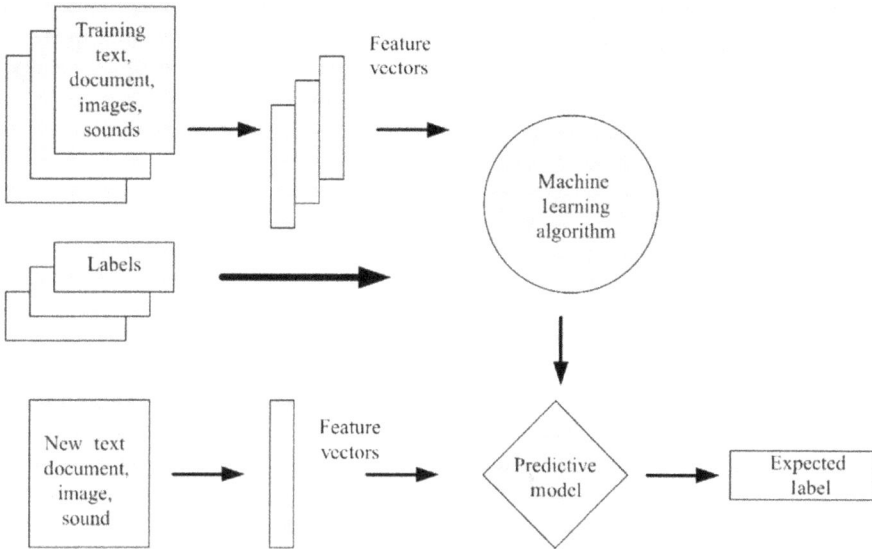

FIGURE 5.6 Functionality of the supervised machine learning [47].

the DNN training approaches are the CNN and RNN. The SVR and SVM are other types of supervised learning algorithms which are used for regression and classification problems, respectively. The goal of the SVR is obtaining a hyperplane in order to have as many data points as possible between the hyperplane's boundaries [45]. The DT algorithm has a tree-like structure with decision nodes and leaves which are continuously divided at decision nodes according to a particular parameter, and the leaves represent the results [48]. The ensemble method is a hybrid of different learning algorithms such as Random Forest, boosting, bagging, and AdaBoost, amongst others, for a more robust and enhanced model's performance than stand-alone algorithms.

Some of the demerits of the supervised learning are as follows. It requires massive volumes of labeled data; this limits such models in situations when it is impractical to produce large-scale labeled datasets [49]. In addition to being laborious and tedious, hand labeling calls for knowledge that is costly and hard to come by. These shortcomings limit the exploitation of the full potentials of the supervised learning in practical applications [14]. An in-depth discussion and more details about these and other supervised learning algorithms are presented in subsequent chapters.

5.5.2 Unsupervised Learning

The unsupervised model is focused on discovering the correlations between individual data points and the hidden pattern in an unlabeled data structure. Generally speaking, an unsupervised algorithm may concomitantly exhibit more than one of these features, and the outcomes of unsupervised learning may then be utilized in supervised learning. The clustering involves the grouping of different data points in a group based on how similar they are without any manual input. The collections of the related data points are called a cluster. The class label (dependent variable) is unknown in unsupervised learning. Unsupervised learning algorithms are mainly used for clustering, association, and dimensionality reduction task. Figure 5.7 represents the functionality of the unsupervised learning algorithms for processing unlabeled data unlike supervised learning in Figure 5.6.

The exceptional ability of the unsupervised learning in finding similarities and differences in data structures is accountable to its fitness as the best candidate for solving problems such as strategies for customer segmentation, cross-selling, exploratory data analysis, and face and image recognition [50]. Unsupervised learning can handle a vast amount of data by automatically extracting inherent feature and pattern in it. Manual labeling is not required for an unsupervised learning task. So before submitting the data to a supervised learning process, unsupervised learning can be employed as a preliminary stage.

These unsupervised learning algorithms search for the hidden trend and pattern in the data in a similar approach as the supervised learning, but the distinction is that the data are not fully understood. For instance, gathering vast amounts of data about a particular disease can aid practitioners in understanding symptom patterns and connecting them to patient result. To classify all the data sources connected to a condition like diabetes would be intensive and time-consuming. Consequently, an unsupervised learning strategy would be a viable alternative to the supervised learning established results since it may take a longer time in this scenario [51].

Clustering is a prominent unsupervised learning task which employs algorithms such as k-means clustering and density-based clustering (DBSCAN). While the DBSCAN algorithm searches clusters based on a continuous zone of high density and a continuous region of low density, the k-means algorithm classifies the points in

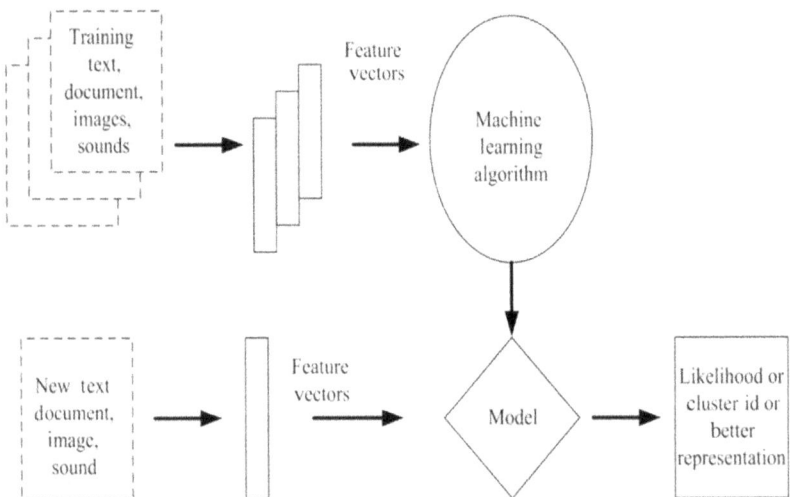

FIGURE 5.7   Approaches of unsupervised learning [47].

the low-density area as outliers and are not assigned to any clusters, and thus assigns all data points to clusters [52]. The K-means clustering is a well-known method that divides M points in N dimensions into K clusters while minimizing the sum of squares inside each cluster. However, the K-means clustering algorithm's fundamental drawback is that K must be known [53]. This challenge has been overcome by using the self-organizing maps (SOMS) and DBSCAN. In order to make comparable input patterns cluster with nearby data, SOMS algorithm uses lateral interaction inside a certain neighborhood [54].

Despite the immense benefits of the unsupervised learning, it is still being faced with some shortfalls, such as (i) computational intensity and complexity owing to the high volume of data being handled by unsupervised learning and the longer time required to process such data, (ii) a high risk of unreliable or inaccurate model outcome as there is no labeled training data. We are often unable to obtain a precise information regarding the output and sorting of the algorithm, (iii) lack of transparency regarding the principles used to cluster data, and (iv) requirement of human involvement in confirming the output variable [35,51,53,55]. An in-depth discussion and more details about these and other unsupervised learning algorithms are presented in subsequent chapters.

5.5.3 Reinforcement Learning

Reinforcement learning is a training approach in ML whose basic principle is based on the reward of desired behaviors, while undesirable actions are penalized. It has the ability to typically perceive and comprehend its surroundings, act, and learn via mistakes [56]. The algorithms allocate a positive value to a beneficial action as a form of motivation to the agent while negative value is allocated to nonbeneficial actions. Thus, the agent learns from its environment and is trained to pursue long-term and optimal overall benefit in order to arrive at the best possible outcome [57]. This is realized through the concepts of dynamical systems theory, utilized to depict the practices of the complex systems over time and typically formulated as a Markov Decision Process (MDP).

The agent is prevented from stagnating on smaller goals by these long-term objectives. Over time, the agent assimilates to focus on the positive goal rather than the negative. Reinforcement learning has different objectives from unsupervised learning. It is focused on creating an appropriate action model that will maximize the agent's overall cumulative reward [58]. The action–reward feedback loop of a typical reinforcement learning is shown in Figure 5.8.

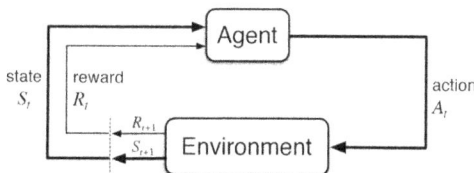

FIGURE 5.8   Framework of reinforcement learning [58].

As long as a distinct reward is available, reinforcement learning can be applied to a circumstance. Reinforcement learning algorithms in enterprise resource management (ERM) can distribute scarce resources across various assignments in as much as there is a broad objective being pursued. In this case, time and resources conservation would be the objectives.

Prominent applications of reinforcement learning are in robotics, gaming, manufacturing, inventory and resource management, and personalized recommendations [56]. Reinforcement learning algorithm are categorized as value-based, policy-based, and actor-critic algorithms [57]. Common examples of reinforcement learning algorithms are Q-learning, SARSA (State-Action Reward-State-Action), Deep Q-Network (DQN), and Deep Deterministic Policy Gradient (DDPG) [58]. Reinforcement learning has been used in few restricted experiments in robotics. Robots based on this type of learning may be able to learn skills that is not possibly exhibited by human teacher, apply previously taught skills to new tasks, or accomplish optimization even in the absence of an analytical formulation. Although it has great potential, reinforcement learning can be challenging to implement and has only a limited number of applications. This form of ML's dependency on environment exploration is one of the challenges to deployment.

5.6   STEPS IN THE MACHINE LEARNING PROCESS

ML processes can be simplified into multiple steps. In this section, we shall discuss how each stage affects ML. These steps are necessary to achieve precise and accurate ML outcomes. These steps ensure that the ML model is properly developed and ready to provide insightful explanations and predictions. Figure 5.9 represents a typical workflow of an ML task.

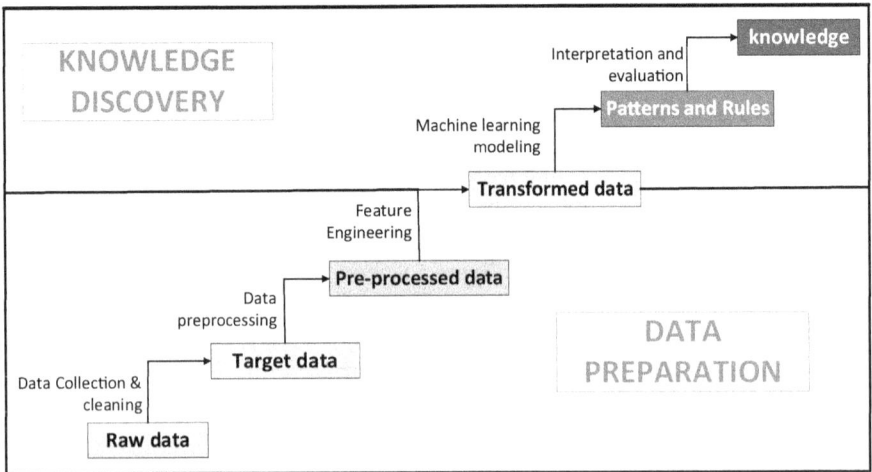

FIGURE 5.9   Typical workflow of a machine learning task [59].

5.6.1 Data Collection

The goal of ML methods is to recognize and mathematically express the patterns seen in data. Consequently, both the quantity and quality of the data employed are crucial for creating effective and usable ML algorithms [7]. It is crucial to gather trustworthy data to ensure that your model accurately identify trend and patterns. Faulty data collection is a major setback to successful ML modeling. Data collection begins with identifying the different data sources. Databases, files, Application Programming Interface (API), and online sources supply ML data. Because these sources' data availability and quality vary, it's important to carefully evaluate them before using them. Common issues encountered with data collection such as bias, inaccuracy, missing data, and imbalance in data are attributed to the involvement of humans. Collecting high-quality data that truly reflect the population being studied is attainable by being aware of these common issues and taking action to mitigate them. Consequently, the precision and dependability of the outcomes will be increased. Any professional in data science can confirm that having too much data is always preferable to having too little data. Thus, when data are insufficient, you might need data augmentation techniques such as latent semantics, data integration, entity augmentation, and synthetic data generators to increase the data size and volume. Thanks to websites like Kaggle, it is now effortless for ML experts to get superior-quality data and the assistance to enable data transformations for a variety of applications. Moreso, data that are not entirely available can be generated manually or automatically using crowdsourcing platforms such as AI-based data collection systems and Amazon Mechanical Turk [60].

5.6.2 Data Pre-Processing

The most crucial and significant factor affecting the generalization capabilities of a supervised ML algorithm is data pre-processing [61]. This step is so important that it represents about 50%–80% of the entire task in classification [62]. Often-times real-life data are unreliable, and devoid of particular trends or pattern. They are also probably full of mistakes, such that they might require being pre-processed into an acceptable format for the algorithm. The information collected in a dataset must not be presumed to be of good quality and is in a satisfactory state. A major competency of ML experts is their ability to thoroughly examine the data and analyze it using the appropriate techniques and tools [10].

This step precedes the training phase, thus signifying its significance. Common pre-processing techniques used by ML modeling are data cleaning, imputations, augmentation, reduction, normalization, and oversampling. The transformation T of a data vector A_{ik} in data pre-processing to give an entirely new process data B_{ij} is shown in Equation 5.5 [63].

$$B_{ij} = TA_{ik} \qquad (5.5)$$

The information in A_{ik} is preserved by B_{ij}, while it gets rid of some of the challenges of the unprocessed data vector A_{ik}, thus presenting B_{ij} as a more significant dataset than A_{ik}. Variables j and k represent the number of extracted features after and before extraction while i is the object number. The pre-processing approach to be adopted depends on the problems identified with the data. For instance, dimensionality reduction might be necessitated for a too large data to enhance its performance [61]. Moreover, if the data are too few such that it contains about 20% of the missing data, the approach is then to get rid of it [64].

The process of data cleansing involves identifying inaccurate or noisy data and either correcting it or eliminating it from the dataset [61]. Data cleaning focuses on locating and replacing records and data that are erroneous, irrelevant, or otherwise noisy. Another pre-processing method is normalization. When there are several features, each one's characteristics could have a different scale, hence normalization is necessary to ensure that they fall in the same scale or the results will be subpar. Thus, to establish that the training data conform to a similar range prior to the training process, we can use a min-max normalization approach in Equation 5.6. Other forms of normalization are z-score and decimal scaling [61]:

$$y_{norm} = \frac{x - x_{min}}{x_{max} - x_{min}} \tag{5.6}$$

where y_{norm} = the normalized data, x = the mean of the variable, x_{min} = minimum variable, and x_{max} = maximum variable.

There are three methods for handling the noise if it persists in class after the loud appearances are already discovered. First, if the model is strong enough to withstand over-fitting, noise can be disregarded. Secondly, it is possible to remove, alter, refine, or relabel noise from the dataset. If the attribute-containing data remains, methods like imputation, filtering, or polishing the incorrect attribute value can reveal other suspicious values and anticipate what needs to be cleaned [61].

Excessively large volume of data resulting into higher number of irrelevant features could sometimes be undesirable for a particular ML application. It is difficult to use high-dimensional data; these models frequently overfit when there aren't large-scale datasets available [14]. In such a situation, it becomes imperative to reduce the initial data using dimensionality reduction which transforms the high dimensionality of the data into a low-dimension while its critical properties are still maintained. Dimensionality reduction approaches such as wavelet transforms, principal component analysis (PCA), feature extraction, or selection have found application. The numerosity reduction could also play a vital role in decreasing the volume of collected data for a smaller representation [61].

5.6.3 SELECTING A MODEL OR ALGORITHM

Your choice of algorithm is contingent on the task you want to carry out. It may be overwhelming to handle all the tasks that ML may assist you with, thus making the choice of appropriate algorithm demanding and as well significant. This step is important because it assists to decide which of the algorithms to select, when to apply them, what parameters to take into account, and the testing approach.

As earlier noted, there is no one-size-fits-for-all rule for selecting the most suitable ML algorithms. The choice of a suitable algorithm relies on factors such as the size, type, and format of the data; model's interpretability and performance; computational speed; time; and cost. The following steps would assist in selecting the right model [65].

1. The correct understanding of your project or task goal is important. What outcome do you desire? Do you desire an intelligent prediction based on historical data, then a supervised training algorithm fits your desire. If you want an image recognition model with low-quality images, a classification algorithm with dimensionality reduction algorithm would be beneficial.
2. Analysis, processing, and annotation of collected data. The determination of the required output precedes the input selection. How does your data look? Is it unprocessed and just collected in raw form from anywhere? Is it unorganized, unclean, and biased? Or do you already have a sizable dataset with annotations? Do you have enough data, or do you need to collect more or perhaps start from scratch? Are you set for the modeling or do you need to spend time getting your data ready for the training?
3. Assess the speed and time of the algorithm's computation. Answering significant questions regarding the speed and time desired helps to select the right algorithm for the task desired. Do you require it quickly at the expense of training and prediction quality? Better training results from more data of greater quality. Can you set aside the time necessary for effective training?
4. A good knowledge of the environment, dimension of the task at hand, and the linearity of the data are significant for algorithm selection. When simplicity and speed is a priority, linear algorithms such as SVM, SVR, and linear regression would be the best bet. However, because they work with linear data, they are not frequently employed for more complicated issues. Linear algorithms might not be adequate for your work if the data are multidimensional, multifarious, and has numerous overlapping associations.
5. The decision of the number of features and parameters, preciseness, and complexity of your proposed model is equally relevant. Remember that an AI model will often perform better and more accurately after a longer period of training. If you have the time to let your model train longer, you can supply more features and parameters for it to consider. Therefore, giving your algorithm additional time to learn could be a wise investment in the accuracy and interpretability of your output in the future.

5.6.4 Training the Model

Once data have undergone several pre-processing stages, the transformed data can be used to train the model. The primary phase in ML modeling is model training, which produces a functioning model that can subsequently be validated, tested, and applied. How well a model performs at the training phase ultimately suggests how well it will perform when integrated into an application. The process of training a ML model entails loading the selected algorithm with the transformed training data so that it can learn. The training process fits the algorithm with the optimal weights and biases such that the loss/error function is minimized. During the training phase, the algorithm updates the weights and biases until pre-determined criteria for stopping the training are satisfied, or else, new weight is relinquished, and a decreasing scalar parameter lowers the training parameters. Equation 5.7 mathematically represents the learning process for optimizing the weight and biases:

$$\theta_{t+1} = \theta_t - \eta . \nabla_{\theta_t} E\left(\theta_t; x^{(i)}; y^{(i)}\right) \tag{5.7}$$

where θ represents the weights and biases, $x^{(i)}$ is the input of the training sample, $y^{(i)}$ is the target label, η is the learning rate, and E is the loss/error function. In supervised learning, the idea of training entails the transformation of the connection between data features and output label into a mathematical expression, while in unsupervised learning, the mathematical relationship within the features themselves are developed [66]. The effectiveness of the training process is contingent on the optimal choice of algorithms, network features and hyper-parameters, and the training data.

5.6.5 MODEL VALIDATION AND EVALUATION

Validation and evaluation are both required to determine whether the ML algorithm's learned model is good or not. An ordinary claim that a function fits exactly for a given set of training data isn't always sufficient and convincing to the data science community and the wide audience. It is imperative to statistically evaluate the performance of the model on a set of testing data. It is necessary to establish that a function fits training data perfectly. Sometimes in training, overfitting causes models to perform best on training data while utterly failing when tested on novel data. While there are many different methods and tools for evaluating models, employing them improperly can lead to an inaccurate assessment of your model's performance. Using the hold-out validation approach, a part of the data is held out for testing the model's performance. A larger amount of the data are used to train the model, and holdout data are utilized to test the model's test metrics [67]. In contrast, when it is impracticable to hold out a fraction of the data specifically for validation reasons due to the limited size of the training dataset, cross-validation is a beneficial technique. Using the k-fold cross validation method, the training dataset is partitioned into k equal folds. Each of these k folds is treated as a holdout dataset, and the remaining $k-1$ folds are used to train the model. There are several performance metrics that are used to statistically establish the goodness of fit of a model. The choice of metrics depends on the tasks, regression, or classification. This will be discussed in detail in the next chapter.

5.6.6 MAKING PREDICTIONS WITH THE MODEL

A model that has been sufficiently and satisfactorily trained and validated can be deployed for a real-life application with a novel dataset. It must be ensured that the model learns satisfactorily instead of memorizing to ensure a good performance at training and avoid failing at the testing phase. A well-trained classification model can effectively classify incoming emails as spam or not, medical images as benign or malignant, and customer's behavior as pushing more sales or not.

5.7 PRACTICAL APPLICATIONS OF MACHINE LEARNING MODELS

Since the dawn of the 21st century, there has been an increased push to incorporate ML into different fields and sectors ranging from medical sciences, social science, engineering, stock market, amongst others. Due to numerous advancements in AI-based decision support approaches over the past few years, this interest has grown. Notably, the impact of the rapid acceptance of these technologies has been particularly widespread in several disciplines, especially science and technology while these tools have been used to delve into new areas which have drawn attention.

The ability of ML to apply higher-dimensional mathematical operations on considerably bigger datasets to unravel complicated, nonlinear relationships distinguishes it from traditional statistical modeling approach and draws on its roots in computer science [7]. The development of statistics-based ML approaches, increased processing power of computers, big data collection and storage have all contributed to an increased interest in ML-based applications. The list of ML applications is inexhaustive. The state-of-the-art algorithms have found wide applications and cannot all be discussed in a chapter. However, to have a brief basic idea of some of its potential benefits in proffering solutions to some real-life problems, we briefly discuss some of the selected applications of ML.

5.7.1 PREDICTIVE ANALYSIS

Both supervised and unsupervised learning are efficient in making intelligent predictions of events. On the basis of labeled data, supervised learning algorithms are used to forecast outcomes. For instance, given a customer's prior credit history, it could be deployed to forecast the risk that they will default on a loan. Owing to the capability of the unsupervised learning to find patterns and connections in unlabeled data, it can perform a task such as identifying a client segment based on their purchasing pattern. The exciting ability of ML in making intelligent predictions have been beneficial to different fields of specialization such as economy and finance, energy sector, manufacturing, construction, and medical field, among others. In building construction, the structural performance of fiber-reinforced concrete prior to the actual construction process was predicted using Extreme Gradient Booting (XGBoot) and Artificial Neural Network (ANN) by Mai et al. [68] and Kilani et al. [69], respectively. Smart grid stability was examined and predicted by Mostafa et al. [70] using predictive ML models for renewable energy management in smart girds. To improve the performance of waste-to-energy thermal plant for optimizing the recovery of inherent energy in the waste resources, the studies by Adeleke et al. [71], and Olatunji et al. [72] developed adaptive neuro-fuzzy inference model (ANFIS, *a hybrid of ANN and fuzzy logic*) to predict the combustion enthalpy of waste and biomass resources, respectively. The quest for a method for valuation that would describe pricing options has spurred the interest of ML applications in the prediction of stock prices of frontier markets [73–77]. In detecting fraudulent activities on credit card transactions, some studies like Afriyie et al. [78] developed DT, RF, and LR for prediction and classification of solutions to credit card transaction frauds.

5.7.2 IMAGE AND VIDEO ANALYSIS

Recently, ML has gained traction as a vital tool for image and video analysis, allowing computers to perform the task of analyzing, comprehending, and interpreting visual information. Classifying an image into one of the many predefined groups, such as landscapes, buildings, or animals, has been made easy with the help of DNN such as CNN which are fitting for gird-like structures applications [79]. ML-based video segmentation is beneficial in different areas of applications such as the analysis of medical image, self-driving cars, augmented reality, and video surveillance [79–81]. The CNN is also versatile for detecting objects and segmentation. This involves identifying and separating objects within an image. Intelligent, autonomous, and smart production and manufacturing systems have benefitted from ML-based image detection and segmentations [82]. An algorithm like the Generative Adversarial Networks (GANs) can be used for image generation and style transfer. This entails producing a brand-new image that is similar to an already-existing one, either by starting from scratch or by incorporating an image's style into another [83–85]. Human motions such as walking, jumping, or sprinting can be conveniently recognized using DNN like convolutional long short-term memory (ConvLSTM) and RNNs [86–89].

5.7.3 MEDICAL DIAGNOSTIC

ML is gaining popularity in medical industry, and one of its most prominent applications is in medical diagnosis. ML has been proposed as a way to enhance suicide risk prediction models that have historically performed badly using traditional approaches [90]. Patient outcomes can be predicted using ML algorithms, and individuals who are at a high risk of developing particular illnesses can be found. This can assist medical professionals with setting treatment priorities and enhancing patient's outcomes. The study by Lebedev et al. [91] developed a DL approach for a digitized prediction of the mental state of a patient. In this study, a deep ANN model to remotely identify emotions was proposed. The task of manually segmenting the size of a brain tumor from 3D Magnetic Resonance Imaging (MRI) volumes takes a lot of time and strongly depends on the operator's skill. DL and computer-aided cancer detection methods have greatly improved this space. Examples of such development are provided in the study by Rao et al. [92], which suggested a modified U-Net structure based on residual networks that uses sub-pixel convolution at the decoder part and periodic shuffling at the encoder section of the original U-Net. ML algorithms can recognize and process images from X-rays, CT images, and MRIs. This helps discover common human heath challenges like cancer, heart disease, and brain abnormalities earlier. ML algorithms can analyze Electronic Health records (EHR) data to uncover patterns and trends that indicate a diagnosis. Drug discovery and personalized medicine may benefit from this. To aid medical professionals in making wise choices and boost diagnostic precision, it can also be incorporated into clinical decision support systems.

5.7.4 NATURAL LANGUAGE PROCESSING

Natural language processing (NLP) uses ML to train computers to process text and other natural language data. This technique has already been used to successfully classify medical text [93]. There are numerous methods for classifying text, including conventional rule-based approaches and DL models like Naive Bayes, SVMs, k-NN, DT, RF, and Gradient Boosting. Kaczmarek [94] utilized the NLP technique of ML to integrate plans for spatial development and implementation by classifying contextual text of plans. Named Entity Recognition (NER) is a branch of NLP that focuses on finding and extracting entities from unstructured text, such as persons, organizations, places, and dates [95]. The typical process of NER involves several steps, including tokenization, part-of-speech (POS) tagging, and entity recognition. In the entity recognition step, ML algorithms are often used to classify each word or token in the text as an entity or not an entity. In an attempt to investigate the susceptibility of urban flood, the study by Fu et al. [96] develops a NER model to extract the site of historical flooding. Machine translation (MT) automates language translation. Due to the internet and the necessity for cross-language communication, NLP research has focused on machine translation [97]. ML-based Neural Machine Translation (NMT) is the most used machine translation method. NMT models learn to map input sequences in one language to output sequences in another by being trained on large parallel corpora of text in two languages [98–100]. Deep neural networks, specifically RNNs or transformer networks, underpin NMT models. Chatbots are computer programs which are built on DL-based methods such as RNN and transformer network and are able to mimic human conversation and perform tasks like answering questions and assisting customers with transactions [101,102]. Due to their 24/7 customer care capabilities and ability to effectively handle a large amount of consumer queries, chatbots have emerged as a crucial tool for businesses and organizations. Large text datasets are used to train these models, which can recognize linguistic patterns and produce coherent and context-sensitive answers.

5.7.5 RECOMMENDER SYSTEM

Another exciting application of ML is the recommender system which gives customized recommendations to users [103]. These recommendations are based on analysis of the user's actions, tastes, and interactions with the system in the past. Recommender systems are used extensively in many fields, including e-commerce, entertainment, and social media, to improve the user experience and increase engagement. Users expect proactive product recommendations from recommender systems, which must be able to respond to shifting user tastes and evolving settings [103–105]. Recommender systems examine the data and generate recommendations using a variety of ML methods, such as matrix factorization [106–108], DL [109–112], and

DTs [113]. The accuracy and effectiveness of the recommender system depend on the quality and quantity of the data used to train the model, as well as the algorithms and techniques used to make the recommendations.

5.7.6 Speech Recognition

The process of turning spoken language into writing is called a fundamental piece of technology for NLP which has a wide range of uses, such as voice-activated virtual assistants, hands-free device control, and automatic transcription of audio and video recordings [114,115]. DNNs are used in speech recognition to create predictions about the most likely transcription by learning patterns in the spectral representations of speech signals [116]. The DNN which is trained using the Connectionist Temporal Classification (CTC) for speech recognition has the ability to handle variable-length input sequences and enables the network to generate predictions at each time step depending on the input and its prior predictions [117]. Spectral representations of the speech signal, such as spectrograms or Mel-frequency cepstral coefficients (MFCCs), are frequently used as the input to a DNN for speech recognition. Large datasets of speech signals and their accompanying transcribed text can be used to train DNNs for speech recognition. The network can learn complex speech signal patterns and their link to transcribed text. Voice-activated virtual assistants, hands-free device control, and audio/video transcription are all possible with the DNN. The DNN can real-time transcribe speech after training. Speech recognition uses probabilistic Hidden Markov Models (HMM) [118,119]. Each observation in a sequence is produced by a hidden state, and HMMs are made to model these sequences of observations. The hidden states in speech recognition often correspond to the underlying phonemes or sub-word units of speech, while the observations are typically spectral representations of the speech signal [118].

5.8   CHALLENGES, LIMITATIONS, AND FUTURE DIRECTIONS OF ARTIFICIAL INTELLIGENCE AND MACHINE LEARNING

The adoption of ML has revolutionized the entire spectrum of different sectors across the globe as a viable and intelligent tool for solving intricate problems which were almost impossible with the traditional approaches. ML techniques has proffered several solutions to long-lived problems through its intelligent data-driven decision-making framework. However, it grapples with several challenges which impedes its full-scale exploration. To fully comprehend the trajectories of practical and innovative applications, it is critical to have a grasp of these limitations. Addressing these challenges will help ML expand and improve its performance, and become more user-friendly. Many ML models are referred to as "black boxes," which makes it challenging to comprehend how they came to a given conclusion. This reduces the public confidence in the systems and hinders their wide adoption. Building trust and confidence in these systems will require methods to make ML models more transparent and understandable. Figure 5.10 represents the future of the AI and ML systems with a focus on the improved interpretability of the systems and models. With the state-of-the-art AI, humans often perform better in communicating their

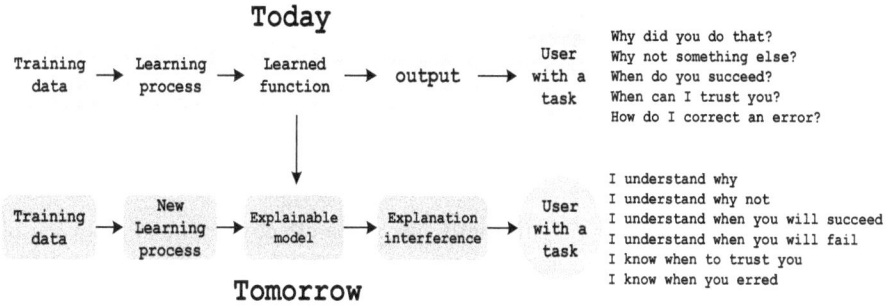

FIGURE 5.10 Future of artificial intelligence and machine learning [120].

thoughts and actions to another human. In order to construct more interpretable models while keeping a high level of learning performance, a set of ML approaches must be developed with more focus on their interpretability [120]. Interpretability, according to Miller [121], is the extent to which individuals can comprehend the rationale behind a decision. This challenge is combated using a rapidly growing concept called Explainable Artificial Intelligence (XAI) which aims to increase the transparency and interpretability of ML models, in order to build trust and confidence in these systems [91,122,123]. There are several approaches to XAI, including saliency map, local interpretable model-agnostic explanations (LIME), and DT. These approaches aim to provide insights into how ML models make decisions, and to identify the key factors and features that are influencing their predictions.

Another giant stride towards overcoming ML hurdles is by improving its large-scale applications and enhancing its privacy and security aspects. This can be achieved by establishing ML training methods which span across vast decentralized devices or data sources. While the conventional approach integrates data onto a centralized server, it often missed out significant and valuable data which results in substantial loss. This is a major training drawback for machine learning. The emergence of federated learning approach has proffered a viable solution by utilizing a decentralized data source approach which offers many benefits such as enhancing its performance and generalizability while allowing models to be trained and updated locally [124]–[126]. Federated learning effectively addresses this drawback of the conventional training methods while reducing the costs associated with collecting, storing, and processing large amounts of data [125]. In addition, it has numerous real-world applications in industries like finance, mobile and IoT devices, and healthcare services..

Significant ML limitations such as bias and fairness have the potential to degrade the effectiveness and dependability of ML models, especially in applications connected to humans. The systematic error or prejudice in the training data, known as bias, leads ML algorithms to generate judgments that are unjust or discriminate against particular groups while fairness is the principle that the decision made by the algorithms is equitable and indiscriminate against individual or groups based on their race, gender, age, or other protected features [127]. It's critical to address bias and fairness in ML not only for ethical reasons but also to enhance model performance and reliability. ML bias mitigation and fairness assurance strategies include data pre-processing, algorithmic fairness requirements, and post-processing techniques [128]. The difficult and ongoing task of addressing bias and fairness in ML systems calls for careful examination of the data, algorithms, and decision-making procedures employed in the systems. We can make sure that ML is used to develop solutions that are moral, just, and dependable and that can have a good impact on society by tackling these issues.

Data quality and quantity significantly affect the effectiveness and precision of ML models. The accuracy, completeness, and relevance of the data used to train ML models are referred to as data quality. Poor data quality can result in unreliable, biased, or inaccurate models that can produce poor performance and inaccurate fore-casts. For instance, it may be challenging for the ML algorithms to produce precise predictions if the training data has missing values, erroneous labels, or inconsistent data types. Data quantity describes the volume of data utilized to train ML models. More data can improve performance, but collecting and analyzing it can be costly and time-consuming. Data collection can often be difficult, especially in sensitive or private applications. ML professionals utilize feature engineering, data augmenta-tion, and transfer learning to solve these difficulties using current data. New ML algorithms that can handle sparse or noisy data are being developed to expand the usage of ML. To overcome these obstacles, significant consideration must be given to data selection, pre-processing, and preparation. In addition, new ML algorithms that can work with sparse or noisy data must be developed.

Scalability must be carefully examined to support the continued development and adoption of ML in various applications. When the data become excessively large, some traditional ML algorithms have limitations in processing such a huge volume of data. For instance, many traditional ML approaches are troublesome for large-scale applications because they have high temporal complexity and require a lot of computer resources to train and evaluate models. Some ML methods cannot handle distributed or parallel computing, which limits their scalability. There is need to cre-ate new algorithms that can handle massive datasets and complex models that are more scalable and efficient. This includes methods for decreasing the computational complexity of current algorithms as well as algorithms created for parallel and dis-tributed processing. It is also becoming simpler to deploy ML models at scale and manage the expanding volume of data created by contemporary applications thanks to new technologies like cloud computing and edge computing.

A significant future prospect of ML-based applications should take into account, from the perspective of biosocial-technical systems, how the various contexts in which human societies learn can influence the use of ML techniques that have been established in other contexts. Fox et al. [129] opined that a broader study of the ethi-cal challenges brought on by ML is required. These can be examined at the level of many societies, each of which holds a distinct view of the connections between bio-logical intelligence and AI. The focus of any ML-related submissions in the future should discuss if and how widely used ML techniques that have an impact on con-cepts like social class and social sustainability are inevitable. Also encouraged are submissions that link ML to collective human learning which ought to take into account nonreinforced learning and reinforcement learning in various types of com-munities across various geographical regions [129].

5.9 CONCLUSION

A transformative impact of ML has been felt across all sectors as it shapes the trajectories of industries towards intelligent data-driven decision making. This chapter examined the interconnectivity and key distinguishing features of AI, ML and DL while exploring their historical context. The chapter proceeded by investigating the algorithms, applications, and approaches of different types of ML alongside their distinctive features. ML has a wide range of applications across numerous industries, such as healthcare, banking, retail, and transportation, among others. The upsurge in volume, veracity and accessibility of data and computational powers has facilitated the widespread use of ML across all organizational spectrums. This has significantly enhanced the competitive edge of several spaces through intelligent data-driven decision-making. Additionally, the chapter examined some of the major limitations of ML applications and some future directions. To fully comprehend the trajectories of practical and innovative applications, it is critical to have a grasp of these limitations. As the discipline progresses, it is probable that novel algorithms will arise and fresh applications will be conceived, resulting in more stimulating progress in the next era.

REFERENCES

[1] J. M. Guerrero, "Chapter 6- Artificial narrow intelligence," in *Mind Mapping and Artificial Intelligence*, J. M. Guerrero, Ed., Academic Press, 2023, pp. 163–185. doi: 10.1016/B978-0-12-820119-0.00010-8.

[2] J. M. Guerrero, "Chapter 7- Artificial general intelligence," In *Mind Mapping and Artificial Intelligence*, J. M. Guerrero, Ed., Academic Press, 2023, pp. 187–201. doi: 10.1016/B978-0-12-820119-0.00009-1.

[3] S. Kumpulainen and V. Terziyan, "Artificial general intelligence vs. industry 4.0: Do they need each other?" *Procedia Computer Science*, vol. 200, pp. 140–150, 2022. doi: 10.1016/j.procs.2022.01.213.

[4] V. Sergievskii, "Super strong artificial intelligence and human mind," *Procedia Computer Science*, vol. 169, pp. 458–460, 2020. doi: 10.1016/j.procs.2020.02.225.

[5] Bernard Marr, "What are the four types of AI?" https://bernardmarr.com/what-are-the-four-types-of-ai/, 2021.

[6] F. Galbusera, G. Casaroli, and T. Bassani, "Artificial intelligence and machine learning in spine research," *JOR Spine*, vol. 2, no. 1, 2019. doi: 10.1002/jsp2.1044.

[7] P. Rattan, D. D. Penrice, and D. A. Simonetto, "Artificial intelligence and machine learning: What you always wanted to know but were afraid to ask," *Gastro Hep Advances*, vol. 1, no. 1, pp. 70–78, 2022, doi: 10.1016/j.gastha.2021.11.001.

[8] D. W. T. H. R. T. G James, *An Introduction to Statistical Learning*. New York: Springer US, 2013.

[9] A. P. King and P. Aljabar, "Machine learning," *Matlab(r) Programming for Biomedical Engineers and Scientists*, pp. 343–372, 2023. doi: 10.1016/B978-0-32-385773-4.00023-X.

[10] C. C. Paul Fergus, *Applied Deep Learning: Tools, techniques and Implementation*, 1st edition. France: Springers, 2022.

[11] IBM, https://www.ibm.com/topics/deep-learning, 2021.

[12] Mathworks, "What is deep learning?" https://www.mathworks.com/discovery/deep-learning.html.

[13] C. Yu, X. Bi, and Y. Fan, "Deep learning for fluid velocity field estimation: A review," *Ocean Engineering*, vol. 271, p. 113693, 2023, doi: 10.1016/j.oceaneng.2023.113693.

[14] F. Maleki, K. Ovens, K. Najafian, B. Forghani, C. Reinhold, and R. Forghani, "Overview of machine learning part 1: Fundamentals and classic approaches," *Neuroimaging Clinics of North America*, vol. 30, no. 4, pp. e17–e32, 2020. doi: 10.1016/j.nic.2020.08.007.

[15] A. Lieto, C. Lebiere, and A. Oltramari, "The knowledge level in cognitive architectures: Current limitations and possible developments," *Cognitive Systems Research*, vol. 48, pp. 39–55, 2018, doi: 10.1016/j.cogsys.2017.05.001.

[16] A. Hoffmann, "Can machines think? An old question reformulated," *Minds Mach (Dordr)*, vol. 20, no. 2, pp. 203–212, 2010, doi: 10.1007/s11023-010-9193-z.

[17] J. K. O'Regan, "How to build a robot that is conscious and feels," *Minds Mach (Dordr)*, vol. 22, no. 2, pp. 117–136, 2012, doi: 10.1007/s11023-012-9279-x.

[18] T. Veale, P. Gervás, and R. Pérez Y Pérez, "Computational creativity: A continuing journey," *Minds Mach (Dordr)*, vol. 20, no. 4, pp. 483–487, 2010, doi: 10.1007/s11023-010-9212-0.

[19] H. T. Tavani, "Can we develop artificial agents capable of making good moral decisions?" *Minds Mach (Dordr)*, vol. 21, no. 3, pp. 465–474, 2011, doi: 10.1007/s11023-011-9249-8.

[20] J. R. Koza, F. H. Bennett, D. Andre, and M. A. Keane, "Automated design of both the topology and sizing of analog electrical circuits using genetic programming," In *Artificial Intelligence in Design '96*, J. S. Gero, and F. Sudweeks, Eds., Springer, South Africa, 1996.

[21] F. C. Pereira, and S. S. Borysov, "Machine learning fundamentals," In *Mobility Patterns, Big Data and Transport Analytics*, Elsevier, pp. 9–29, 2018. doi: 10.1016/B978-0-12-812970-8.00002-6.

[22] A. M. Turing, *Computing Machinery and Intelligence*, England, 1950.

[23] N. J. Nilsson, *The Quest for Artificial Intelligence*, Cambridge, UK: Cambridge University Press, 2009.

[24] Joseph Weizenbaum, "ELIZA A computer program for the study of natural language communication between man and machine," *Communications of the ACM*, vol. 9, no. 1, pp. 36–45, 1966.

[25] D. J. State, and R. Lawrence, "Chess 4.5-the Northwestern University chess program," In *Chess Skill in Man and Machine*, P. W. Frey, Ed., New York: Springer, pp. 80–103, 1983.

[26] A. L. Samuel, "Some studies in machine learning using the game of checkers," *IBM Journal*, vol. 3, no. 3, pp. 210–229, 1959.

[27] G. Wiederhold, and J. McCarthy, "Arthur Samuel: Pioneer in machine learning," *IBM Journal of Research and Development*, vol. 36, no. 3, pp. 329–331, 1992, doi: 10.1147/rd.363.0329.

[28] D. Hillis *et al.*, "In honor of Marvin Minsky's contributions on his 80th birthday," *AI Magazine (Journal)*, Washington DC, pp. 103–110, 2007.

[29] A. L. Fradkov, "Early history of machine learning," *IFAC-PapersOnLine*, vol. 53, no. 2, pp. 1385–1390, 2020. doi: 10.1016/j.ifacol.2020.12.1888.

[30] M. Minsky, and S. Papert, "Perceptron: An introduction to computational geometry," *The MIT Press, Cambridge, Expanded Edition*, vol. 19, no. 88, p. 2, 1969.

[31] F. Rosenblatt, "The perceptron: A probabilistic model for information storage and organization in the brain," *Psychological Reviews*, vol. 65, no. 6, pp. 386–408, 1958.

[32] J. J. Hopfield, "Neural networks and physical systems with emergent collective computational abilities (associative memory/parallel processing/categorization/content-addressable memory/fail-soft devices)," 1982. [Online]. Available: https://www.pnas.org.

[33] K. Fukushima, *Biological Cybernetics Neocognitron: A Self-organizing Neural Network Model for a Mechanism of Pattern Recognition Unaffected by Shift in Position*, South Africa: Springer Link, 1980.

[34] R. Dechter, "Learning while searching in constraint-satisfaction-problems*," [Online]. Available: www.aaai.org.

[35] Santiago Valdarrama, "Considerations when choosing a machine learning model," 2021, https://towardsdatascience.com/considerations-when-choosing-a-machine-learning-model-aa31f52c27f3.

[36] Jorge Garza-Ulloa, *Applied Biomedical Engineering Using Artificial Intelligence and Cognitive Model*, 1st edition. London: Elsevier, 2022.

[37] Ramya Vidiyala, "How to select the right machine learning algorithm," https://towardsdatascience.com/how-to-select-the-right-machine-learning-algorithm-b907a3460e6f, 2020.

[38] D. Paudel, A. de Wit, H. Boogaard, D. Marcos, S. Osinga, and I. N. Athanasiadis, "Interpretability of deep learning models for crop yield forecasting," *Computers and Electronics in Agriculture*, vol. 206, 2023, doi: 10.1016/j.compag.2023.107663.

[39] O. Adeleke, S. Akinlabi, T. C. Jen, and I. Dunmade, "A machine learning approach for investigating the impact of seasonal variation on physical composition of municipal solid waste," *Journal of Reliable Intelligent Environments*, 2022, doi: 10.1007/s40860-021-00168-9.

[40] P. A. Adedeji, S. A. Akinlabi, N. Madushele, and O. O. Olatunji, "Neuro-fuzzy resource forecast in site suitability assessment for wind and solar energy: A mini review," *Journal of Cleaner Production*, vol. 269, p. 122104, 2020.

[41] K. P. Murphy, *Probabilistic Machine Learning: An Introduction*. Cambridge: MIT Pres," 2022.

[42] N. Kühl, M. Goutier, L. Baier, C. Wolff, and D. Martin, "Human vs. supervised machine learning: Who learns patterns faster?" *Cognitive Systems Research*, vol. 76, pp. 78–92, 2022, doi: 10.1016/j.cogsys.2022.09.002.

[43] K. Duraisamy, G. Iaccarino, and H. Xiao, "Turbulence modeling in the age of data," *Annual Review of Fluid Mechanics*, vol. 51, pp. 357–377, 2019.

[44] S. L. Brunton, B. R. Noack, and P. Koumoutsakos, "Machine learning for fluid mechanics," *Annual Review of Fluid Mechanism*, vol. 52, pp. 477–508, 2020.

[45] L. Zhou, Y. Song, W. Ji, and H. Wei, "Machine learning for combustion," *Energy and AI*, vol. 7, 2022, doi: 10.1016/j.egyai.2021.100128.

[46] H. G. W. RJ. Rumelhart DE, "Learning representations by backpropagating errors," *Nature*, vol. 323, pp. 533–536, 1986.

[47] I. Kumar, S. P. Singh, and Shivam, "Chapter 26- Machine learning in bioinformatics," In *Bioinformatics*, D. B. Singh and R. K. Pathak, Eds., Academic Press, pp. 443–456, 2022. doi: 10.1016/B978-0-323-89775-4.00020-1.

[48] J. R. Quinlan., "Induction of decision trees," *Machine Learning*, vol. 1, pp. 81–106, 1986.

[49] I. Goodfellow, Y. Bengio, and A. Courville, *A Deep Learning*. Cambridge, MA: MIT Press, 2016.

[50] R. Polikar, "Ensemble learning," *Ensemble Machine Learning: Methods and Applications*, pp. 1–34, 2012.

[51] J. Hurwitz, and D. Kirsch, *Machine Learning for Dummies*, IBM Limited Edition, New Jersey: John Wiley and Sons, Inc., 2018.

[52] C. Han, Z. He, and A. J. W. Toh, "Pairs trading via unsupervised learning," *European Journal of Operational Research*, 2022, doi: 10.1016/j.ejor.2022.09.041.

[53] N. C. Caballé, J. L. Castillo-Sequera, J. A. Gómez-Pulido, J. M. Gómez-Pulido, and M. L. Polo-Luque, "Machine learning applied to diagnosis of human diseases: A systematic review," *Applied Sciences (Switzerland)*, vol. 10, no. 15, 2020. doi: 10.3390/app10155135.

[54] T. Kohonen, "The self-organizing map," *Neurocomputing*, vol. 21, no. 1–3, 1998.

[55] J. Wang, and F. Biljecki, "Unsupervised machine learning in urban studies: A systematic review of applications," *Cities*, vol. 129, 2022, doi: 10.1016/j.cities.2022.103925.

[56] J. M. Carew, "Reinforcement learning. In-depth guide to machine learning in the enterprise. Techtarget," https://www.techtarget.com/searchenterpriseai/definition/reinforcement-learning, 2023.

[57] Í. Elguea-Aguinaco, A. Serrano-Muñoz, D. Chrysostomou, I. Inziarte-Hidalgo, S. Bøgh, and N. Arana-Arexolaleiba, "A review on reinforcement learning for contact-rich robotic manipulation tasks," *Robotics and Computer-Integrated Manufacturing*, vol. 81, 2023. doi: 10.1016/j.rcim.2022.102517.

[58] S. Bhatt, "Reinforcement learning 101. Towards data science," 2018, https://towardsdatascience.com/reinforcement-learning-101-e24b50e1d292.

[59] Z. Wang, L. Xia, H. Yuan, R. S. Srinivasan, and X. Song, "Principles, research status, and prospects of feature engineering for data-driven building energy prediction: A comprehensive review," *Journal of Building Engineering*, vol. 58, 2022. doi: 10.1016/j.jobe.2022.105028.

[60] Yuliia Kniazieva, "What is data collection in machine learning?" https://labelyourdata.com/articles/data-collection-methods-AI#:~:text=Simply%20put%2C%20data%20collection%20is,fed%20into%20an%20ML%20model, 2022.

[61] K. Maharana, S. Mondal, and B. Nemade, "A review: Data pre-processing and data augmentation techniques," *Global Transitions Proceedings*, vol. 3, no. 1, pp. 91–99, 2022, doi: 10.1016/j.gltp.2022.04.020.

[62] A. Kadhim, "An evaluation of preprocessing techniques for text classification pattern recognition view project improvement text classification using log(TF-IDF) with K-NN algorithm view project." [Online]. Available: https://sites.google.com/site/ijcsis/.

[63] A. Famili, W.-M. Shen, R. Weber, and E. Simoudis Famili, "Data preprocessing and intelligent data analysis," *Intelligent Data Analysis*, vol. 1, no. 1–4, pp. 3–23, 1997.

[64] J. F. Davis, M. J. Piovoso, K. A. Hoo, and B. R. Bakshi, "Process data analysis and interpretation," *Advances in Chemical Engineering*, vol. 25, pp. 1–103, 1999.

[65] Iryna Sydorenko, "How to choose the right machine learning algorithm: A pragmatic approach," 2021, https://labelyourdata.com/articles/how-to-choose-a-machine-learning-algorithm.

[66] David Weedmark, "Machine learning model training: What it is and why it's important," 2021, https://www.dominodatalab.com/blog/what-is-machine-learning-model-training.

[67] M. Mohammed, M. B. Khan, and E. B. M. Bashie, *Machine Learning: Algorithms and Applications*. CRC Press, 2016. doi: 10.1201/9781315371658.

[68] H. V. T. Mai, M. H. Nguyen, and H. B. Ly, "Development of machine learning methods to predict the compressive strength of fiber-reinforced self-compacting concrete and sensitivity analysis," *Construction and Building Materials*, vol. 367, 2023, doi: 10.1016/j.conbuildmat.2023.130339.

[69] A. J. Kilani, O. Adeleke, and C. A. Fapohunda, "Application of machine learning models to investigate the performance of concrete reinforced with oil palm empty fruit brunch (OPEFB) fibers," *Asian Journal of Civil Engineering*, vol. 23, no. 2, pp. 299–320, 2022, doi: 10.1007/s42107-022-00424-0.

[70] N. Mostafa, H. S. M. Ramadan, and O. Elfarouk, "Renewable energy management in smart grids by using big data analytics and machine learning," *Machine Learning with Applications*, vol. 9, p. 100363, 2022, doi: 10.1016/j.mlwa.2022.100363.

[71] O. Adeleke, S. Akinlabi, T. C. Jen, P. A. Adedeji, and I. Dunmade, "Evolutionary-based neuro-fuzzy modelling of combustion enthalpy of municipal solid waste," *Neural Computing and Applications*, vol. 2, 2022, doi: 10.1007/s00521-021-06870-2.

[72] O. Olatunji, S. Akinlabi, N. Madushele, and P. A. Adedeji, "Estimation of municipal solid waste (MSW) combustion enthalpy for energy recovery," *EAI Endorsed Transactions on Energy Web*, vol. 19, no. 23, pp. 1–9, 2019, doi: 10.4108/eai.11-6-2019.159119.

[73] R. Chowdhury, M. R. C. Mahdy, T. N. Alam, G. D. al Quaderi, and M. Arifur Rahman, "Predicting the stock price of frontier markets using machine learning and modified black-scholes option pricing model," *Physica A: Statistical Mechanics and its Applications*, vol. 555, 2020, doi: 10.1016/j.physa.2020.124444.

[74] M. A. Khattak, M. Ali, and S. A. R. Rizvi, "Predicting the European stock market during COVID-19: A machine learning approach," *MethodsX*, vol. 8, 2021, doi: 10.1016/j.mex.2020.101198.

[75] Y. Han, J. Kim, and D. Enke, "A machine learning trading system for the stock market based on N-period min-max labeling using XGBoost," *Expert Systems with Applications*, vol. 211, 2023, doi: 10.1016/j.eswa.2022.118581.

[76] P. Chhajer, M. Shah, and A. Kshirsagar, "The applications of artificial neural networks, support vector machines, and long-short term memory for stock market prediction," *Decision Analytics Journal*, vol. 2, 2022, doi: 10.1016/j.dajour.2021.100015.

[77] M. Bansal, A. Goyal, and A. Choudhary, "Stock market prediction with high accuracy using machine learning techniques," *Procedia Computer Science*, vol. 215, pp. 247–265, 2022, doi: 10.1016/j.procs.2022.12.028.

[78] J. K. Afriyie *et al.*, "A supervised machine learning algorithm for detecting and predicting fraud in credit card transactions," *Decision Analytics Journal*, vol. 6, 2023, doi: 10.1016/j.dajour.2023.100163.

[79] G. Balachandran and J. V. G. Krishnan, "Machine learning based video segmentation of moving scene by motion index using IO detector and shot segmentation," *Image and Vision Computing*, vol. 122, 2022, doi: 10.1016/j.imavis.2022.104443.

[80] M. Kawka, T. MH. Gall, C. Fang, R. Liu, and L. R. Jiao, "Intraoperative video analysis and machine learning models will change the future of surgical training," *Intelligent Surgery*, vol. 1, pp. 13–15, 2022, doi: 10.1016/j.isurg.2021.03.001.

[81] S. S. Harakannanavar, S. R. Sameer, V. Kumar, S. K. Behera, A. v Amberkar, and V. I. Puranikmath, "Robust video summarization algorithm using supervised machine learning," *Global Transitions Proceedings*, vol. 3, no. 1, pp. 131–135, 2022, doi: 10.1016/j.gltp.2022.04.009.

[82] L. Malburg, M. P. Rieder, R. Seiger, P. Klein, and R. Bergmann, "Object detection for smart factory processes by machine learning," *Procedia Computer Science*, pp. 581–588, 2021, doi: 10.1016/j.procs.2021.04.009.

[83] Z. Zhang, H. Zhang, J. Wang, Z. Sun, and Z. Yang, "Generating news image captions with semantic discourse extraction and contrastive style-coherent learning," *Computers and Electrical Engineering*, vol. 104, 2022, doi: 10.1016/j.compeleceng.2022.108429.

[84] K. Panwar, S. Kukreja, A. Singh, and K. K. Singh, "Towards deep learning for efficient image encryption," *Procedia Computer Science*, vol. 218, pp. 644–650, 2023, doi: 10.1016/j.procs.2023.01.046.

[85] M. V. da Silva, L. H. F. P. Silva, J. D. D. Junior, M. C. Escarpinati, A. R. Backes, and J. F. Mari, "Generating synthetic multispectral images using neural style transfer: A study with application in channel alignment," *Computers and Electronics in Agriculture*, vol. 206, p. 107668, 2023, doi: 10.1016/j.compag.2023.107668.

[86] T. v. Nguyen and B. Mirza, "Dual-layer kernel extreme learning machine for action recognition," *Neurocomputing*, vol. 260, pp. 123–130, 2017, doi: 10.1016/j.neucom.2017.04.007.

[87] A. Iosifidis, A. Tefas, and I. Pitas, "Dynamic action recognition based on dynemes and extreme learning machine," *Pattern Recognition Letters*, vol. 34, no. 15, pp. 1890–1898, 2013, doi: 10.1016/j.patrec.2012.10.019.

[88] G. Varol, and A. A. Salah, "Efficient large-scale action recognition in videos using extreme learning machines," *Expert Systems with Applications*, vol. 42, no. 21, pp. 8274–8282, 2015, doi: 10.1016/j.eswa.2015.06.013.

[89] X. Chen, and M. Koskela, "Skeleton-based action recognition with extreme learning machines," *Neurocomputing*, vol. 149, no. Part A, pp. 387–396, 2015, doi: 10.1016/j.neucom.2013.10.046.

[90] K. Kusuma *et al.*, "The performance of machine learning models in predicting suicidal ideation, attempts, and deaths: A meta-analysis and systematic review," *Journal of Psychiatric Research*, vol. 155, pp. 579–588, 2022. doi: 10.1016/j.jpsychires.2022.09.050.

[91] C. (Abigail) Zhang, S. Cho, and M. Vasarhelyi, "Explainable artificial intelligence (XAI) in auditing," *International Journal of Accounting Information Systems*, vol. 46, 2022, doi: 10.1016/j.accinf.2022.100572.

[92] K. R. Pedada, A. Bhujanga Rao, K. K. Patro, J. P. Allam, M. M. Jamjoom, and N. A. Samee, "A novel approach for brain tumour detection using deep learning based technique," *Biomed Signal Process Control*, vol. 82, 2023, doi: 10.1016/j.bspc.2022.104549.

[93] J. M. Inglis, S. Bacchi, A. Troelnikov, W. Smith, and S. Shakib, "Automation of penicillin adverse drug reaction categorisation and risk stratification with machine learning natural language processing," *Journal of Medical Informatics*, vol. 156, 2021, doi: 10.1016/j.ijmedinf.2021.104611.

[94] I. Kaczmarek, A. Iwaniak, A. Świetlicka, M. Piwowarczyk, and A. Nadolny, "A machine learning approach for integration of spatial development plans based on natural language processing," *Sustainable Cities and Society*, vol. 76, 2022, doi: 10.1016/j.scs.2021.103479.

[95] N. V. Patil, "An emphatic attempt with cognizance of the marathi language for named entity recognition," *Procedia Computer Science*, vol. 218, pp. 2133–2142, 2023, doi: 10.1016/j.procs.2023.01.189.

[96] S. Fu, H. Lyu, Z. Wang, X. Hao, and C. Zhang, "Extracting historical flood locations from news media data by the named entity recognition (NER) model to assess urban flood susceptibility," *Journal of Hydrology*, vol. 612, 2022, doi: 10.1016/j.jhydrol.2022.128312.

[97] A. v Hujon, T. D. Singh, and K. Amitab, "Transfer learning based neural machine translation of english-khasi on low-resource settings," *Procedia Computer Science*, vol. 218, pp. 1–8, 2023, doi: 10.1016/j.procs.2022.12.396.

[98] M. Brour and A. Benabbou, "ATLASLang NMT: Arabic text language into Arabic sign language neural machine translation," *Journal of King Saud University - Computer and Information Sciences*, vol. 33, no. 9, pp. 1121–1131, 2021, doi: 10.1016/j.jksuci.2019.07.006.

[99] S. R. Laskar, B. Paul, P. Pakray, and S. Bandyopadhyay, "English-assamese multimodal neural machine translation using transliteration-based phrase augmentation approach," *Procedia Computer Science*, vol. 218, pp. 979–988, 2023, doi: 10.1016/j.procs.2023.01.078.

[100] S. K. Sheshadri, D. Gupta, and M. R. Costa-Jussà, "A voyage on neural machine translation for indic languages," *Procedia Computer Science*, vol. 218, pp. 2694–2712, 2023, doi: 10.1016/j.procs.2023.01.242.

[101] L. Gkinko and A. Elbanna, "The appropriation of conversational AI in the workplace: A taxonomy of AI chatbot users," *International Journal of Information Management*, 2022, doi: 10.1016/j.ijinfomgt.2022.102568.

[102] S. Han and M. K. Lee, "FAQ chatbot and inclusive learning in massive open online courses," *Computers & Education*, vol. 179, 2022, doi: 10.1016/j.compedu.2021.104395.

[103] X. Chen, L. Yao, J. McAuley, G. Zhou, and X. Wang, "Deep reinforcement learning in recommender systems: A survey and new perspectives," *Knowledge-Based Systems*, p. 110335, 2023, doi: 10.1016/j.knosys.2023.110335.

[104] B. Walek and P. Fajmon, "A hybrid recommender system for an online store using a fuzzy expert system," *Expert Systems with Applications*, vol. 212, 2023, doi: 10.1016/j.eswa.2022.118565.

[105] M. Kuanr and P. Mohapatra, "Outranking relations based multi-criteria recommender system for analysis of health risk using multi-objective feature selection approach," *Data & Knowledge Engineering*, vol. 145, 2023, doi: 10.1016/j.datak.2023.102144.

[106] X. Ran, Y. Wang, L. Y. Zhang, and J. Ma, "A differentially private matrix factorization based on vector perturbation for recommender system," *Neurocomputing*, vol. 483, pp. 32–41, 2022, doi: 10.1016/j.neucom.2022.01.079.

[107] A. Pujahari and D. S. Sisodia, "Item feature refinement using matrix factorization and boosted learning based user profile generation for content-based recommender systems," *Expert Systems with Applications*, vol. 206, 2022, doi: 10.1016/j.eswa.2022.117849.

[108] M. H. Aghdam, "A novel constrained non-negative matrix factorization method based on users and items pairwise relationship for recommender systems," *Expert Systems with Applications*, vol. 195, 2022, doi: 10.1016/j.eswa.2022.116593.

[109] M. Ahmadian, S. Ahmadian, and M. Ahmadi, "RDERL: Reliable deep ensemble reinforcement learning-based recommender system," *Knowledge-Based Systems*, vol. 263, 2023, doi: 10.1016/j.knosys.2023.110289.

[110] S. Lee and D. Kim, "Deep learning based recommender system using cross convolutional filters," *Information Sciences*, vol. 592, pp. 112–122, 2022, doi: 10.1016/j.ins.2022.01.033.

[111] N. Heidari, P. Moradi, and A. Koochari, "An attention-based deep learning method for solving the cold-start and sparsity issues of recommender systems," *Knowledge-Based Systems*, vol. 256, 2022, doi: 10.1016/j.knosys.2022.109835.

[112] M. Dong, F. Yuan, L. Yao, X. Wang, X. Xu, and L. Zhu, "A survey for trust-aware recommender systems: A deep learning perspective," *Knowledge-Based Systems*, vol. 249, 2022, doi: 10.1016/j.knosys.2022.108954.

[113] Y. Ho Cho, J. K. Kim, and S. H. Kim, "A personalized recommender system based on web usage mining and decision tree induction," [Online]. Available: www.elsevier.com/locate/eswa.

[114] P. Zhang, Y. Huang, C. Yang, and W. Jiang, "Estimate the noise effect on automatic speech recognition accuracy for mandarin by an approach associating articulation index," *Applied Acoustics*, vol. 203, 2023, doi: 10.1016/j.apacoust.2023.109217.

[115] C. Guerrero Flores, G. Tryfou, and M. Omologo, "Cepstral distance based channel selection for distant speech recognition," *Computer Speech & Language*, vol. 47, pp. 314–332, 2018, doi: 10.1016/j.csl.2017.08.003.

[116] L. Sun, B. Zou, S. Fu, J. Chen, and F. Wang, "Speech emotion recognition based on DNN-decision tree SVM model," *Speech Communication*, vol. 115, pp. 29–37, 2019, doi: 10.1016/j.specom.2019.10.004.

[117] X. Kang, H. Huang, Y. Hu, and Z. Huang, "Connectionist temporal classification loss for vector quantized variational autoencoder in zero-shot voice conversion," *Digital Signal Processing: A Review Journal*, vol. 116, 2021, doi: 10.1016/j.dsp.2021.103110.

[118] C. Champion and S. M. Houghton, "Application of continuous state Hidden Markov Models to a classical problem in speech recognition," *Computer Speech & Language*, vol. 36, pp. 347–364, 2016, doi: 10.1016/j.csl.2015.05.001.

[119] B. Mouaz, B. H. Abderrahim, and E. Abdelmajid, "Speech recognition of Moroccan dialect using hidden Markov models," *Procedia Computer Science*, pp. 985–991, 2019. doi: 10.1016/j.procs.2019.04.138.

[120] Z. Shao, R. Zhao, S. Yuan, M. Ding, and Y. Wang, "Tracing the evolution of AI in the past decade and forecasting the emerging trends," *Expert Systems with Applications*, vol. 209, 2022. doi: 10.1016/j.eswa.2022.118221.

[121] T. Miller, "Explanation in artificial intelligence: Insights from the social sciences," *Artificial Intelligence*, vol. 267, pp. 1–38, 2019. doi: 10.1016/j.artint.2018.07.007.

[122] A. B. Haque, A. K. M. N. Islam, and P. Mikalef, "Explainable Artificial Intelligence (XAI) from a user perspective: A synthesis of prior literature and problematizing avenues for future research," *Technological Forecasting and Social Change*, vol. 186, 2023, doi: 10.1016/j.techfore.2022.122120.

[123] M. Langer *et al.*, "What do we want from explainable artificial intelligence (XAI)? - A stakeholder perspective on XAI and a conceptual model guiding interdisciplinary XAI research," *Artificial Intelligence*, vol. 296, 2021, doi: 10.1016/j.artint.2021.103473.

[124] Y. Chen, L. Liang, and W. Gao, "Non trust detection of decentralized federated learning based on historical gradient," *Engineering Applications of Artificial Intelligence*, vol. 120, p. 105888, 2023, doi: 10.1016/j.engappai.2023.105888.

[125] I. Ullah, U. U. Hassan, and M. I. Ali, "Multi-level federated learning for industry 4.0- A crowdsourcing approach," *Procedia Computer Science*, vol. 217, pp. 423–435, 2023, doi: 10.1016/j.procs.2022.12.238.

[126] Y. Chen, L. Liang, and W. Gao, "Non trust detection of decentralized federated learning based on historical gradient," *Engineering Applications of Artificial Intelligence*, vol. 120, p. 105888, 2023, doi: 10.1016/j.engappai.2023.105888.

[127] D. Pessach and E. Shmueli, "Improving fairness of artificial intelligence algorithms in privileged-group selection bias data settings," *Expert Systems with Applications*, vol. 185, 2021, doi: 10.1016/j.eswa.2021.115667.

[128] A. Ashokan and C. Haas, "Fairness metrics and bias mitigation strategies for rating predictions," *Information Processing & Management*, vol. 58, no. 5, 2021, doi: 10.1016/j.ipm.2021.102646.

[129] C. G.-B. Stephen Fox, "Machine learning in society: Technology in society briefing," *Technology in Society*, vol. 72, p. 102147, 2023.

6 Supervised Learning

6.1 INTRODUCTION

The previous chapter introduced the preliminary information to properly understand the concept of machine learning and its types, algorithms, and applications. In this chapter, we want to delve deeper into understanding of a technique in machine learning, which is supervised learning. We present an overview of its types and components, algorithms, and common practices. Worthy of note is the fact that the information provided in this chapter about supervised learning is not exhaustive. The reader is advised to consult specialized machine learning textbooks for more information beyond what is provided in this chapter.

In supervised learning, algorithms are used to discover a relationship between input variables (features) and output variables (labels) based on a labeled training dataset. Using input attributes, supervised learning aims to create a model that can precisely predict outputs for novel data. A labeled dataset with input variables (features) and associated output variables is used to train the model (labels). The model learns a function that maps the inputs to the outputs using the relationships between the input and output variables. Then, using this function, predictions for fresh, unseen data can be made.

Utilizing the statistical relationship between the input and output, supervised learning models adjust model parameters in response to the discrepancy between the actual result and the prediction [1]. Supervised learning is a potent method for resolving issues involving classification or prediction based on input data. It is widely employed in a variety of industries, and as artificial intelligence technology develops, so do its applications.

6.2 TYPES OF SUPERVISED MACHINE LEARNING

The types and algorithms of the common types of supervised learning is illustrated in Figure 6.1 and briefly discussed as follows.

6.2.1 REGRESSION ANALYSIS

A supervised learning method called regression is used to forecast a continuous output value. Regression analysis with a continuous objective variable links input and output variables. Finance, economics, physics, engineering, and others employ regression. Regression analysis may simulate any input–output relationship, linear or not. Try estimating a house's price based on its size, location, and other factors. To predict this, you may use a regression model. This requires compilation of a dataset of homes with information on their capacity, locations, and other characteristics

DOI: 10.1201/9781003346234-8

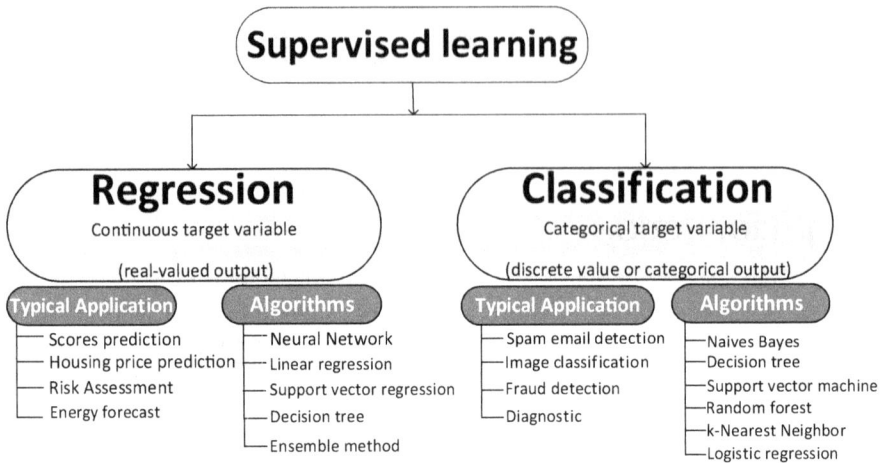

FIGURE 6.1   Types and algorithms of supervised learning.

in addition to their sale prices. This dataset would subsequently be used to train a regression model. As soon as the model has been trained, it can be deployed to predict the anticipated cost of a new home depending on its size, location, and other characteristics. The studies by Adetunji et al. [2], Kusan et al. [3] and Gerek [4] were carried out to achieve this aim using random forest, fuzzy logic, and adaptive neuro-fuzzy inference system (ANFIS), respectively.

The study by Adeleke et al. [5] investigated the relationship between the meteorological parameters and electrical energy consumption at the students' residence at the University of Johannesburg, South Africa. The study developed a neuro-fuzzy model, which is a hybrid of artificial neural network (ANN) and fuzzy logic to predict the electrical energy consumption of the campus residence based on weather parameters such as temperature, humidity, and windspeed. The study is significant for creating an intelligent and workable data-driven decision-making approach to manage the campus's electrical energy consumption in the case study. A similar study explored the regression analysis task of the supervised machine learning to develop a viable system for an on-line and offline monitoring of waste-to-energy (WTE) thermal plants. This study developed an ANFIS model optimized with evolutionary algorithms to estimate the combustion enthalpy of municipal solid waste (MSW) based on its physical composition [6]. The list of the real-work practical applications of the regression analysis of the supervised learning to solve several linear and nonlinear problems is inexhaustive, while the literature is replete with several studies that have deployed these algorithms across several fields for intelligent decision-making, planning, and designs.

6.2.2   Classification

Classification is a type of supervised learning used to predict a categorical output variable. This suggests two or more classes: male or female, yes or no, males or women, 0 or 1, black or white, etc. Classification seeks a link between input factors and output variables, usually shown as a judgment border on a graph. Many issues, like email spam detection, sentiment analysis, image categorization, etc., can be modeled using classification [7]. Common classification tasks are as follows.

6.2.2.1 Binary Classification

Binary classification predicts one of two outputs from the input. The two outcomes are called good and negative. A spam filtering system might produce "spam" or "not spam" from an email. When employing binary classification, the machine learning system recognizes input data patterns related to each class. After training on labeled samples, the method can predict new, unlabeled samples. Support vector machines, decision trees, and logistic regression are popular binary classification techniques.

6.2.2.2 Multi-Class Classification

Multi-class classification divides input into more than two classes. The input can be classified into many categories. For example, a collection of animal photographs with species tags. "Dog," "cat," "bird," "fish," and others may apply here. Classification challenges aim to identify each image's species. Multi-class classification uses logistic regression, decision trees, Support Vector Machine (SVMs), and neural networks. The dataset size, feature count, and classification task difficulty affect algorithm selection.

6.2.2.3 Multi-Label Classification

In multi-label classification, each input can be classified under more than one class or label. This implies that different labels may be simultaneously applied to one input [8,9]. The goal of the multi-label classification task would be to predict the genres that best describe each movie review. A variety of techniques, including classifier chains, binary relevance, and label powersets, can be used to conduct multi-label classification.

6.3 COMMON SUPERVISED LEARNING ALGORITHMS

We shall discuss supervised learning algorithms in the following sections. Understanding these algorithms is crucial for selecting the optimum algorithm and improving model performance. The task to be carried out whether regression or classification determines the algorithm to select.

6.3.1 LINEAR REGRESSION

Linear regression is a prominent predictive analysis technique in machine learning. It is a statistical approach for depicting the relationship between two variables, with the objective of identifying the optimal linear model that characterizes this association. It seeks for the line of greatest fit that most closely approximates the connection between the variables. The determination of the line is achieved by minimizing the sum of the squared discrepancies between the anticipated values and the actual values. Regression analysis may be used to quantify the magnitude of an independent variable's influence over a dependent variable. It can also be applied to project effects or change's influence. Regression analysis gives insights into the understanding of the trend in the variation of the independent variable when one or more independent variables change [10]. Linear regression models offer a simple understandable mathematical technique that can produce predictions. It is represented mathematically in Equation 6.1:

$$y = b_0 + b_1 x \tag{6.1}$$

where y is the dependent variable, x is the independent variable, b_0 is the y-intercept (constant term), while b_1 is the slope (regression coefficient). Equation 6.2 describes a multiple linear regression which encompasses several explanatory variables to predict an outcome. In real life scenarios, several variables determine the turnout of events, thus multiple linear regression finds more applications in real life context.

$$y = b_0 + b_1 x_1 + b_2 x_2 + \dots b_n x_n \tag{6.2}$$

where y is the dependent variable, x_1, x_2,....x_n are the independent variables, b_0 is the y-intercept (constant terms), while b_1, b_2, ...b_n are the slope (regression coefficients).

The Least Squares method is a frequently used method in linear regression to minimize the variance between data points [1]. It fits data by minimizing the sum of squares of residuals vis-à-vis, r_1 and r_2, thus helping to search for optimal parameters of the linear regression model. Finding the regression coefficient values that reduce the mean squared error (MSE) between the observed and predicted values of the dependent variable is the aim of Least Mean Squares. Equation 6.3 is used in Least Mean Squares to iteratively update the regression coefficients:

$$b(t+1) = b(t) + \alpha * r(t) * x(t) \tag{6.3}$$

In this context, $x(t)$ represents the values of the independent variable at iteration t. The residual at iteration t, denoted as $r(t)$, is defined as the discrepancy between the actual and predicted. The current estimate of the regression coefficients at iteration t is represented by $b(t)$. Additionally, the learning rate α is a parameter that governs the magnitude of the update step. The objective of linear regression is to determine the line of best fit that minimizes the sum of squared residuals, representing the discrepancy between observed and anticipated values. Additionally, finding an optimal value of projection with minimal error function is a major focus of linear regression. The error function in linear regression tasks is measured by a vital metric known as the R^2. When making a comparison between observed and anticipated values, a higher R^2 value signifies a strong correspondence, whereas a lower R^2 value implies a worse fit. Linear regression is a machine learning-based statistical approach that finds relevance in data analytics and interpretations.

6.3.2 LOGISTIC REGRESSION

Logistic regression also known as logit model is another significant supervised machine learning algorithm which finds application in classification and predictive analysis task. It is a statistical method used for binary classification problems and estimating the probability of an event. A primary distinctive feature of the linear regression and logistic regression is their respective outputs. While linear regression processes a continuous variable spanning over a wide range, logistic regression computes probability of an instance belonging to a certain group. Furthermore, linear regression seeks for a linear best fit while logistic regression finds S-curve. Moreover, performance of the linear regression is estimated using a least square estimation while logistic regression employs maximum likelihood for its performance estimations. Maximum likelihood estimate seeks to identify the model parameter values that maximize the probability of witnessing the data given the model. A logistic function as represented in Equation 6.4 is used to represent the connection between the independent factors and the dependent variable [11,12];

$$F(x) = \frac{1}{1+e^{-x}} = \frac{e^x}{e^x+1} \tag{6.4}$$

Since the outcome is often depicted in the form of probability, the dependent variable is expressed in the range 0 and 1. However, it involves a categorical or continuous independent variable. The logistic transformation of the unbounded linear equation $p(x)$ of the probability is given in Equation 6.5 while Equation 6.6 represents the resulting solution for $p(x)$. [11,12]. This is a logarithm function which maps the predicted value to a probability, encompassing a real value and threshold value of 0 and 1.

$$\log\frac{p(x)}{1-p(x)} = \alpha_0 + \alpha.x \tag{6.5}$$

$$p(x) = \frac{e^{\alpha_0+\alpha}}{e^{\alpha_0+\alpha+1}} \tag{6.6}$$

Now, we can reduce the misclassification rate by predicting $y = 1$ when $p \geq 5$ and $y = 0$ when $p < 0.5$, while classes in this case are 1 and 0 [13]:

$$L(\alpha_0,\alpha) = \prod_{i=1}^{n} p(x_i)^{y_i}(1-p(x_i))^{1-y_i} \tag{6.7}$$

Once the model parameters have been determined, predictions based on fresh data can be made. By differentiating the Equation 6.7 with respect to various parameters and setting it to zero, we may determine the Maximum likelihood estimation. For instance, the derivative with regard to a_j, one of the parameter alpha's components, is given by Equation 6.8 [14]:

$$l(\alpha_0,\alpha) = \sum_{i=1}^{n} -\log 1 + e^{\alpha_0+\alpha} + \sum_{i=0}^{n} y_i(\alpha_0 + \alpha.x_i) \tag{6.8}$$

Logistic regression's key benefit is its capacity to model the relationship between dependent and independent variables, even in the face of intricate and nonlinear correlations. In addition, logistic regression is reasonably simple to comprehend because each independent variable's influence on the projected probability of the positive class can be determined using the independent variable's coefficients. Logistic regression assumes linearity. Logistic regression assumes linearity between independent factors and dependent variables. To account for nonlinear interactions, the model may need to be altered.

6.3.3 Neural Networks

A form of supervised machine learning algorithm known as a neural network is based on the structure and operation of the human brain [15]. These algorithms are made to identify data patterns and predict or decide based on those patterns. Neural network is one of the most crucial technologies in contemporary artificial intelligence, which is essential to applications including computer vision, natural language processing, and autonomous systems.

A neural network is made of artificial neurons which are a group of interconnected processing nodes. Figure 6.2 compares the biological neurons and the design of the artificial neuron designs. Each neuron processes input from other neurons and outputs to other neurons. Each neuron's processing and inputs and outputs are mathematical vectors. A neuron with label i gets input $x_i(t)$ from a precursor neuron with an activation $U_i(t)$ whose neuron's status relies on discreet time parameter. A default threshold value of θ_i is always set, except altered by the learning process. The new activation in the current neuron at a time $t+1$ from precursor neuron's activation $U_i(t)$ is computed using an activation function f as shown in Equation 6.9:

$$U_i(t+1) = f\left(U_i(t),\ x_i(t),\ \theta_i\ \right) \tag{6.9}$$

The output of the activation of the current neuron is computed using an output function f_{out} (identity function) as shown in Equation 6.10:

$$y_i(t) = f_{out}\left(U_i(t)\right) \tag{6.10}$$

A neural network typically consists of three types of layers: input, hidden, and output. The input data must be changed by the hidden layers into a format that is better suited for the output layer. The network's ultimate output is produced by the output layer. While serving as the network's input interface, an input neuron has no predecessor, likewise an output neuron has no successor and as a result acts as the network's output interface.

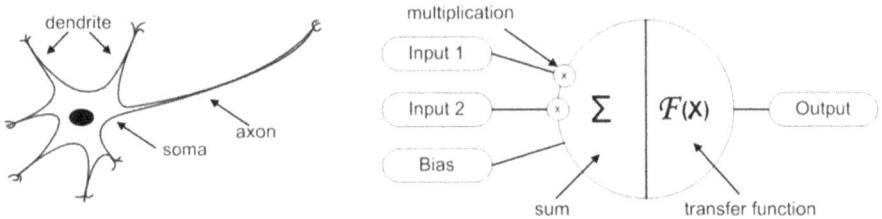

FIGURE 6.2   (a) Biological neurons and (b) artificial neurons [16].

A neural network's core consists of a collection of weights and biases that control how the neurons behave. The biases modify the activation of the neurons, while the weights regulate the strength of the link between neurons. The design of the network and the manner in which input data are processed are both governed by the weights and biases combination. Using a method known as backpropagation, the weights between neurons are modified during the training phase. The network's output error computed is used to modify weights in backpropagation. This operation is repeated several times to improve network performance and eliminate errors. Generally, the neural network architecture can be simplified mathematically as in Equation 6.11:

$$y(t) = F\left(\sum_{i=0}^{m} w_i(t) \cdot x_i(t) + b\right) \tag{6.11}$$

where $w_i(t)$ is the weight value in discreet time t, b is the bias, and F is the transfer function.

The common types of neural networks are discussed as follows.

6.3.3.1 Feedforward Neural Network

The feedforward neural network is a type of neural network with a unidirectional flow of information from input nodes through the hidden nodes and finally to the output nodes. It is also called Multi-Later Perceptron (MLP). In a feedforward neural network, there are connections between every neuron in the layer above it but none within the layer itself. Weights control the strength of the connections between neurons, and they are modified throughout the learning process to enhance the network's functionality. There are no restrictions on the number of layers, kind of transfer function employed in each artificial neuron, or the number of connections between each artificial neuron. A single perceptron, which is the simplest feed-forward ANN and can only learn linear separable problems, is used in this system as shown in Figure 6.3 [16]. The cost function is a critical factor in a feedforward neural network. The adjustment of the weights and bias is aimed at reducing the error (cost function) between the output given by the network and the target outcome Backpropagation, a method for accomplishing this, entails sending the error backward across the network and modifying the weights as necessary.

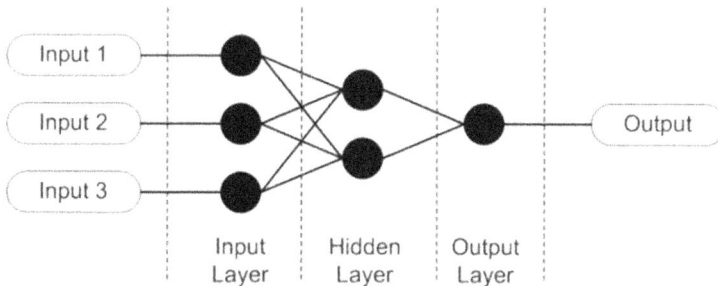

FIGURE 6.3   Topology of the feed forward neural network [16].

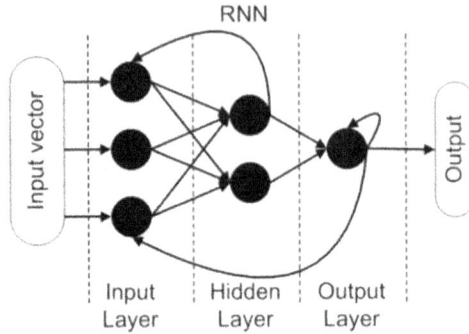

FIGURE 6.4   Topology of the recurrent neural network [16].

6.3.3.2   Recurrent Neural Network

A Recurrent Neural Network (RNN) is a kind of ANN that utilizes feedback connections to handle sequential data. RNNs allows a loop-form flow of information. Its internal state or "memory" enables them to recall the input and thus able to process sequences of input data. Without restrictions on back loops, it is comparable to a feed-forward neural network. In these circumstances, information is no longer only sent in one direction but also in reverse, making nodes interconnections to form a cycle [16]. An RNN's architecture as shown in Figure 6.4 typically consists of a number of recurrent layers, with several memory cells or "hidden units" included in each layer. The network is able to keep a representation of the current context in the input sequence because these hidden units are updated at each time step based on the current input and the prior hidden state [17,18]. RNNs are applicable in scenarios where a sequential dataset is present, and the temporal dynamics linking the data have greater signigicance than the spatial characteristics of any individual frame [19]. For training an RNN, the weights of the network are changed to minimize some loss function through, backpropagation through time (BPTT), a type of backpropagation that takes into account the temporal dependencies in the data [20]. Common examples of RNN are Long Short-Term Memory (LSTM) networks, Hierarchical RNNs, Bidirectional RNNs, and Gated Recurrent Units (GRUs).

6.3.3.3   Convolutional Neural Network

ANNs have a limited application when the inputs are videos and images. Instead, we can utilize a deep neural network such as Convolutional Neural Network (CNN) specially designed for such task [21]. CNNs are modeled after the cerebral cortex's primary visual cortex comprising six layers which are in charge of extracting various features such as edges, shapes, color, etc. [22]. CNNs are used to selectively focus on distinct sections of an image in order to extract features that are pertinent to the job at hand, as opposed to fully linked layers in classic neural networks that treat each input feature equally. The basic architecture of CNN comprises several layers such as convolutional layers, pooling layers, and fully connected layers as shown in Figure 6.5. A series of feature maps that highlight various facets of the input image are created by each convolutional layer by applying a set of filters to the image.

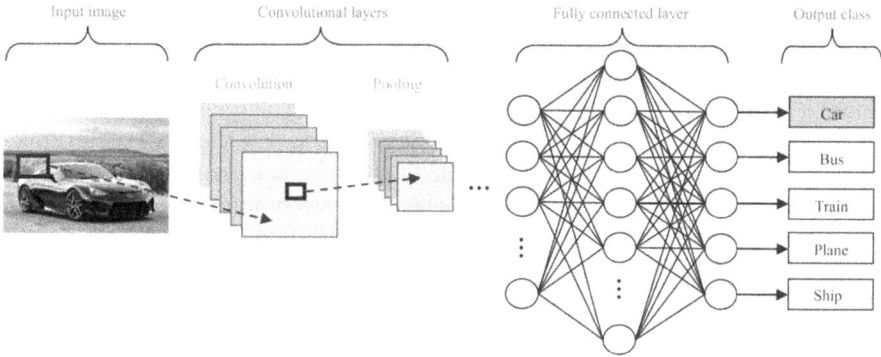

FIGURE 6.5 Framework of convolutional neural network [25].

The feature maps are subsequently down sampled by pooling layers, which lowers their dimensionality while maintaining crucial characteristics. Features that are extracted from the convolutional and pooling layer features are combined by fully connected layers into a set of outputs that are appropriate for the task such as image classification [23]. In order to reduce any loss function during training, the CNN modifies the weights of the filters, often through backpropagation [24].

6.3.4 SUPPORT VECTOR MACHINE

The support vector machine (SVM) is a class of supervised learning algorithms used to determine the best boundary between two classes. SVMs, which are based on statistical learning frameworks, are one of the most reliable prediction techniques [26]. An SVM training algorithm creates a model that categorizes fresh samples to one of two categories based on a collection of training examples, making it a non-probabilistic binary linear classifier. In order to maximize the distance between the two categories, SVM maps training examples to points in space. Then, depending on the gap side where they are positioned, new examples are projected to fit into one of the categories by being mapped into that same space. The basic focus of the SVM is finding the hyperplane that maximizes the margin between the two classes, i.e., the distance between the nearest data points of each class and the hyperplane [27]. The margin is defined as the distance between the hyperplane and the closest data points of each class. This hyperplane is known as the optimal boundary, and the data points closest to it are called support vectors.

Early in the 1960s, a Russian mathematician and computer scientist named Vladimir Vapnik originally proposed the idea of structural risk reduction for pattern recognition, which is when SVMs initially came into existence [28–30]. The theory and method for SVMs in its current version were created in the 1990s by Vapnik and his colleagues at AT&T Bell laboratories in the early 1990s, and they later rose to become one of the most used machine learning techniques [30]. SVM techniques for regression problems were established in 1995 and are now referred to as Support Vector Regression (SVR). Instead of classifying the data, SVR algorithms seek to identify the hyperplane that best fits the data. SVM algorithms have increased in popularity with the development of powerful computing systems, and the creation of new kernel functions has allowed for the solution of increasingly challenging problems. The versatility and efficacy of the method are demonstrated by the history of SVMs, which have played a key role in the advancement of machine learning algorithms.

SVMs are used for both regression and classification problem. SVMs can effectively execute nonlinear classification in addition to linear classification and handle challenges such as nonlinear, high dimension, and local minimum [26]. The regression part of the SVM is known as support vector regression (SVR), which has demonstrated greater performance due to its innate capacity to avoid the overfitting challenge faced in regression and increased response approximation [31]. The objective of the classification task of the SVM is to divide two groups by a function which is inferred from the examples that are provided, while the classifier performs well on unseen instance cases as shown in Figure 6.6a [32]. Although there are numerous alternative linear classifiers that may segregate the data, just one maximizes the margin. The optimal separating hyperplane is the linear classifier.

The mathematical framework of the SVM is the optimization problem which can be formulated as a quadratic programming problem and can be solved using optimization algorithms such as gradient descent or the primal-dual method. Let us consider the task of splitting a set of training vectors into two distinct classes $\{(x_1, y_1), (x_2, y_2).......(x_n, y_n)\}$, while $x \in R^n$ and $y \in (+1, -1)$ alongside a hyperplane $w.x + b = 0$. If there is no error in the separation and the closest vector is as close to the hyperplane as possible, an optimal separation of the collection of vectors is said to occur by the hyperplane as shown in Figure 6.6b. When the margin $2/w$ is maximized, we obtain the optimal hyperplane. The approach can otherwise be depicted by minimizing $w/2$ giving the expression in Equation 6.12:

$$\text{Min.} \ \frac{w^2}{2} \ \text{st.,} \ y_i \left[w.x_i + b \right] \geq \tag{6.12}$$

Beyond the linear classification techniques, the SVM technique could be utilized for nonlinear classification by using the kernel trick methods for inherently charting the

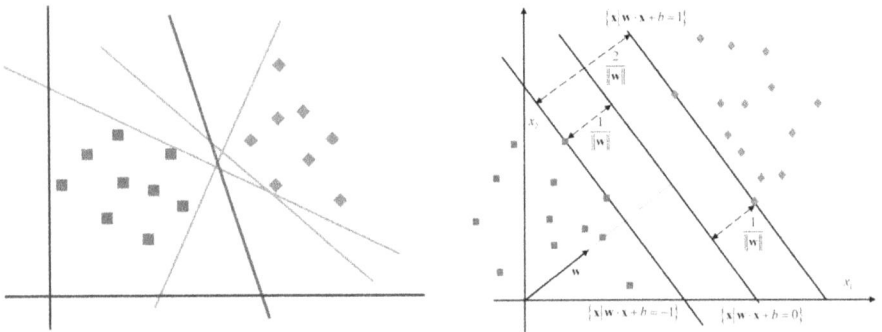

FIGURE 6.6   (a) Separation of hyperplanes for two-class data and (b) optimal hyperplanes to separate two-class data [32].

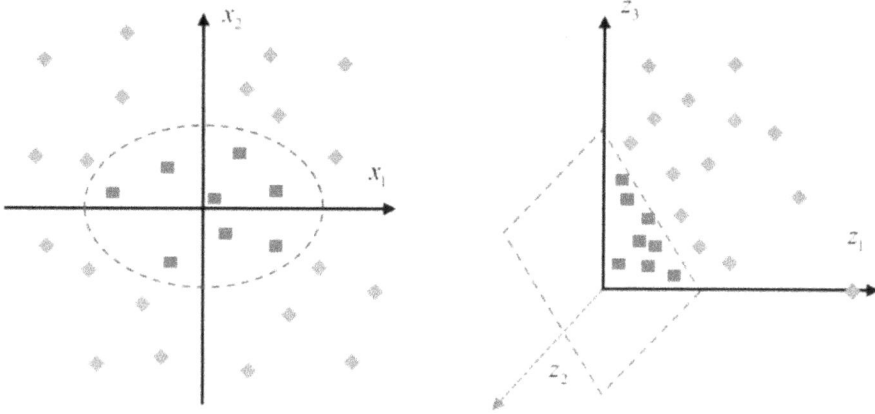

FIGURE 6.7   Input space mapping into high-dimensional feature space [32].

inputs into high-dimensional features space. Figure 6.7 illustrates an instance, in which the mapping $z_1 = x_i^2$, $z_2 = \sqrt{2}\, x_1 x_2$, $z_3 = x_2^2$ is used to transform a two-dimensional input space (x_1, y_1) into a three-dimensional feature space (z_1, z_2, z_3). According to Figure 6.7, it can be observed that the two-class data exhibit linear inseparability inside the initial input space. However, it is noteworthy that these classes may be effectively differentiated within a three-dimensional feature space.

The SVM can be deployed for a regression task. Considering the same set of training vector $\{(x_1, y_1), (x_2, y_2)\ldots\ldots(x_n, y_n)\}$, while $x \in R^n$. A point in the space of R^n is mapped to the space R by using a regression function, f, represented in Equation 6.13:

$$F = \left\{ f(x, w),\ w \in \Lambda | f : R^n \to R \right\} \tag{6.13}$$

where w represents an unidentified parameter vector to be decided and while Λ represents a set of parameters. The regression function is further expressed in terms of the minimum anticipated risk factor and error function, e, depicted in Equation 6.14:

$$R(f) = \int e\big(y - f(x, w)\big) dP(x, y) \tag{6.14}$$

There are various processes involved in training an SVM model, such as data preprocessing, feature selection, and model selection. The data must be cleaned and standardized during data preprocessing to get rid of any outliers or missing values. The process of choosing the features that are most important to the classification of the data is then carried through feature selection. Further to this, selecting the suitable kernel function and customizing the SVM model's hyperparameters is a critical step. Many different applications of SVM, amongst others, include text classification, picture classification, and bioinformatics. SVMs have been used in text classification to divide publications into various groups according to their content and to classify images into different classes based on their visual features. Bioinformatics has made an extensive use of the SVM to forecast protein–protein interactions and categorize proteins into several functional groups.

6.3.5 Decision Tree

Another prominent machine learning approach for classification and regression of tasks is decision trees. They are a kind of supervised learning algorithm that forecasts a target variable's value based on a number of input features. Decision trees divide a dataset into ever-smaller subsets, until arriving at a final classification for a particular observation [33]. A decision tree is a model that resembles hierarchical tree-like structures and has leaf nodes as well as decision nodes. Each leaf node represents a class label, and each decision node represents a test on an attribute. With regard to the target variable, a decision tree aims to divide the data into subsets that are as homogeneous as possible [34].

Due to their transparent decision-making processes and easily comprehensible categorization outcomes, decision trees have emerged as being a widely used classification design [35,36]. This makes them helpful for outlining a decision's justification to stakeholders who aren't technically inclined. In addition to this, other benefits of the decision tree classifier over other classifiers include [37–40]:

- i. It can handle high-dimensional data without the need for domain knowledge. It is capable of handling missing values and both categorical and numerical data.
- ii. It is simple and quick.
- iii. It has good accuracy.
- iv. It can produce rules for classification that are easy to grasp.
- v. It is a versatile technique that may be utilized to solve a wider range of problems in both classification and regression tasks.

Figure 6.8 represents a decision with nodes and edges. While edges result from a split to the next node, nodes divide based on the value of a particular property. The internal nodes, sometimes referred to as decision nodes, are fed by the root node's

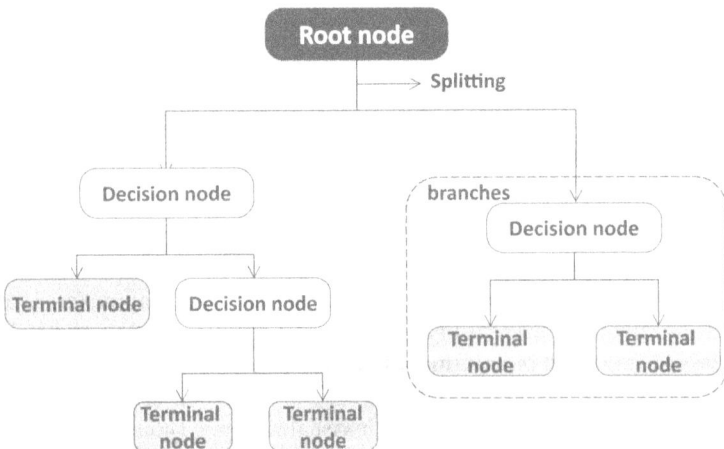

FIGURE 6.8   Decision tree structure.

outgoing branches. Both node types undertake assessments based on the available attributes to create homogenous subsets, which are represented by leaf nodes or terminal nodes.

We begin by creating a decision tree at the root node, which stands in for the entire dataset. Then, we select the attribute that divides the data into two subsets most effectively. Up until leaf nodes that represent a single class or a range of target variable values, we continue to recursively partition the subsets into smaller subsets. By using a greedy search to find the ideal split points inside a tree, decision tree learning uses a divide and conquer technique. Once most or all records have been categorized under particular class labels, this dividing procedure is then repeated in a top-down, recursive fashion [34]. Although there are other approaches to choose the optimal attribute at each node, the Gini impurity and information gain methods are the two that are most frequently used as a splitting criterion in decision tree models. They aid in assessing the effectiveness of each test condition and its capacity to categorize samples into a group.

Decision tree algorithms are as follows:

  i. **ID3 (Iterative Dichotomiser 3):** ID3 is a classic decision which works by selecting the attribute that best splits the data based on the information gain criterion [41,42].
 ii. **C4.5:** This algorithm is an improvement over the ID3 algorithm. It uses a similar splitting criterion based on information gain, but it can handle both categorical and continuous data. C4.5 can also handle missing values and can prune the tree to prevent overfitting [43].
iii. **Classification and Regression Trees (CART):** CART can handle categorical and numerical data, and as a splitting criterion, it employs the MSE or the Gini impurity. CART can be applied to tasks requiring classification and regression [44–46].
 iv. **Multivariate Adaptive Regression Splines (MARS):** A decision tree method that is especially effective for regression applications is the MARS. To do this, piecewise linear models are fitted to the data, and the attribute with the greatest fit at each split is chosen [47,48].
  v. **Chi-square Automatic Interaction Detector (CHAID):** Decision tree algorithms like CHAID are best suited for categorical data. The attribute that gives the greatest significant split is chosen after using the chi-square statistic to verify the input features' independence from the target variable [49–52].

6.3.6 RANDOM FOREST

A popular ensemble learning algorithm in machine learning is random forest. It is a potent method for resolving binary and multi-class classification and regression problems and is a member of the decision tree class [1]. A decision tree–based technique called random forest makes use of numerous decision trees to increase the model's precision and generalizability. A number of decision trees are trained using various subsets of the training data as part of an ensemble learning technique [53]. When building each decision tree, a subset of features is randomly chosen, hence

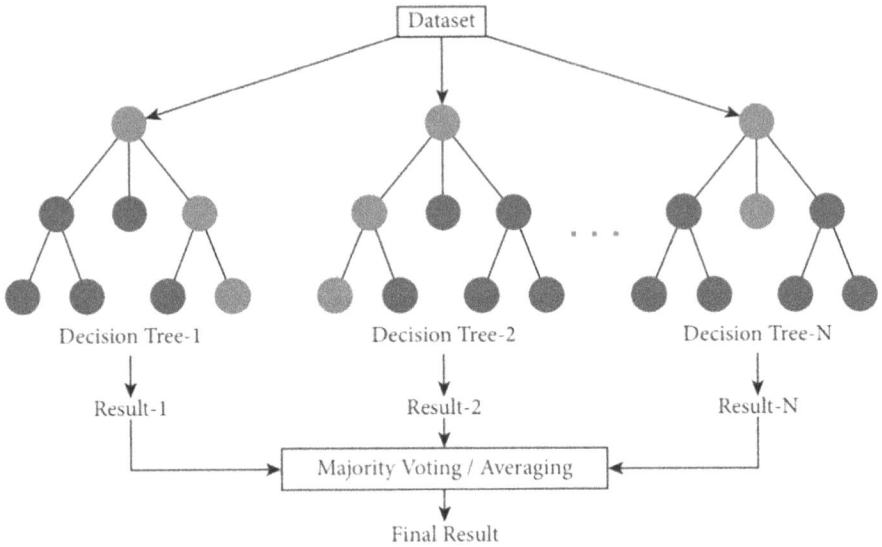

FIGURE 6.9   Vote-based random forest [55].

the term "Random" in the name random forest. The building of numerous prede-termined decision trees serves as the foundation for random forests. The result in classification is the mode of the classes, whereas the output in regression is the mean forecast of the individual trees [54]. The random forest approach used today builds a set of decision trees with controlled variance using bagging and a random selection of features. Figure 6.9 illustrates the functionality of the random forest classifier and the final class from decision trees.

There are three key hyperparameters for random forest algorithms that must be set prior to training. Node size, tree count, and sampled feature count are a few of them. After that, regression or classification issues can be resolved using the random forest classifier.

Each decision tree in the ensemble that makes up the random forest method is built of a data sample taken from a training set with replacement known as the boot-strap sample. One-third of the training dataset, called the out-of-bag (oob) sample, is designated as test data sample [56]. The dataset is subsequently given a second ran-domization injection by feature bagging, increasing dataset diversity and decreasing decision tree correlation. The prediction will be determined differently depending on the problem. The mean of the individual decision trees will be estimated for the regression job, and for the classification task, the predicted class will be determined by a majority vote, or the most common categorical variable. The prediction is then finalized by cross-validation using the OOB sample.

The way random forest creates predictions is by creating a lot of decision trees and combining their outputs. The algorithm operates in the following steps [57]:

i. Select a random subset of features from the input features.
ii. Build a decision tree using the selected features and a random subset of the training data.

iii. Repeat steps 1 and 2 multiple times to create a forest of decision trees.
iv. Predict the output by aggregating the results of all decision trees in the forest.

Bagging, often referred to as bootstrap aggregating, is the process of choosing a random subset of features and a random subset of the training data. By building numerous models that are trained on various subsets of the data, bagging aids in the reduction of overfitting [53].

Random forest has the following drawbacks [54,58,59]:

i. For large datasets, random forest can be sluggish, especially when the number of features is high.
ii. Understanding how each decision tree contributes to the final forecast in a random forest model is challenging.
iii. If there are too many trees in the forest, random forest may overfit.

6.3.7 NAIVE BAYES

Naive Bayes is a widely used machine learning algorithm for classification problems based on the Bayes theorem, a probabilistic method. The Bayes theorem enables you to calculate the likelihood of an event based on past occurrences [1]. According to Bayes' theorem, the likelihood of an event occurring given some previous knowledge is equal to the product of the event's likelihood and the prior knowledge's likelihood given the event has occurred. "Naive" because it thinks input features are independent. Represented in Figure 6.10 is the structure of a hierarchical Naive Bayes model. As a probabilistic model, given an instance that needs to be classified, and represented by a vector $x = (x_i......x_n)$ with some n attributes, the model sets probability $p(C_k \mid x_1....x_n)$ for each k potential class C_k. If there are many features (n) or if a feature can have several values, creating a model using probability tables is impractical. Reconstructing the model makes it more manageable. The probability is therefore broken down as in Equation 6.15 using Bayes' theorem:

$$p(C_k \mid x) = \frac{p(C_k)\, p(x \mid C_k)}{p(x)} \qquad (6.15)$$

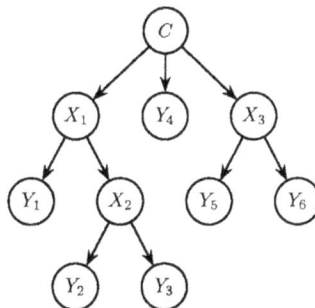

FIGURE 6.10   Hierarchical Naive Bayes model [60].

Thus, a classifier can be constructed based on this Bayes' theorem as follows:

$$\hat{y} = \underset{k \in (1....K)}{argmax} \; p(C_k) \prod_{i=1}^{n} p(x_i \mid C_k) \qquad (6.16)$$

Naive Bayes works in the following steps:

  i. Determine each class's previous probability. This is the likelihood that each class will occur before any input is provided.
  ii. Determine the probability of every input feature given every class. This represents the likelihood that each input feature will occur given the class.
  iii. Add the likelihood of each input feature given each class to the previous probability of each class.
  iv. To determine the posterior probability of each class given the input, normalize the probabilities.

The class with the highest posterior probability is what Naive Bayes produces as its result.

6.3.8 k-Nearest Neighbor

k-Nearest Neighbor (k-NN) algorithm is a non-parametric classifier in supervised learning which leverages proximity to classify or predict the grouping of a given data point. In this context, the target variable is categorical in situations involving classification, while the input data are categorized by the k-nearest neighbor class that appears most frequently in the k-NN algorithm. The input data will be classified as belonging to class A; for instance, if the k-NN method chooses three neighbors, two of whom are in class A and one is in class B [61]. In the context of classification issues, the assignment of a class label is determined by a "majority vote" mechanism. This entails selecting the label that is more commonly seen among the surrounding data points. The target variable is continuous when there is regression problem. A key variable in the k-NN algorithm is the value of k which establishes how many neighbors should be taken into account when producing a prediction. The elbow method or cross-validation can determine k by employing subsets of the training data as test sets to evaluate the model. Based on the average of the k-nearest neighbors, the k-NN method predicts the target value of the input data. The anticipated target value, for instance, will be equal to the average of the target values of the five neighbors chosen by the k-NN algorithm [62].. By graphing the model's performance against various values of k, the elbow technique determines the value of k at which the performance begins to plateau. A classification of a brand-new observation $(k = 1)$ using KNN is shown in Figure 6.11.

6.4 COMMON CHALLENGES OF SUPERVISED LEARNING

Although supervised learning has significantly advanced a number of industries, including healthcare, banking, and engineering, it is not without its drawbacks,

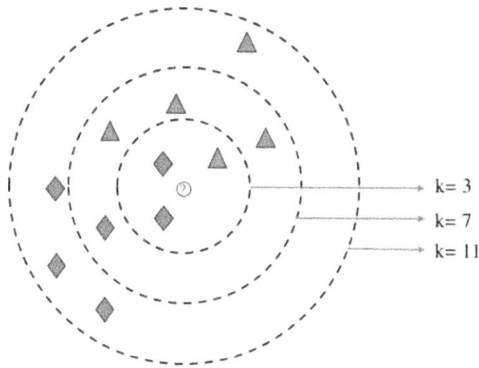

FIGURE 6.11 K-Nearest Neighbour (KNN) classification approach [63].

which should not be disregarded or dismissed. We can create better models and more precise forecasts while being aware of their ramifications if we are aware of these constraints. Challenges with data quality, bias and discrimination, overfitting, and limited generalization are some of these shortfalls which are discussed as follows.

6.4.1 OVERFITTING AND UNDERFITTING

The complexity of the model is one design decision we must make when selecting a machine learning model. The quantity of parameters that need to be optimized in order to train the model is what we mean by complexity. The model is possibly more powerful, the more parameters there are. Yet, we must be cautious of any risks that can be associated with models containing lots of parameters. Underfitting and overfitting are two pertinent concepts that relates to model's parameters and complexity [64]. When a model is overly complicated and learns the noise in the training data rather than the underlying patterns, this is known as overfitting. Because of this, the model does well with training data but poorly with novel, untried data. This is due to the model's inability to generalize successfully to new data because it has learned the training data so well that it has effectively memorized it [65].

Overfitting usually occurs when a model is too complex for the training dataset. The model contains too many degrees of freedom, allowing it to account for data noise and randomness. Overfitting can also occur when a model is trained too many times and memorizes the training data instead of learning the patterns. Underfitting happens when a model is overly simplistic and unable to discern the fundamental trends in the data. In essence, the model performs poorly on both the training and validation datasets because it is not complicated enough to understand the correlations between the input and output variables. A linear model can match a nonlinear data connection. Underfitting can also occur when the model is not trained long enough or has insufficient data.

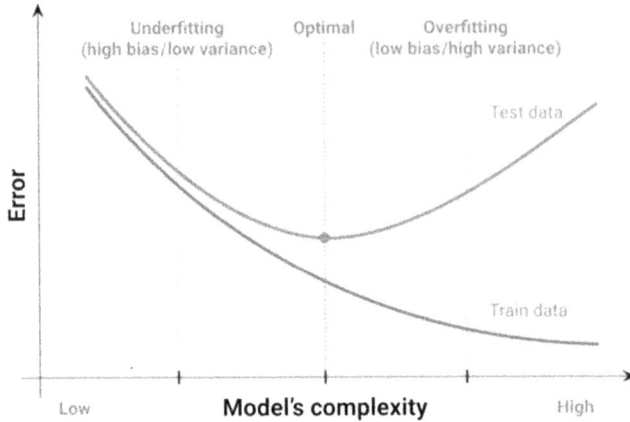

FIGURE 6.12   Learning curve for model training and testing based on their complexity [64].

The task of finding an optimal model is achieved at a mid-point between underfitted and overfitted points as shown in the learning curve in Figure 6.12 depicting the error in training and testing phase based on the model's complexity [64].

A simple regression analysis for demonstrating overfitting and underfitting is presented in Figure 6.13. In all three charts, the same data points are displayed. The graphs in Figure 6.13a show a first-order polynomial (straight line) fit the data. The model underfits the data because it cannot reflect the variability in the data. In Figure 6.13b, we have fifth-order polynomial (overfitting). The model's additional parameters make it evidently strong enough to fit the data. The model might not, however, generalize very well. For instance, there would likely be huge mistakes if we were to assess how well it fits at other x locations. This is known as "overfitting the data". Figure 6.13c shows a third-order polynomial which in this instance most likely indicates a satisfactory fit. The curve is likely to generalize well since it fits the known data well without overfitting to the data noise [66]. Although this is a simple illustration, this sample applies to complex machine learning models.

In order to prevent overfitting and underfitting, it's crucial to select a model with the right level of complexity and train it on the right quantity of data. By encouraging the model to have simpler parameter values, regularization strategies, like adding a penalty term to the loss function, can help to prevent overfitting. By evaluating the model on various subsets of the data, cross-validation like leave-one-out cross-validation (LOOCV) and k-fold cross-validation can also aid in the diagnosis and prevention of overfitting [67]. In k-fold cross-validation, the model is trained and tested on k equal-sized subsets of the dataset.

6.4.2 Class Imbalance

With supervised learning, each example has a corresponding class or label because the dataset has been labeled. For instance, each case can be classified as either positive or negative in a binary classification task. Class imbalance happens when one class has a disproportionately low number of samples compared to the other classes,

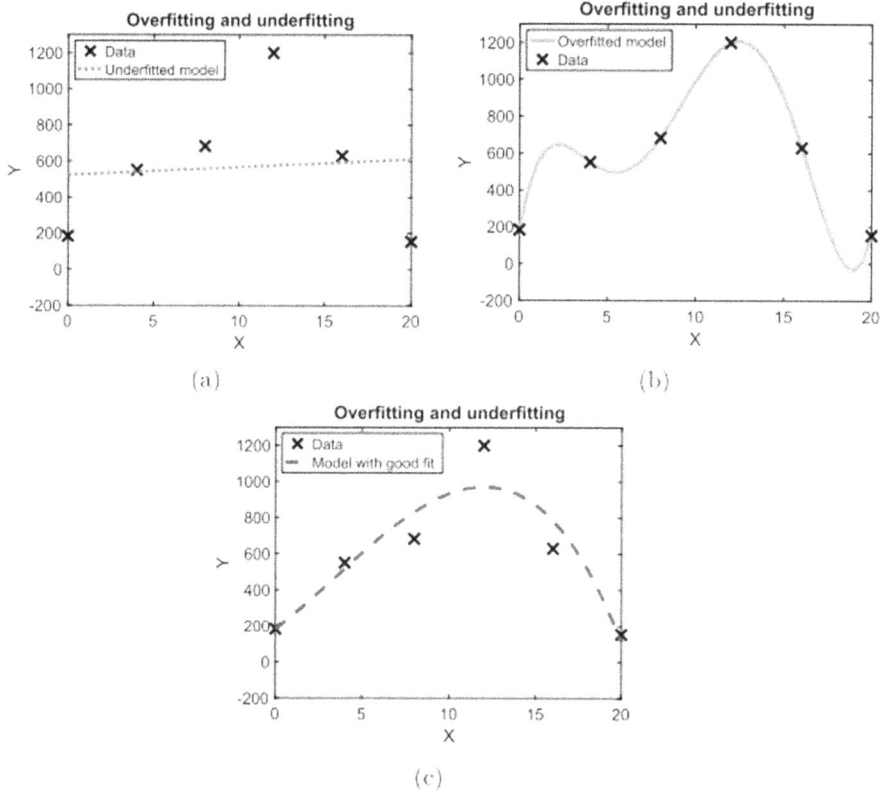

FIGURE 6.13 Illustration of (a) underfitting, (b) overfitting, and (c) good fit [66].

making it more challenging for the machine learning algorithm to identify patterns and forecast outcomes [68]. Take a binary classification problem with 90% of examples in class A and 10% in class B. Class B dominates this dataset. If we create a binary classification model to predict whether a transaction is fraudulent and only 1% is fraudulent, the dataset is highly skewed. Building precise and efficient machine learning models requires addressing the class imbalance. The following are some strategies for addressing the disparity between classes:

6.4.2.1 Data Resampling

One of the most common methods to address class imbalance is to resample the data to balance the number of observations in each class [69]. This can be done either by oversampling the minority class or under sampling the majority class. Oversampling techniques include random oversampling, where the minority class is replicated to match the number of samples in the majority class, and Synthetic Minority Oversampling Technique (SMOTE), which generates new synthetic minority class samples based on the k-nearest neighbors of the minority class samples. Under sampling techniques involve randomly selecting a subset of the majority class samples to match the number of samples in the minority class [70].

6.4.2.2   Cost-Sensitive Learning

To do this, the learning algorithm must be changed to increase the cost of misclassifying instances of the minority class. This can be accomplished by giving each class weights based on how frequently they occur or by altering the loss function to give the minority class greater weight [71,72].

6.4.2.3   Anomaly Detection

Anomaly detection techniques can be used to identify and isolate the minority class from the majority class, allowing for targeted analysis and classification of the minority class [73].

6.4.2.4   Algorithm-Specific Techniques

Some algorithms have built-in methods for addressing class imbalance. For example, decision trees can use cost-sensitive learning by assigning weights to each class, and SVMs can use class-weighting or use a kernel function that is sensitive to the minority class.

Figure 6.14 diagrammatically represents some of the approaches adopted for mitigating the challenge of class imbalance.

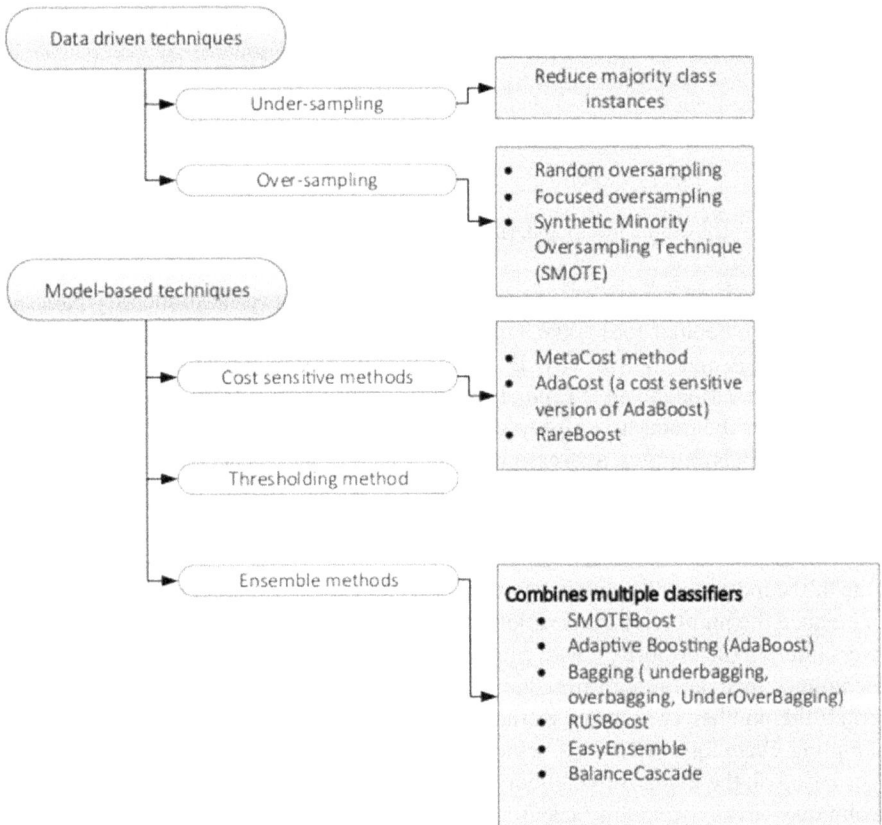

FIGURE 6.14   Overview of approaches in minimizing the effect of class imbalance [74].

6.4.3 DATA AVAILABILITY AND QUALITY

Supervised machine learning grapples with these two significant hurdles [75]. To create a robust supervised learning model, a reasonable amount of data is a vital requirement. However, acquiring such a dataset can be time-consuming and costly. Poor data quality inhibits the accuracy and decision making capability of machine learning models. It might not always be possible to gather enough labeled data to solve a particular issue. When working with newer technology or in specialized industries where data are limited, this task can be more challenging. The correctness and dependability of the labeled data used to train a supervised learning model can have a significant impact on the model's performance, which is why data quality is equally crucial. Bad data might hinder a model's capacity to generalize to new data and lead to erroneous model predictions. Incomplete or missing data, labeling errors, and data bias may affect data quality. Data availability and quality can be improved by using data augmentation to increase and improve the dataset [76]. Techniques for data augmentation include adding noise to the data, flipping or rotating photos, and altering the brightness or contrast of the images [77]. Using pre-trained models or transfer learning strategies that can benefit from currently available labeled datasets to enhance the model's performance on fresh data is an additional alternative. Moreover, techniques like active learning or semi-supervised learning can be employed to maximize the use of scant labeled data when training supervised learning models.

6.4.4 CHALLENGES WITH FEATURE ENGINEERING

The act of choosing and converting raw data features into a format that machine learning algorithms can use is known as feature engineering. Effective feature engineering can lead to improved model performance and better insights into the underlying patterns in the data. However, feature engineering can present several challenges in supervised learning. Some researchers assert that the primary goal of feature engineering is to improve machine learning by optimizing the representation of the feature space [73]. A crucial phase in supervised learning is feature engineering because the effectiveness of the final model depends heavily on the caliber and applicability of the features utilized. To reduce data dimensionality, eliminate redundancy, and reduce noise, feature engineering also includes feature selection and extraction. Effective feature engineering requires domain knowledge, an understanding of the problem being solved, and careful consideration of factors such as data distribution, missing data, and feature interactions. Common feature-engineering-related challenges in supervised machine learning are handling missing data, including redundant features in feature selections, computational complexities amongst others.

6.5 COMMON SUPERVISED MACHINE LEARNING OPERATIONS

To use supervised learning effectively, we must understand model creation and refinement. This section discusses the main supervised machine learning techniques and their importance for model accuracy. From feature selection through regularization, performance evaluation, and hyper-parameter tuning, we will examine machine learning practitioners' best practices for building robust models.

6.5.1 FEATURE SELECTION

Data gathered from real-world context typically contains a significant number of attributes and features [80]. High dimensional data requires more processing power and storage space. In addition, bad performance of the classification model is caused by noisy, redundant, and irrelevant features in these data. The choice of an effective feature subset, minimize the dimensionality of the data, cut down the training time and making the model interpretation simpler, feature selection become necessitated as it purges redundant, irrelevant, and noisy features from the original feature space [81]. It is crucial to remember that feature selection may not be required or advantageous without exception because certain models may perform better with more features, and some features may be crucial when combined with other features. There is no one-size-fits-all approach to feature selection, thus the way to use will depend on the particular issue and dataset at hand. To get the best results in practice, a combination of techniques gives a satisfactory outcome. The No Free Lunch (NFL) Theorem [82] states that no single feature selection strategy can perform satisfactorily on datasets generated in different contexts. Although a single feature selection technique has some advantages, it also has some drawbacks. Sequel to this, Wu et al. [81] suggested an ensemble method of feature selection to overcome this challenge.

When there are multiple input features or variables that might not all be pertinent to the goal variable(s) being forecasted, evolutionary multi-objective optimization (EMO) can be utilized as a feature selection strategy. EMO-based approach is especially well suited for handling feature selection tasks because it can identify a number of nondominated solutions (feature subsets) with the trade-off between various competing objectives in a single run [80]. The categories of feature selection methods in supervised learning is summarized as follows and illustrated in Figure 6.15.

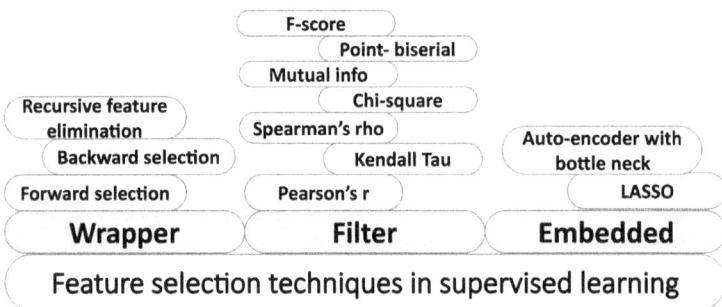

FIGURE 6.15   Feature selection techniques.

6.5.1.1 Wrapper Technique

This method utilizes a model to score many feature subsets before choosing the best one. A model is trained on each new subset, and its performance is then assessed on a hold-out set. Features subset that produces the best model performance is chosen [83]. Wrapper approaches typically offer the best feature set for the particular model type selected, which is a significant advantage. Common feature selection approaches in this category are forward and backward selection and recursive feature elimination.

6.5.1.2 Filter Technique

The filter technique is a simpler and easier substitute for wrapper technique. They statistically establish the importance of each feature with the model's target and rank them accordingly using metrics like correlation or mutual information [84,85]. In addition to being quicker than wrappers, filter techniques are also more general because they don't overfit to any one algorithm because they are model-agnostic. They are also rather simple to understand: if a feature has no statistical relevance to the aim, it is eliminated [83]. Common feature selection approaches in this category includes the chi-square, Pearson's correlation, Spearman's rank correlation, Kendall rank correlation, mutual information, Point-biserial correlation, and ANOVA F-score.

6.5.1.3 Embedded Technique

With the embedded technique, feature selection is included into the model construction process. The goal is to achieve the best of both methods by combining filter speed with the ability to obtain the ideal subset for a given model, much like from a wrapper. Examples are the LASSO and auto-encoder with a bottleneck.

6.5.2 REGULARIZATION

The term "regularization" describes methods for calibrating machine learning models to reduce the adjusted loss function, avoid overfitting or underfitting, and obtain an optimal model. It achieves this by instituting additional constraints or penalties on their parameters during training. Figure 6.16 illustrates the idea of regularization on an over-fitted model. Keep in mind that regularization does not minimize training

FIGURE 6.16 Overfitted model regularization [86].

error; rather, it seeks to reduce test or generalization error. The gap between train-ing and test error is reduced by minimizing the test error, which also reduces model overfitting.

6.5.2.1 L1 (LASSO) Regularization

This method involves adding a penalty term to the model's cost function that is pro-portional with the total of the parameters' absolute values. As a result, the model is encouraged to have sparse weights, which means that some of the weights will be driven to zero, making the model simpler. Equation 6.17 represents the L1 regular-ization for a linear regression model with a modified cost function:

$$\text{Cost } (w) = RSS \ (w) + \lambda * \left(\text{Sum of absolute values of weights} \right)$$

$$= \sum_{i=1}^{n} \left\{ y_i - \sum_{j=0}^{M} w_j x_{ij} \right\}^2 + \lambda \sum_{j=o}^{M} w_{ij} \tag{6.17}$$

6.5.2.2 L2 (Ridge) Regularization

The proposed approach entails augmenting the cost function of the model by includ-ing a penalty term that is directly proportional to the square of the overall parameter count. Consequently, the model is incentivized to possess reduced weights, perhaps leading to a more streamlined model. The utilisation of this technique successfully mitigates the issue of overfitting and further enhances the interpretability of the model. Equation 6.18 denotes the modified cost function employed in linear regres-sion models that incorporate both ridge and LASSO regularizations [87]:

$$\text{cost} = \sum_{i=0}^{n} \left\{ y_i - \sum_{j=0}^{M} x_{ij} w_j \right\}^2 + \lambda \sum_{j=o}^{M} w_j^2 \tag{6.18}$$

6.5.2.3 Elastic Net Regularization

The objective of linear regression is to identify the set of coefficients that minimize the sum of squared errors between the predicted and observed values. In this objec-tive function, elastic net regularization introduces a penalty term that combines L1 (LASSO) and L2 (ridge) regularization. The elastic net technique can handle corre-lated predictor variables better than L1 or L2 alone since it uses both L1 and L2 regu-larization. It is especially helpful for high-dimensional datasets with many features since it enables the selection of pertinent predictors and the suppression of ones that are irrelevant. Elastic net provides a solution to the problem in equation for a range of α-values strictly between 0 and 1 and a nonnegative λ [88]:

$$\min_{\beta_0 \beta} \left(\frac{1}{2N} \sum_{i=1}^{N} \left(y_i - \beta_0 - x_i^T \beta \right)^2 + \lambda P_\alpha (\beta) \right) \tag{6.19}$$

where

$$P_\alpha(\beta) = \frac{1-\alpha}{2}\beta_2^2 + \alpha\beta_1 = \sum_{j=1}^{P}\left(\frac{1-\alpha}{2}\beta_j^2 + \alpha|B_j|\right) \qquad (6.20)$$

When $\alpha = 1$, elastic net and LASSO are the same. Elastic net approaches ridge regression as α decreases toward 0. The penalty term $P_\alpha(\beta)$ interpolates between the L1 norm of β and the squared L2 norm of β for other values of α [87].

6.5.3 Model Performance Evaluation

The evaluation of model's performance on a certain dataset is a critical stage in supervised machine learning. It aims to discover any problems or limits with the model that needs to be fixed as well as how effectively it can generalize to new, untested data. Every machine learning workflow should consider model generalization after developing a model from scratch. We need to know if it works to trust its projections. These issues can be overcome by evaluating a machine learning model. The techniques used for assessing the performance of supervised learning are described as follows.

6.5.3.1 Train-Test Split (Hold-Out) Technique

This approach splits the dataset into training and testing sets. Holdout evaluation tests a model with data different from its training data. This evaluates learning efficacy impartially. The training dataset trains the model, while the testing dataset evaluates it. The hold-out method created a training and testing set [89]. Depending on dataset size and task complexity, the split is 70:30, 80:20, or 90:10. The training set trains the model, whereas the testing set evaluates it. The train-test split technique's merits include simplicity, speed, minimal computational resources, and instantaneous model performance estimation. If the testing set is small or imbalanced, it may have large variance [90].

6.5.3.2 Cross-Validation Technique

Cross validation evaluates a supervised machine learning model's performance by dividing the given dataset into numerous folds [91]. When the dataset is tiny or the model is prone to overfitting, this method is especially helpful. Cross-validation helps tune hyperparameters and estimate model performance by comparing different configurations. Popular cross-validation approaches include the following:

i. **k-fold cross-validation:** In this approach, the dataset is divided into k folds of equal size. The model is tested on the last fold after being tested on $k-1$ folds. Each fold serves as the testing set once during this process, which is performed k times. To estimate the model's performance, the performance metrics are averaged over the k folds [92,93].
ii. **Stratified k-fold cross-validation:** When the dataset is unbalanced or the class distribution is not uniform, stratified k-fold cross-validation is utilized.

With this method, each fold has a balanced representation of each class because of the way the folds are made [94].

iii. **Leave one-out cross-validation:** A special example of k-fold cross-validation, where k is the number of samples in the dataset, is LOOCV. Every sample in LOOCV serves as the testing set, while the remaining samples are utilized to train the model. The performance indicators are averaged across the samples after repeating this procedure for each sample [95–97].

iv. **Time-series cross-validation:** When the dataset is sorted by time, time series cross-validation is preferred. The dataset is divided into training and testing sets using this method, with the testing set containing only future data points. This makes sure that the model's performance is measured by how well it can forecast upcoming data points [98].

6.5.3.3   Evaluation Metrics

One of the most important factors in determining whether a supervised learning algorithm is successful, accurate, reliable, and eligible is performance metrics. These metrics evaluate a model's ability to predict incoming data. Classification, regression, grouping, and anomaly detection problems determine performance metrics. Commonly used performance metrics for classification tasks include accuracy, precision, F1-score, false-positive rate (FPR), and specificity or recall. For regression tasks, metrics such as MSE, mean absolute error (MAE), correlation co-efficient (R^2), mean absolute deviation (MAD), variance accounted for (VAR), root mean bias error (rMBE), and mean absolute percentage error (MAPE), amongst others, are often used [6,99]. Beyond the type of problem, the selection of performance indicators is influenced by the objectives of the task and the demands of the stakeholders. Table 6.1 summarizes some of the metrics used in supervised learning. To understand metrics used in the classification problem, we must define the different kinds of results we always obtain:

i. **True Positive (TP):** model outcome accurately predicts that an observation is a positive class.
ii. **True Negative (TN):** model outcome accurately predicts that an observation is a negative class.
iii. **False positive (FP):** model outcome wrongly predicts that an observation is a positive class.
iv. **False negative (FN):** model outcome wrongly predicts that an observation is a negative class.

6.5.4   HYPERPARAMETER TUNING

In machine learning, hyperparameters are parameters that the user directly defines to regulate the learning process. They are also referred to as settings for the model that are not directly learned from the data during training, such as learning rate, regularization strength, or number of hidden layers. These hyperparameters are set before model learning to improve it [97]. Hyperparameter tuning optimizes machine

TABLE 6.1

Performance Evaluation Metrics for Supervised Learning

Task	Performance Metrics	Mathematical Expression
Classification	Accuracy	$\dfrac{(TP+TN)}{(TP+FP+TN+FN)}$
	Precision	$\dfrac{TP}{(TP+FP)}$
	Specificity	$\dfrac{TN}{(TN+FP)}$
	Recall/sensitivity	$\dfrac{TP}{(TP+FN)}$
	False-positive rate (FPR)	$\dfrac{FP}{(FP+TN)}$
	F1-score	$2 \times \dfrac{(\text{Precision*Recall})}{(\text{Precision} + \text{Recall})}$
Regression	Co-efficient of determination (R^2)	$1 - \left(\dfrac{\frac{1}{N}\sum_{k=1}^{N}(\widehat{y_k} - y_k)^2}{\sum_{k=1}^{N}(\widehat{y_k} - y_k)^2} \right)$
	Mean absolute deviation (MAD)	$\dfrac{1}{N}\sum_{k=1}^{N}|y_k - \bar{y}|$
	Mean square error (MSE)	$\dfrac{\sum_{k=1}^{N}\left[y_k - \widehat{y_k} \right]}{N}$
	Mean absolute percentage error (MAPE)	$\dfrac{1}{N}\sum_{k=1}^{N}\left| \dfrac{y_k - \widehat{y_k}}{y_k} \right| \times 100\%$
	Root mean bias error (rMBE)	$\dfrac{1}{N}\sum_{k=1}^{N}\left(\dfrac{\widehat{y_k} - y_k}{y_k} \right)$

NB: N, number of testing data sample; y_k, actual observation; $\widehat{y_k}$, predicted value; $\bar{y}$, mean observation.

learning model hyperparameters. These hyperparameters affect how well the model learns from data and generalizes to new, unknown data; thus, they must be fine-tuned to build a high-performing machine learning model. Iteratively changing a variety of hyperparameters and assessing the model's performance on a validation set are the usual methods for hyperparameter tuning [100]. Many methods, including grid search, random search, and Bayesian optimization, can automate the hyperparameter tuning process. These methods are briefly described as follows.

6.5.4.1   Grid Search

In this method, each hyperparameter is given a set of values, and the model's performance is assessed for each and every combination of these values. It entails specifying a range of values for every hyperparameter and thoroughly looking through all viable value combinations [101]. The combination that yields the highest performance for the model after training and evaluation is chosen. Although this approach can be computationally expensive, particularly for models with many hyperparameters, it ensures that the best values will be found within the defined search space.

6.5.4.2   Random Search

Random search is a more efficient alternative to grid search. This approach involves assessing the model's performance after randomly selecting some hyperparameters from a predetermined range or distribution [102]. When the search space is large and complex, this method is computationally more effective than grid search.

6.5.4.3   Bayesian Optimization

This advanced method uses probabilistic models to find the best hyperparameters. New hyperparameter configurations are assessed, and the model's predictions determine the next set to evaluate [103]. Bayesian optimization is often more efficient than grid search and random search, especially when the search space is large.

6.5.4.4   Genetic Algorithms

Genetic algorithms are optimization techniques inspired by the process of natural selection. In hyperparameter tuning, the configurations with the best performance are chosen to "reproduce" and produce a new generation of hyperparameter configurations from a population of hyperparameter configurations that are generated at random [99]. This procedure is repeated until the ideal hyperparameters are identified.

6.5.4.5   Gradient-Based Optimization

Gradient-based hyperparameter optimization uses gradient descent or other optimization methods [104]. When the search space is continuous and the hyperparameters are interdependent, this strategy can be useful.

Hyperparameter tuning is crucial since it can considerably enhance a machine learning model's performance. Tuning the hyperparameters can guarantee that the model is operating at its best for the particular task because the best hyperparameters can change depending on the dataset and the particular problem being addressed [100]. A machine learning model may underfit or overfit the data if the hyperparameters are not tuned properly, which will lead to subpar performance on new, untried data. There are many hyperparameters in supervised learning models. Some of the prominently used hyperparameters in supervised machine learning are briefly discussed in the following:

i. **Learning rate:** The optimization algorithm's learning rate defines the step size for each iteration. It regulates how quickly the model picks up new information from training data. A low learning rate indicates a slow convergence while a high learning rate indicates a high convergence but at the expense of accuracy [105].

ii. **Epoch number:** The epoch number establishes how many times the training algorithm will run over the full training dataset [106]. The performance
of the model can be significantly impacted by the choice of the epoch number. Underfitting may be attributed to a low epoch number while a high
epoch number is accountable for overfitting [107].

iii. **Number of hidden layers:** The number of hidden layers in a deep learning model determines its depth. A more sophisticated model may possess
enhanced capabilities in discerning nuanced patterns within data, although
it may also have a higher susceptibility to overfitting [107].

iv. **Number of neurons in each hidden layer:** Each hidden layer's neurons
determine the model's width. Larger models may capture more complex
data patterns, but they may also overfit [107].

v. **Activation function:** In order to add nonlinearity to the model, the activation
function is used. Depending on the data and the job, several activation functions, such as sigmoid, ReLU, and tanh, may perform better or worse [108].

vi. **Dropout rate:** To forestall overfitting in deep learning models, dropout is a
regularization strategy commonly used [109]. The likelihood that each neuron will be lost during training is determined by the dropout rate. Although
a larger dropout rate might minimize overfitting, it might also reduce the
model's precision.

vii. **Batch size:** The batch size decides how many samples are utilized to iteratively update the model's parameters [110]. While a bigger batch size might
hasten convergence, it also might make overfitting more likely.

6.6 CONCLUSION

Supervised learning represents a basic and transformative paradigm in machine
learning space. In this chapter, we presented an overview of the concept of supervised
machine learning, which is a subset of machine learning in which the algorithm is
trained on labeled data or input data that have a known goal value or output. It makes
intelligent decisions by either classifying a categorical data or predicting a numerical
output. This chapter provided insights into the distinctive features, types and applications of different supervised machine learning spectrum while establishing the importance of right choice of features and data representation. Further to this, we examined
common supervised learning tasks while establishing the significance of performance
evaluation of machine learning model. Some of the critical hyper-parameters which
are significant to its robustness were examined and evaluation metrics used for establishing their reliability were discussed. Despite the immense benefits of supervised
machine learning in solving complex real-life problems, it grapples with some challenges which impedes its full-scale exploration. This chapter identified some of the
potential drawbacks of supervised learning, including overfitting and the requirement
for an adequate amount of labeled data, while we present practical solutions for overcoming these issues, such as regularization, cross-validation, and model selection. To
shape the trajectories of machine learning applications, a proper understanding of its
features, types, algorithms, and approaches in addressing some of its challenges is
pertinent.

REFERENCES

[1] C. C. Paul Fergus, *Applied Deep Learning: Tools, Techniques and Implementation*, 1stFrist edition. France: Springers, 2022.

[2] A. B. Adetunji, O. N. Akande, F. A. Ajala, O. Oyewo, Y. F. Akande, and G. Oluwadara, "House price prediction using random forest machine learning technique," *Procedia Computer Science*, pp. 806–813, 2021. doi: 10.1016/j.procs.2022.01.100.

[3] H. Kuşan, O. Aytekin, and I. Özdemir, "The use of fuzzy logic in predicting house selling price," *Expert Systems with Applications*, vol. 37, no. 3, pp. 1808–1813, 2010, doi: 10.1016/j.eswa.2009.07.031.

[4] I. H. Gerek, "House selling price assessment using two different adaptive neuro-fuzzy techniques," *Automation in Construction*, vol. 41, pp. 33–39, 2014, doi: 10.1016/j.autcon.2014.02.002.

[5] T.-C. J. Oluwatobi Adeleke, "Prediction of electrical energy consumption in university campus residence using FCM-clustered neuro-fuzzy model," In *ASME 2022 International Mechanical Engineering Congress and Exposition*, Columbus, OH: ASME, 2022.

[6] O. Adeleke, S. Akinlabi, T. C. Jen, P. A. Adedeji, and I. Dunmade, "Evolutionary-based neuro-fuzzy modelling of combustion enthalpy of municipal solid waste," *Neural Computing and Applications*, vol. 2, 2022, doi: 10.1007/s00521-021-06870-2.

[7] I. H. Sarker, "Machine learning: Algorithms, real-world applications and research directions," *SN Computer Science*, vol. 2, no. 3, 2021. doi: 10.1007/s42979-021-00592-x.

[8] Y. Liu and F. bin Zheng, "Object-oriented and multi-scale target classification and recognition based on hierarchical ensemble learning," *Computers and Electrical Engineering*, vol. 62, pp. 538–554, 2017, doi: 10.1016/j.compeleceng.2016.12.026.

[9] N. John, R. Surya, R. Ashwini, S. Sachin Kumar, and K. P. Soman, "A low cost implementation of multi-label classification algorithm using mathematica on Raspberry Pi," *Procedia Computer Science*, pp. 306–313, 2015. doi: 10.1016/j.procs.2015.02.025.

[10] Statistics Solutions, "What is linear regression," Retrieved from https://www.statisticssolutions.com/free-resources/directory-of-statistical-analyses/what-is-linear-regression/, 2013.

[11] K. Rai, "The math behind logistic regression," https://medium.com/analytics-vidhya/the-math-behind-logistic-regression-c2f04ca27bca#:~:text=Maths%20behind%20Logistic%20Regression&text=To%20address%20this%20problem%2C%20let,solving%20for%20p(x)%3A, 2020.

[12] M. Mohammadi, A. A. Atashin, and D. A. Tamburri, "From ℓ1 subgradient to projection: A compact neural network for ℓ1-regularized logistic regression," *Neurocomputing*, vol. 526, pp. 30–38, 2023, doi: 10.1016/j.neucom.2023.01.021.

[13] T. Ning, Y. Yang, and Z. Du, "Quantum kernel logistic regression based Newton method," *Physica A: Statistical Mechanics and Iits Applications*, vol. 611, 2023, doi: 10.1016/j.physa.2023.128454.

[14] S. Zhao and C. F. Parmeter, "The 'wrong skewness' problem: Moment constrained maximum likelihood estimation of the stochastic frontier model," *Economics Letters*, vol. 221, 2022, doi: 10.1016/j.econlet.2022.110901.

[15] A. Khan, S. Hayat, Y. Zhong, A. Arif, L. Zada, and M. Fang, "Computational and topological properties of neural networks by means of graph-theoretic parameters," *Alexandria Engineering Journal*, vol. 66, pp. 957–977, 2023, doi: 10.1016/j.aej.2022.11.001.

[16] A. Krenker, J. Bešter, and A Kos, *Introduction to the Artificial Neural Networks, Artificial Neural Networks - Methodological Advances and Biomedical Applications*, ISBN: 978-953-307-243-2, London: InTech, 2011.

[17] R. Khanduzi and A. K. Sangaiah, "An efficient recurrent neural network for defensive Stackelberg game," *Journal of Computational Science*, vol. 67, p. 101970, 2023, doi: https://doi.org/10.1016/j.jocs.2023.101970.

[18] C. G. S. Capanema, G. S. de Oliveira, F. A. Silva, T. R. M. B. Silva, and A. A. F. Loureiro, "Combining recurrent and graph neural networks to predict the next place's category," *Ad Hoc Networks*, vol. 138, p. 103016, 2023, doi: https://doi.org/10.1016/j.adhoc.2022.103016.

[19] R. Khaldi, A. El Afia, R. Chiheb, and S. Tabik, "What is the best RNN-cell structure to forecast each time series behavior?" *Expert Systems with Applications*, vol. 215, p. 119140, 2023, doi: https://doi.org/10.1016/j.eswa.2022.119140.

[20] T. P. Lillicrap and A. Santoro, "Backpropagation through time and the brain," *Current Opinion in Neurobiology*, vol. 55, pp. 82–89, 2019. doi: 10.1016/j.conb.2019.01.011.

[21] K. Fukushima and S. Miyake, "Neocognitron: A self-organizing neural network model for a mechanism of visual pattern recognition," In *Competition and Cooperation in Neural Nets*, S. Amari and M. A. Arbib, Eds., Berlin, Heidelberg: Springer, 1982, pp. 267–285.

[22] K. Grill-Spector and R. Malach, "The human visual cortex," *Annual Review of Neuroscience*, vol. 27, no. 1, pp. 649–677, 2004, doi: 10.1146/annurev.neuro.27.070203.144220.

[23] B. Walter, "Analysis of convolutional neural network image classifiers in a hierarchical max-pooling model with additional local pooling," *Journal of Statistical Planning and Inference*, vol. 224, pp. 109–126, 2023, doi: 10.1016/j.jspi.2022.11.001.

[24] J. He, X. Wang, Y. Song, and Q. Xiang, "A multiscale intrusion detection system based on pyramid depthwise separable convolution neural network," *Neurocomputing*, vol. 530, pp. 48–59, 2023, doi: 10.1016/j.neucom.2023.01.072.

[25] R. Zenghui and W. Waseem, "Deep convolutional neural networks for image classification: a comprehensive review," *Neural Computation*, vol. 29, pp. 2352–2449, 2017.

[26] A. Mat Deris, A. Mohd Zain, and R. Sallehuddin, "Overview of support vector machine in modeling machining performances," *Procedia Engineering*, pp. 308–312, 2011. doi: 10.1016/j.proeng.2011.11.2647.

[27] B. Samanta, K. R. Al-Balushi, and S. A. Al-Araimi, "Artificial neural networks and support vector machines with genetic algorithm for bearing fault detection," *Engineering Applications of Artificial Intelligence*, vol. 16, no. 7–8, pp. 657–665, 2003, doi: 10.1016/j.engappai.2003.09.006.

[28] C. Cortes, V. Vapnik, and L. Saitta, *Support-Vector Networks Editor*, Dordrecht, Netherlands: Kluwer Academic Publishers, 1995.

[29] B. E. Boser, I. M. Guyon, and V. N. Vapnik, *A Training Algorithm for Optimal Margin Classifiers*, New York: ACM Digital library, 1992.

[30] V. Vapnik, *The Nature of Statistical Learning Theory*, New York: Springer, 1995.

[31] A. J. Smola, B. Sch¨olkopf, and S. Sch¨olkopf, *A Tutorial on Support Vector Regression *,*" Dordrecht, Netherlands: Kluwer Academic Publishers, 2004.

[32] A. Roy and S. Chakraborty, "Support vector machine in structural reliability analysis: Aa review," *Reliability Engineering & System Safety*, p. 109126, 2023, doi: 10.1016/j.ress.2023.109126.

[33] X. Han, X. Zhu, W. Pedrycz, and Z. Li, "A three-way classification with fuzzy decision trees," *Applied Soft Computing*, vol. 132, 2023, doi: 10.1016/j.asoc.2022.109788.

[34] IBM, "What is a decision tree," https://www.ibm.com/topics/decision-trees#:~:text=A%20 decision%20tree%20is%20a,internal%20nodes%20and%20leaf%20nodes.

[35] C. C. Wu, Y. L. Chen, Y. H. Liu, and X. Y. Yang, "Decision tree induction with a constrained number of leaf nodes," *Applied Intelligence*, vol. 45, no. 3, pp. 673–685, 2016, doi: 10.1007/s10489-016-0785-z.

[36] F. E. B. Otero, A. A. Freitas, and C. G. Johnson, "Inducing decision trees with an ant colony optimization algorithm," *Applied Soft Computing Journal*, vol. 12, no. 11, pp. 3615–3626, 2012, doi: 10.1016/j.asoc.2012.05.028.

[37] Y. Cai, H. Zhang, Q. He, and J. Duan, "A novel framework of fuzzy oblique decision tree construction for pattern classification," *Applied Intelligence*, vol. 50, no. 9, pp. 2959–2975, 2020, doi: 10.1007/s10489-020-01675-7.

[38] A. Isazadeh, F. Mahan, and W. Pedrycz, "MFlexDT: Mmulti flexible fuzzy decision tree for data stream classification," *Soft Computing*, vol. 20, no. 9, pp. 3719–3733, 2016, doi: 10.1007/s00500-015-1733-2.

[39] A. Frini, A. Guitouni, and J. M. Martel, "A general decomposition approach for multi-criteria decision trees," *European Journal of Operational Research*, vol. 220, no. 2, pp. 452–460, 2012, doi: 10.1016/j.ejor.2012.01.032.

[40] E. Borgonovo and M. Marinacci, "Decision analysis under ambiguity," *European Journal of Operational Research*, vol. 244, no. 3, pp. 823–836, 2015, doi: 10.1016/j.ejor.2015.02.001.

[41] S. Yang, J. Z. Guo, and J. W. Jin, "An improved Id3 algorithm for medical data classification," *Computers and Electrical Engineering*, vol. 65, pp. 474–487, 2018, doi: 10.1016/j.compeleceng.2017.08.005.

[42] F. Arif, N. Suryana, and B. Hussin, "Cascade quality prediction method using multiple PCA+ID3 for multi-stage manufacturing system," *IERI Procedia*, vol. 4, pp. 201–207, 2013, doi: 10.1016/j.ieri.2013.11.029.

[43] S. Moral-García, C. J. Mantas, J. G. Castellano, and J. Abellán, "Using credal C4.5 for calibrated label ranking in multi-label classification," *International Journal of Approximate Reasoning*, vol. 147, pp. 60–77, 2022, doi: 10.1016/j.ijar.2022.05.005.

[44] M. G. bin Md Ghazi, L. C. Lee, A. S. B. Samsudin, and H. Sino, "Evaluation of ensemble data preprocessing strategy on forensic gasoline classification using untargeted GC-MS data and classification and regression tree (CART) algorithm," *Microchemical Journal*, vol. 182, 2022, doi: 10.1016/j.microc.2022.107911.

[45] E. J. Shim et al., "An MRI-based decision tree to distinguish lipomas and lipoma variants from well-differentiated liposarcoma of the extremity and superficial trunk: Classification and regression tree (CART) analysis," *European Journal of Radiology*, vol. 127, 2020, doi: 10.1016/j.ejrad.2020.109012.

[46] G. S. Kori and M. S. Kakkasageri, "Classification and regression tree (CART) based resource allocation scheme for wireless sensor networks," *Computer Communications*, vol. 197, pp. 242–254, 2023, doi: 10.1016/j.comcom.2022.11.003.

[47] C. S. Smith and J. P. Schwieterman, "Using multivariate adaptive regression splining (MARS) to identify factors affecting the performance of dock-based bikesharing: The case of Chicago''s Divvy system," *Research in Transportation Economics*, vol. 89, 2021, doi: 10.1016/j.retrec.2021.101032.

[48] A. H. Naser, A. H. Badr, S. N. Henedy, K. A. Ostrowski, and H. Imran, "Application of multivariate adaptive regression splines (MARS) approach in prediction of compressive strength of eco-friendly concrete," *Case Studies in Construction Materials*, vol. 17, 2022, doi: 10.1016/j.cscm.2022.e01262.

[49] M. Hiranuma, D. Kobayashi, K. Yokota, and K. Yamamoto, "Chi-square automatic interaction detector decision tree analysis model: Predicting cefmetazole response in intra-abdominal infection," *Journal of Infection and Chemotherapy*, vol. 29, no. 1, pp. 7–14, 2023, doi: 10.1016/j.jiac.2022.09.002.

[50] A. Meydanlioglu et al., "Prevalence of obesity and hypertension in children and determination of associated factors by CHAID analysis," *Archives de Pediatrie*, vol. 29, no. 1, pp. 30–35, 2022, doi: 10.1016/j.arcped.2020.10.017.

[51] E. L. Murphy and C. M. Comiskey, "Using chi-squared automatic interaction detection (CHAID) modelling to identify groups of methadone treatment clients experiencing significantly poorer treatment outcomes," *Journal of Substance Abuse Treatment*, vol. 45, no. 4, pp. 343–349, 2013, doi: 10.1016/j.jsat.2013.05.003.

[52] J. M. Muñoz-Rodríguez, C. P. Alonso, T. Pessoa, and J. Martín-Lucas, "Identity profile of young people experiencing a sense of risk on the internet: A data mining application of decision tree with CHAID algorithm," *Computers & Education*, p. 104743, 2023, doi: 10.1016/j.compedu.2023.104743.

[53] W. Gao, F. Xu, and Z. H. Zhou, "Towards convergence rate analysis of random forests for classification," *Artificial Intelligence*, vol. 313, 2022, doi: 10.1016/j.artint.2022.103788.

[54] M. Minnoor and V. Baths, "Diagnosis of breast cancer using random forests," *Procedia Computer Science*, vol. 218, pp. 429–437, 2023, doi: 10.1016/j.procs.2023.01.025.

[55] M. Y. Khan, A. Qayoom, M. S. Nizami, M. S. Siddiqui, S. Wasi, and S. M. K. U. R. Raazi, "Automated prediction of good dictionary examples (GDEX): A comprehensive experiment with distant supervision, machine learning, and word embedding-based deep learning techniques," *Complexity*, vol. 2021, 2021, doi: 10.1155/2021/2553199.

[56] IBM, "What is random forest?" https://www.ibm.com/topics/random-forest#:~:text=Random%20forest%20is%20a%20commonly,both%20classification%20and%20regression%20problems.

[57] G. Cattani, "Combining data envelopment analysis and random forest for selecting optimal locations of solar PV plants," *Energy and AI*, vol. 11, 2023, doi: 10.1016/j.egyai.2022.100222.

[58] M. A. Ganaie, M. Tanveer, P. N. Suganthan, and V. Snasel, "Oblique and rotation double random forest," *Neural Networks*, vol. 153, pp. 496–517, 2022, doi: 10.1016/j.neunet.2022.06.012.

[59] Z. Xia and K. Stewart, "A counterfactual analysis of opioid-involved deaths during the COVID-19 pandemic using a spatiotemporal random forest modeling approach," *Health Place*, vol. 80, p. 102986, 2023, doi: 10.1016/j.healthplace.2023.102986.

[60] N. L. Zhang, T. D. Nielsen, and F. V. Jensen, "Latent variable discovery in classification models," *Artificial Intelligence in Medicine*, vol. 30, no. 3, pp. 283–299, 2004, doi: 10.1016/j.artmed.2003.11.004.

[61] M. M. dos Santos Freitas *et al.*, "KNN algorithm and multivariate analysis to select and classify starch films," *Food Packag Shelf Life*, vol. 34, 2022, doi: 10.1016/j.fpsl.2022.100976.

[62] M. Fopa, M. Gueye, S. Ndiaye, and H. Naacke, "A parameter-free KNN for rating prediction," *Data & Knowledge Engineering*, vol. 142, 2022, doi: 10.1016/j.datak.2022.102095.

[63] W. Zhang, X. Chen, Y. Liu, and Q. Xi, "A distributed storage and computation k-nearest neighbor algorithm based cloud-edge computing for cyber-physical-social systems," *IEEE Access*, vol. 8, pp. 50118–50130, 2020, doi: 10.1109/ACCESS.2020.2974764.

[64] F. C. Pereira and S. S. Borysov, "Machine learning fundamentals," in *Mobility Patterns, Big Data and Transport Analytics: Tools and Applications for Modeling*, Elsevier, 2018, pp. 9–29. doi: 10.1016/B978-0-12-812970-8.00002-6.

[65] J. L. Andrews, "Addressing overfitting and underfitting in Gaussian model-based clustering," *Computational Statistics & Data Analysis*, vol. 127, pp. 160–171, 2018, doi: 10.1016/j.csda.2018.05.015.

[66] A. P. King and P. Aljabar, "Machine learning," in *Matlab(r) Programming for Biomedical Engineers and Scientists*, Elsevier, 2023, pp. 343–372. doi: 10.1016/B978-0-32-385773-4.00023-X.

[67] S. Shakil, D. Arora, and T. Zaidi, "Feature identification and classification of hand based biometrics through ensemble learning approach," *Measurement: Sensors*, vol. 25, 2023, doi: 10.1016/j.measn.2022.100593.

[68] P. Baro and M. D. Borah, "A factor based multiple imputation approach to handle class imbalance," *Procedia Computer Science*, vol. 218, pp. 103–112, 2023, doi: 10.1016/j.procs.2022.12.406.

[69] R. M. Pereira, Y. M. G. Costa, and C. N. Silla, "Toward hierarchical classification of imbalanced data using random resampling algorithms," *Information Sciences*, vol. 578, pp. 344–363, 2021, doi: 10.1016/j.ins.2021.07.033.

[70] Q. Chen, Z. L. Zhang, W. P. Huang, J. Wu, and X. G. Luo, "PF-SMOTE: A novel parameter-free SMOTE for imbalanced datasets," *Neurocomputing*, vol. 498, pp. 75–88, 2022, doi: 10.1016/j.neucom.2022.05.017.

[71] S. Fu, Y. Tian, J. Tang, and X. Liu, "Cost-sensitive learning with modified stein loss function," *Neurocomputing*, vol. 525, pp. 57–75, 2023, doi: 10.1016/j.neucom.2023.01.052.

[72] Y. Ding, M. Jia, J. Zhuang, and P. Ding, "Deep imbalanced regression using cost-sensitive learning and deep feature transfer for bearing remaining useful life estimation," *Applied Soft Computing*, vol. 127, 2022, doi: 10.1016/j.asoc.2022.109271.

[73] N. Wang *et al.*, "Search-based cost-sensitive hypergraph learning for anomaly detection," *Information Sciences*, vol. 617, pp. 451–463, 2022, doi: 10.1016/j.ins.2022.07.029.

[74] P. A. Adedeji, O. O. Olatunji, N. Madushele, and A. O. Ajayeoba, "Soft computing in renewable energy system modeling," in *Design, Analysis and Applications of Renewable Energy Systems*, Elsevier, 2021, pp. 79–102. doi: 10.1016/B978-0-12-824555-2.00026-5.

[75] S. Srivastava, R. N. Shah, C. Teodoriu, and A. Sharma, "Impact of data quality on supervised machine learning: Case study on drilling vibrations," *Journal of Petroleum Science and Engineering*, vol. 219, 2022, doi: 10.1016/j.petrol.2022.111058.

[76] F. Dornaika, D. Sun, K. Hammoudi, J. Charafeddine, A. Cabani, and C. Zhang, "Object-centric contour-aware data augmentation using superpixels of varying granularity," *Pattern Recognition*, p. 109481, 2023, doi: https://doi.org/10.1016/j.patcog.2023.109481.

[77] F. Garcea, A. Serra, F. Lamberti, and L. Morra, "Data augmentation for medical imaging: A systematic literature review," *Computers in Biology and Medicine*, vol. 152, p. 106391, 2023, doi: 10.1016/j.compbiomed.2022.106391.

[78] A. Thakkar and K. Chaudhari, "Predicting stock trend using an integrated term frequency-inverse document frequency-based feature weight matrix with neural networks," *Applied Soft Computing Journal*, vol. 96, 2020, doi: 10.1016/j.asoc.2020.106684.

[79] N. N. Amir Sjarif, N. F. Mohd Azmi, S. Chuprat, H. M. Sarkan, Y. Yahya, and S. M. Sam, "SMS spam message detection using term frequency-inverse document frequency and random forest algorithm," in *Procedia Computer Science*, Elsevier B.V., 2019, pp. 509–515. doi: 10.1016/j.procs.2019.11.150.

[80] P. Wang, B. Xue, J. Liang, and M. Zhang, "Feature selection using diversity-based multi-objective binary differential evolution," *Information Sciences*, vol. 626, pp. 586–606, 2023, doi: 10.1016/j.ins.2022.12.117.

[81] T. Wu, Y. Hao, B. Yang, and L. Peng, "ECM-EFS: An ensemble feature selection based on enhanced co-association matrix," *Pattern Recognition*, p. 109449, 2023, doi: 10.1016/j.patcog.2023.109449.

[82] R. Polikar, "Ensemble based systems in decision making," *IEEE Circuits and Systems Magazine*, vol. 6, no. 3. pp. 21–44, 2006. doi: 10.1109/MCAS.2006.1688199.

[83] Michał Oleszak, "Feature selection methods and how to choose them. Neptune AI," 2023, https://neptune.ai/blog/feature-selection-methods.

[84] M. Chemmakha, O. Habibi, and M. Lazaar, "Improving machine learning models for malware detection using embedded feature selection method," in *IFAC-PapersOnLine*, Elsevier B.V., 2022, pp. 771–776. doi: 10.1016/j.ifacol.2022.07.406.

[85] K. Robindro, U. B. Clinton, N. Hoque, and D. K. Bhattacharyya, "JoMIC: A joint MI-based filter feature selection method," *Journal of Computational Mathematics and Data Science*, vol. 6, p. 100075, 2023, doi: 10.1016/j.jcmds.2023.100075.

[86] Mayank Banoula, "The best guide to regularization in machine learning," 2023, https://www.simplilearn.com/tutorials/machine-learning-tutorial/regularization-in-machine-learning.

[87] Samet Girgin, "Day-32 regularization in machine learning-1," 2019, https://medium.com/pursuitnotes/day-32-regularization-in-machine-learning-1-58e02add851a.

[88] MathWorks, "Lasso and elastic net," https://www.mathworks.com/help/stats/lasso-and-elastic-net.html.

[89] J. J. Salazar, L. Garland, J. Ochoa, and M. J. Pyrcz, "Fair train-test split in machine learning: Mitigating spatial autocorrelation for improved prediction accuracy," *Journal of Petroleum Science and Engineering*, vol. 209, 2022, doi: 10.1016/j.petrol.2021.109885.

[90] G. Mezzadri, T. Laloë, F. Mathy, and P. Reynaud-Bouret, "Hold-out strategy for selecting learning models: Application to categorization subjected to presentation orders," *Journal of Mathematical Psychology*, vol. 109, 2022, doi: 10.1016/j.jmp.2022.102691.

[91] S. de Bruin, D. J. Brus, G. B. M. Heuvelink, T. van Ebbenhorst Tengbergen, and A. M. J. C. Wadoux, "Dealing with clustered samples for assessing map accuracy by cross-validation," *Ecological Informatics*, vol. 69, 2022, doi: 10.1016/j.ecoinf.2022.101665.

[92] H. L. Vu, K. T. W. Ng, A. Richter, and C. An, "Analysis of input set characteristics and variances on k-fold cross validation for a recurrent neural network model on waste disposal rate estimation," *Journal of Environmental Management*, vol. 311, 2022, doi: 10.1016/j.jenvman.2022.114869.

[93] J. Li *et al.*, "Quantum k-fold cross-validation for nearest neighbor classification algorithm," *Physica A: Statistical Mechanics and iIts Applications*, vol. 611, 2023, doi: 10.1016/j.physa.2022.128435.

[94] R. Priyadarshini, H. Joardar, S. K. Bisoy, and T. Badapanda, "Crystal structural prediction of perovskite materials using machine learning: A comparative study," *Solid State Communications*, vol. 361, 2023, doi: 10.1016/j.ssc.2022.115062.

[95] R. Kelter, "Bayesian model selection in the M-open setting - Approximate posterior inference and subsampling for efficient large-scale leave-one-out cross-validation via the difference estimator," *Journal of Mathematical Psychology*, vol. 100, 2021. doi: 10.1016/j.jmp.2020.102474.

[96] Z. Shao and M. J. Er, "Efficient leave-one-out cross-validation-based regularized extreme learning machine," *Neurocomputing*, vol. 194, pp. 260–270, 2016, doi: 10.1016/j.neucom.2016.02.058.

[97] T. T. Wong, "Performance evaluation of classification algorithms by k-fold and leave-one-out cross validation," *Pattern Recognition*, vol. 48, no. 9, pp. 2839–2846, 2015, doi: 10.1016/j.patcog.2015.03.009.

[98] C. Bergmeir, R. J. Hyndman, and B. Koo, "A note on the validity of cross-validation for evaluating autoregressive time series prediction," *Computational Statistics & Data Analysis*, vol. 120, pp. 70–83, 2018, doi: 10.1016/j.csda.2017.11.003.

[99] O. Adeleke, S. Akinlabi, T. C. Jen, and I. Dunmade, "A machine learning approach for investigating the impact of seasonal variation on physical composition of municipal solid waste," *Journal of Reliable Intelligent Environments*, 2022, doi: 10.1007/s40860-021-00168-9.

[100] G. Rajendra Kannammal, P. Sivamalar, P. Santhi, T. Vetriselvi, V. Kalpana, and T. M. Nithya, "Prediction of quality in production using optimized Hyper-parameter tuning based deep learning model," *Materials Today: Proceedings*, vol. 69, pp. 703–709, 2022, doi: https://doi.org/10.1016/j.matpr.2022.07.133.

[101] G. Li, W. Wang, W. Zhang, Z. Wang, H. Tu, and W. You, "Grid search based multi-population particle swarm optimization algorithm for multimodal multi-objective optimization," *Swarm and Evolutionary Computation*, vol. 62, 2021, doi: 10.1016/j.swevo.2021.100843.

[102] W. Gao, R. Fan, R. Huang, Q. Huang, W. Gao, and L. Du, "Augmented random search based inter-area oscillation damping using high voltage DC transmission," *Electric Power Systems Research*, vol. 216, 2023, doi: 10.1016/j.epsr.2022.109063.

[103] M. J. Begall, A. M. Schweidtmann, A. Mhamdi, and A. Mitsos, "Geometry optimization of a continuous millireactor via CFD and Bayesian optimization," *Computers & Chemical Engineering*, vol. 171, p. 108140, 2023, doi: 10.1016/j.compchemeng.2023.108140.

[104] A. A. Ewees, F. H. Ismail, and A. T. Sahlol, "Gradient-based optimizer improved by slime mould algorithm for global optimization and feature selection for diverse computation problems," *Expert Systems with Applications*, vol. 213, 2023, doi: 10.1016/j.eswa.2022.118872.

[105] X. Hu, S. Wen, and H. K. Lam, "Dynamic random distribution learning rate for neural networks training," *Applied Soft Computing*, vol. 124, 2022, doi: 10.1016/j.asoc.2022.109058.

[106] O. G. Ajayi and J. Ashi, "Effect of varying training epochs of a faster region-based convolutional neural network on the accuracy of an automatic weed classification scheme," *Smart Agricultural Technology*, vol. 3, p. 100128, 2023, doi: https://doi.org/10.1016/j.atech.2022.100128.

[107] O. Adeleke, S. A. Akinlabi, T. Jen, and I. Dunmade, "Application of artificial neural networks for predicting the physical composition of municipal solid waste : An assessment of the impact of seasonal variation," *Waste Management & Research*, vol. 39, no. 8, pp. 1058–1068 2021, doi: 10.1177/0734242X21991642.

[108] S. Yu, H. Li, X. Chen, and D. Lin, "Multistability analysis of quaternion-valued neural networks with cosine activation functions," *Applied Mathematics and Computation*, vol. 445, p. 127849, 2023, doi: https://doi.org/10.1016/j.amc.2023.127849.

[109] Y. Chen and Z. Yi, "Adaptive sparse dropout: Learning the certainty and uncertainty in deep neural networks," *Neurocomputing*, vol. 450, pp. 354–361, 2021, doi: https://doi.org/10.1016/j.neucom.2021.04.047.

[110] I. Kandel and M. Castelli, "The effect of batch size on the generalizability of the convolutional neural networks on a histopathology dataset," *ICT Express*, vol. 6, no. 4, pp. 312–315, 2020, doi: https://doi.org/10.1016/j.icte.2020.04.010.

7 Unsupervised Learning

7.1 INTRODUCTION

Going forward from Chapter 6 where we discussed supervised learning, in this current chapter we will discuss an alternative machine learning approach, which is unsupervised learning. Unsupervised learning, unlike supervised learning, finds structure and patterns without labels, making it useful for handling and gaining insights from large, complex datasets where it may be difficult or impossible to manually classify all the data. Big data has made unsupervised learning crucial in natural language processing, computer vision, and anomaly detection. Now, the machine learning model is being trained using the unlabeled input data. It will first analyze the raw data to identify any hidden patterns in the data. Once the appropriate algorithm is deployed, the data objects are split into groups based on how similar and different the objects are [1]. In real-world machine learning solutions, unsupervised learning has benefits and drawbacks. Clustering, dimensionality reduction, and association rule learning are introduced in this chapter. We'll also look at some other important unsupervised learning algorithms, their benefits, as well as drawbacks and applications to gain knowledge and improve decision-making. This chapter on unsupervised learning will help us understand about machine learning and how it can solve difficulties in atomic layer deposition (ALD) and thin film technologies. Machine learning improves ALD processes. Most of these strategies require labeled data, which may be difficult to obtain. Thus, this chapter's knowledge can advance ALD and accelerate the development of more effective and efficient methods.

7.2 UNSUPERVISED LEARNING TECHNIQUES

As illustrated in Figure 7.1, this section examines some unsupervised learning techniques, typical applications, and corresponding algorithms.

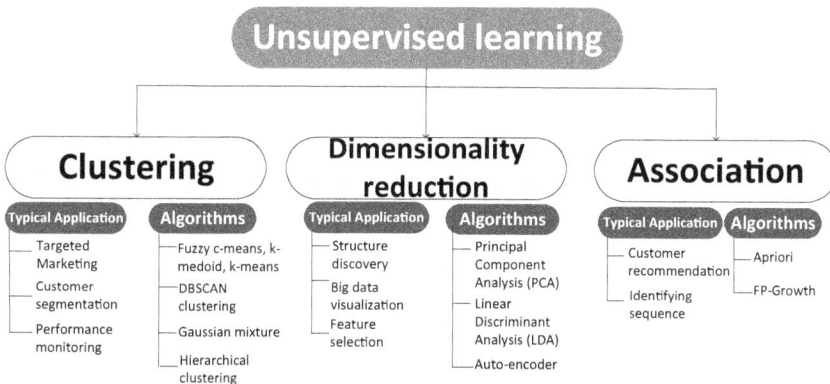

FIGURE 7.1 Unsupervised learning techniques and algorithms.

DOI: 10.1201/9781003346234-9

7.2.1 CLUSTERING

Clustering can be referred to as a technique for arranging the data points into various clusters made up of related data points [2–4]. The items with potential resemblances continue to be in a group that shares little to no characteristics with another group. It accomplishes this by identifying comparable patterns in the unlabeled dataset, such as shape, size, color, and behavior, and then classifying the data according to the presence or absence of these patterns. Each cluster or group is given a cluster-ID after using this clustering technique, which machine learning systems can employ to streamline the processing of big and complicated datasets. The full feature set for an example can now be reduced to its cluster-ID. Clustering is effective when a complicated example is represented by a simple cluster-ID [1]. Most clustering algorithms start by specifying a distance or similarity metric and then iteratively clustering comparable locations. The created clusters should have low inter-cluster similarity (items in different clusters are dissimilar) and high intra-cluster similarity [2]. Several sectors like marketing, medicine, computer vision, and natural language processing find a wide use for clustering tasks such as market division, statistical evaluation of data, assessment of social networks, segmentation of images, anomaly detection, and medical imaging, amongst others. For instance, clustering can be used to group genes in biological studies that have comparable expression patterns or to divide customers into groups with similar purchase habits. The classification algorithm and clustering share certain similarities, but the type of dataset we are utilizing is different. We use the labeled dataset for classification while the unlabeled dataset is used for clustering. As illustrated in Figure 7.2, some of the techniques in clustering are, but not limited to, the following [5]:

7.2.1.1 Centroid-Based Clustering

Centroid-based clustering is an iterative clustering process where the clusters are produced by the proximity of data points to the cluster centroid [6,7]. The centroid is constructed in this case such that the least distance is obtained between the data points and the center [8]. A central vector is used to represent clusters in centroid-based clustering, which might not actually be a part of the data. It's crucial to have prior knowledge of the dataset because these models require that the number of clusters needed be stated upfront. These models iteratively search for the regional optimum. The most important yet challenging step in the clustering approach is determining the number of clusters in advance. It is a widely used clustering strategy for surfacing and optimizing huge datasets, notwithstanding the disadvantage. By using

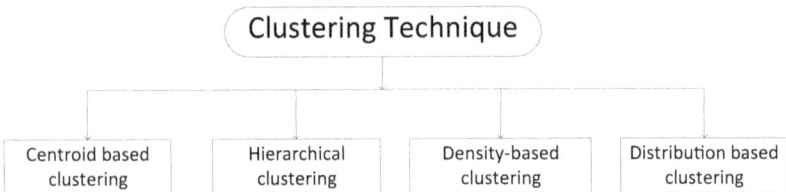

FIGURE 7.2 Types of clustering techniques.

centroid-based clustering, non-hierarchical groupings are created from the data. The feature of determining the distance measure between the clusters and the characteristic centroids is the fundamental component of all centroid-based algorithms.

The distance measure is often calculated using the common metrics such as Manhattan distance, Minkowski distance, or Euclidean distance. The steepest descent method, the mean, and the median are the approaches deployed for computing the distance measures in the Minkowski, Euclidean, and Manhattan distances, respectively. These algorithms for iterative clustering draw the idea of similarity from how near an observation of data is to the centroid of the clusters [9]. Because the optimization problem is NP-hard, solutions are frequently approximated over several trials [10]. The typical approximations method is the k-means algorithm that often runs numerous times with various random initializations in an effort to locate a local optimum. While using k-means clustering, we aim to maximize the distance between the centroids of distinct clusters while minimizing the distance between a cluster's centroid and other points within it [11].

7.2.1.2 Hierarchical Clustering

Hierarchical clustering is another technique of clustering, which builds a hierarchy of clusters according to how related objects are. The approach is useful when there is an inherent hierarchical structure in the data as well as when there is no prior knowledge of the cluster number [12]. It is a notably valuable tool for analyzing and visualizing the structure of complicated datasets and locating natural groupings or specific groups inside the data. Each data point is initially treated as an independent cluster, and from there, the algorithm iteratively aggregates the nearest clusters until a stopping requirement is met [13]. A dendrogram, which displays the relationships between the clusters, can be used to illustrate the resulting hierarchy.

The diagonal values are zeros since a cluster of values against itself is always 0, the values on the diagonal are all zeros. The additional numbers represent the separation between each cluster. For instance, the distance between clusters 1 and 2 is 3 and indicates a closer association; however, clusters 1 and 5 have such a distance between them of considerably greater magnitude, indicating that the clusters are farther away. To create a dendrogram, you will need to use the proximity matrix and the resulting clusters. The dendrogram displays the sequence in which the clusters were combined, and each node's height corresponds to the separation between the clusters at that particular level of the hierarchy.

Hierarchical clustering can be categorized into two:

   i. Agglomerative (bottom-up) Hierarchical Clustering
  ii. Divisive (top-down) Hierarchical Clustering

The former begins with a single dataset and sequentially combines nearby points in line with a set rule till all data points are clustered into a single class. The latter, however, involves initially treating the entire dataset as a whole and then segmenting it in accordance with predetermined principles till all datasets are isolated from one another [14]. Both techniques are essentially reverse processes, and the dendrogram produced using the same criteria is identical. Agglomerative clustering provides a clustering sequence with a decreasing cluster count at each iteration. A single cluster is produced by combining two clusters into one at each level from the preceding clusters while divisive algorithms operate in the reverse manner, that is, they produce a clustering sequence of m at each stage [15]. At each stage, the clustering is produced by splitting a single cluster into two components from the prior one. We will dwell much on agglomerative clustering, being the most commonly used techniques in hierarchical clustering. The most important agglomerative algorithm representatives are single and full connection algorithms. These are the best algorithms for recovering large and small clusters. Agglomerative algorithms, like AGNE, treat each point as a separate cluster and repeatedly merge them to form a hierarchy [12]. The split algorithm, like DIANA, considers every point as a cluster and splits them into smaller clusters iteratively [16]. Since CURE extracts a predetermined number of evenly distributed points from each cluster to serve as the cluster's representative points, it can identify nonspherical clusters [17].

7.2.1.3   Density-Based Clustering

Density-based clustering groups data points by proximity in a high-density area [18–20]. Density-based clustering is useful for asymmetric or uneven datasets since it can locate clusters of any size and shape. Density-based clustering uses density. It operates by finding high-density locations, which are places where many data points are congregated close to one another. Then, pockets of lower density, often referred to as noise or outliers, separate these zones from one another. The recognition and categorization of items with comparable densities is done using density-based clustering algorithms. This method divides data collection by point density. Low-density data points are clustered together, whereas high-density data points are clustered separately [18].

Density-based methods typically involve two key steps. Using data from the immediate neighborhood, a suitable technique is utilized to calculate each data point's density in the first stage. The following stage involves identifying and merging comparable data points in denser regions to create clusters. A data point's density is determined in the first stage using information about its nearby neighbors. The accuracy of density computation typically reflects the correctness of closeness estimation.

The Density-Based Spatial Clustering of Applications with Noise (DBSCAN) and fast search and find of density peaks (FDP) algorithm are the most widely used density-based clustering algorithms [21]. The DBSCAN algorithm specifies two parameters, namely, ε and MinPts. The minimal number of points necessary to establish a dense zone is MinPts, and ε is the greatest distance between two points before they can be regarded as belonging to the same cluster. A border point is a point that is less than MinPts neighbors away from a core point but is still within distance of the core point. Noise points are locations with less than MinPts neighbors within and are not a part of any dense region [22]. The idea of the noise, core, and border object is depicted in Figure 7.3. A density core-based clustering technique with dynamic scanning radius (DCNaN) was proposed to replace the need for fixed global parameter setting following the success of DBSCAN and FDP

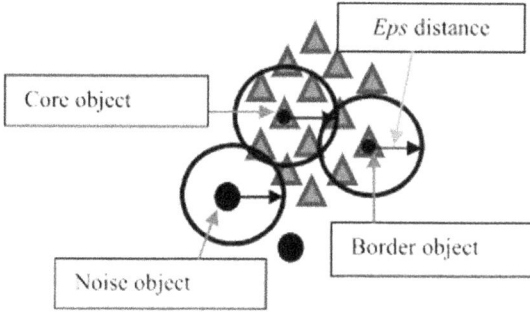

FIGURE 7.3   Core, noise, and border objects with MinPts = 3 [23].

[20]. A Fast Density Peaks Clustering Algorithm with Sparse Search (FSDPC) is an alternative approach, which was developed to address the FDP's disadvantage of increased computing cost [19].

Density-based clustering has several advantages over k-means and hierarchical clustering:

i. It may discover clusters of any form or size, unlike k-means clustering, which assumes spherical and equal-sized clusters.
ii. As noise points are identified as such and excluded from any cluster, it is robust to noise and outliers.
iii. It does not need the number of clusters beforehand, unlike k-means clustering.
iv. It is computationally efficient because it just needs to calculate distances between close sites.

7.2.1.4   Distribution-Based Clustering

The distribution-based clustering algorithm, sometimes referred to as model-based clustering, locates clusters in a dataset based on the data's underlying probability distribution. The algorithm operates by assigning data points to clusters based on how likely they are to belong to each distribution and modeling the probability distribution of the data [24]. The capacity of distribution-based clustering to handle datasets with complex distributions is one of its key benefits. Distribution-based clustering may locate clusters of any distribution, unlike k-means, which assumes Gaussian and similar-sized clusters [25].

GMM is the most used distribution-based clustering algorithm. GMM assumes data points are formed by a combination of Gaussian distributions, each corresponding to a cluster [26]. The Gaussian distribution's mean and covariance are among the parameters of the method that are initially initialized at random. The algorithm then assigns each data point to the cluster with the highest probability [27]. Next, the Gaussian distribution parameters are changed using the allotted data points and repeated until convergence. The Dirichlet Process Gaussian Mixture Model (DPGMM) is another distribution-based clustering method. Similar to GMM, but with an unspecified number of clusters, is DPGMM [28]. The procedure begins by presuming the existence of a prior distribution over the number of clusters and then allocates data points to clusters on the basis of the probability that each distribution is a fit for each individual data point [29]. The parameters of the Gaussian distributions are then automatically changed based on the data to reflect the number of clusters. The potency of the model-based clustering as a viable tool for data analysis has been attributed to its capability to handle complex distribution and assign number of clusters automatically [30]. However, it is not without its own demerits, which are computational intensity and requirement for advance knowledge of the probability distribution of the data, which might negatively impact the datasets with unknown distributions [30].

7.2.2 DIMENSIONALITY REDUCTION

A dataset's dimensionality refers to how many properties, features, or input variables it has. This could be two-dimensional or three-dimensional or more attributes [31]. The features of real-world datasets are numerous as well; it is thus difficult to comprehend that such datasets' observations are located in high-dimensional space. Dimensionality reduction is the technique of decreasing the number of features in data while retaining the same or more variance as is feasible in the original data [32]. Dimensionality reduction is a potent unsupervised learning technique that may be used to streamline complex datasets and enhance the reliability of clustering. We may obtain an equivalent of the dataset using fewer attributes by reducing noise and ambiguity in the dataset through dimensionality reduction.

There are primarily two categories of dimensionality reduction techniques which lower the number of dimensions in distinct approaches. One approach eliminates the redundant characteristics from the dataset and only preserves the most crucial information while the set of features receives no transformation [32]. Examples of this approach include backward elimination, forward selection, and random forests. The alternative approach uncovers a collection of novel attributes while it is transformed appropriately. This approach can be further categorized into linear and nonlinear techniques [33]. These linear methods work with linear data but fail with nonlinear data because they project the data onto a new set of orthogonal basis vectors that capture the highest variance.. Nonlinear methods to map data to a lower-dimensional space retain the data structure. The nonlinear approaches comprise algorithms such as Kernel Principal Component Analysis (PCA), t-distributed Stochastic Neighbor Embedding (t-SNE), Multidimensional Scaling (MDS), and Isometric mapping (Isomap) [31,33].

One of the most popular methods for reducing linear dimensions in unsupervised learning is PCA [33]. The orthogonal basis vectors with the highest variance in the data are identified by PCA as the major components of the data [32]. The first principal component represents the biggest variation, and the subsequent principal components are orthogonal to the first and represent the remaining variance. The next section will explain this algorithm. While Singular Value Decomposition (SVD) is a closely comparable method to PCA, it decomposes the data matrix into a product of three matrices instead of identifying the primary components of the data: a left singular matrix, a diagonal singular value matrix, and a right singular matrix [34–36]. SVD are used for matrix completion, low-rank approximation, and cooperative filtering.

A nonlinear dimensionality reduction technique called manifold learning trains a low-dimensional embedding of the data while maintaining its local neighborhood framework [37,38]. The visualization, clustering, and outlier detection of data can all be accomplished using a variety of learning techniques, including t-SNE [39,40], Isomap [41], and locally linear embedding (LLE) [33,42]. By keeping the pairwise distances between the data points, t-SNE is particularly helpful for visualizing high-dimensional data in two or three dimensions. In order to learn a compacted way to represent the information in a lower-dimensional space, autoencoders use neural networks that reduce the reconstruction error between the input and output data. Autoencoders can compress, identify anomalies, and extract features. Variational autoencoders (VAEs) learn a probabilistic representation of input to create new data samples [43,44].

7.2.3 ASSOCIATION RULE LEARNING

In huge datasets stored in different forms of databases, association rules are "if-then" statements that serve to illustrate the possibility of associations between data elements [45]. It is frequently used to discover unnoticed links or hidden connections between variables. The method in association learning is predicated on the idea of locating common patterns or itemset in a dataset. A frequent itemset is a group of items that appear in a dataset together more frequently than other groups of items. In order to predict the recurrence of one item based on the presence of another, association rule learning algorithms construct rules based on these frequently occurring itemset. If you wish to find rules that broadly characterize your data, such as "those who buy X also tend to buy Y," you have an association rule learning problem. In order to unveil relevant relationships between features, association rules are created. We will employ the following measures, which also constitute the fundamental steps in association learning, to choose the most intriguing rules from among the many potential rules:

   i. **Support**: Finding a common itemset is the first step in learning association rules. Support is a metric for determining how frequently a set of objects appear in a dataset. The number of transactions that contain an itemset divided by the total number of transactions in the dataset is referred to as the itemset's support. The support reflects how often the item emerges in the dataset. For instance, the level of product popularity in a store. The support for the pairing of A and B would be, $-P(AB)$ or $P(A)$ for Single A [45].
  ii. **Confidence**: This creates rules using the frequently occurring itemset from the first step. Confidence measures a rule's antecedent (if) and consequent (then) relationships. Divide the total number of transactions by the number that contains both the antecedent and the consequent. Confidence reflects how often the rule is correct. The rule is reliable. How often do people buy

toothpaste and toothbrushes? Confidence also represents the likelihood of the consequent given the antecedent, $-P(B|A) = P(AB)/P(A)$ [45].

iii. **Lift**: Lift, which accounts for the antecedent and consequent's separate occurrence rates, measures the intensity of a rule's link. Lift is the difference between the product of the supports of itemsets that only contain the antecedent and the consequent and the support of the itemset that contains both. How viable is it to buy another product while regulating the first's popularity? When a lift score is near 1, the antecedent and consequent are autonomous and do not affect each other. If the Lift score is more than 1, the antecedent and consequent are interconnected and positively affect each other. The antecedent and consequent are substitutes if the lift score is less than 1 [45].

Some of the prominent applications of association rule learning includes: (i) Market Basket Analysis uses association learning to discover goods that are commonly bought together and to create tailored marketing campaigns and the improvement of product placement in retail establishments. (ii) Health care can find patterns and connections between symptoms, diseases, and therapies using association learning, thus improving the outcomes for patients by using this information to create more efficient treatment strategies. (iii) It can be used in the finance industry to find links and patterns in financial transactions as well as to spot fraud. These data can be utilized to create fraud detection systems that are more effective and to stop financial crimes.

Considering a data sample on supermarket transaction data comprising 30 days of grocery transactions extracted from R library and reported by Analyn [46], the network diagram in Figure 7.4 illustrates the association between particular items under consideration. Red circles indicate greater lift, whereas larger circles indicate greater support. Common algorithms used for association learning are Apriori algorithm, FP-growth, and Eclat algorithm, amongst others. The Apriori algorithm is predicated on the notion of identifying frequent itemsets and producing association rules from them. It begins by locating all frequently occurring itemsets, and then, by combining them, it develops rules [47,48]. The approach of the FP-growth algorithm is based on the FP-Tree, a data structure that compresses data and improves the efficiency of mining frequent itemsets. It traverses the FP-Tree to produce rules [49].

7.3   COMMON UNSUPERVISED LEARNING ALGORITHMS

Without a map or a guidebook, picture yourself wandering in a strange city. You're pacing around, taking everything in, trying to make sense of the noise and turmoil. You suddenly begin to see connections and patterns in the language, the people, and the architecture. You start to put related items together and become aware of any outliers that don't follow the pattern. You should feel proud of using unsupervised learning, which computers use to find patterns in data without explicit supervision. From detecting fraud to predicting market fluctuations, unsupervised learning helps us understand the world around us. In this subsection, we shall examine some of the prominent algorithms used in unsupervised learning. These algorithms are based on

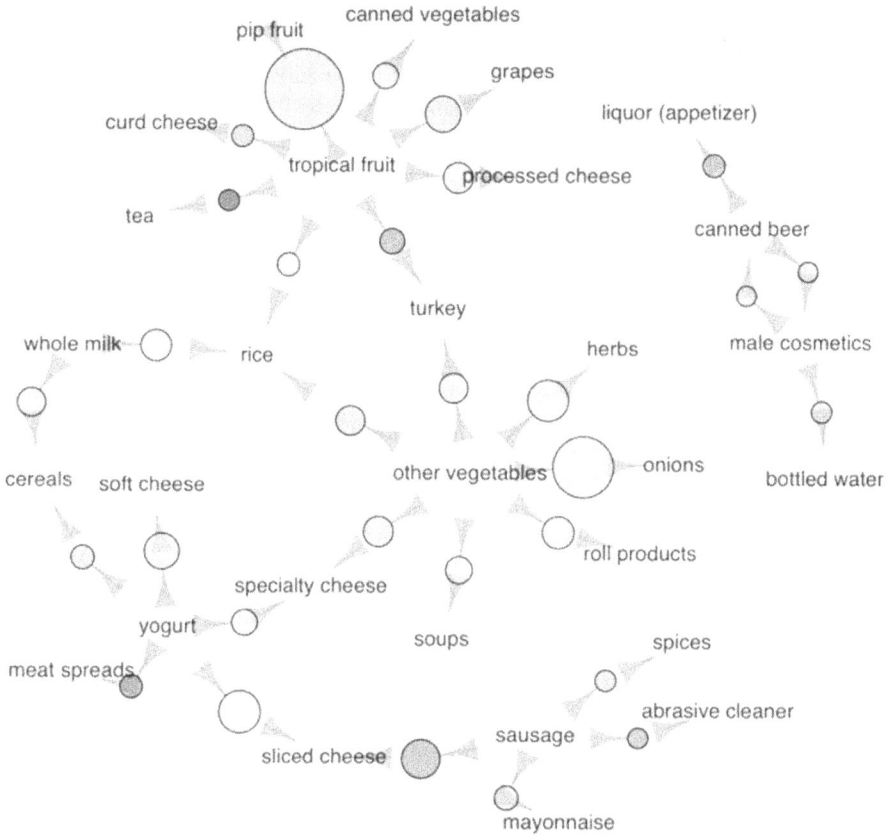

FIGURE 7.4 Illustration of association between selected items [46].

the category of unsupervised learning techniques discussed in the previous subsection. Unsupervised learning algorithm can range from clustering methods that group comparable data points to dimensionality reduction strategies that simplify huge and complicated datasets. Unsupervised learning is used in anomaly detection, photo identification, market segmentation, and consumer profiling. Because it lacks explicit feedback, it requires thorough data analysis and algorithm selection.

7.3.1   k-MEANS CLUSTERING ALGORITHMS

K-means clustering is one of the most widely used algorithms for centroid-based clustering [50]. Using a user-defined parameter called k, it divides a set of n data points into k clusters. The technique operates by first selecting k initial centroids at random from the data points, allocating each subsequent data point to the closest centroid, and then recalculating the centroids of the clusters that result [11]. Up until convergence, when no more data points are assigned to new clusters, the process is repeated. It remains one of the primary methods used today in data clustering–related applications [51–53].

While using k-means clustering, we aim to maximize the distance between the centroids of distinct clusters while minimizing the distance between a cluster's centroid and other points within it [11]. The steps in k-means algorithm operations as firstly proposed by Stuart Lloyd [50] are summarized as follows [11,54]:

i. Specify "k" as the arbitrary clusters. Each point should be assigned at random to one of the clusters.
ii. Estimate the distance between each observation and each cluster centroid and assign it to the cluster with the smallest distance.
iii. Recalculate the k centers. Use the mean vector of the cluster's points, to determine the cluster centroid for each cluster.
iv. Repeat steps ii and iii until no clusters are changing any more.

The areas of application of k-means are but not limited to the following: crime prediction and detection of fraud [55,56], cyber profiling [57], document processing [58], customer segmentation [53], and drone networks [59], amongst others. The extensive use of the k-means clustering in most clustering problems has been attributed to its simplicity, flexibility with huge datasets, and strong generalization properties [60]. k-means can be used in the early phases of machine learning tasks to develop a thorough grasp of your data [54]. However, k-means are not ideal for all applications due to a variety of disadvantages. Among these are the realities that the value of k-means must be known beforehand; thus, the initial cluster centers chosen are not always ideal, and the influence of noise results in a drop in accuracy [11]. Although these drawbacks haven't led to the abandonment of clustering, serious users must be ready to design algorithms that are better suited to their data than those that are commercially accessible. According to Nayini et al. [60], few other limitations of the k-means algorithm are as follows:

i. The Euclidean distance has a constraint that restricts its use to numerical data alone. All data types should be supported by a clustering method in its optimal condition.
ii. The choice of cluster centers and the value of k have significant effect on the clustering outcomes.
iii. Due to the possibility of algorithms based on square-error convergent to local minimum, different beginning partitions can result in different final clustering. This is particularly accurate if clusters are not properly separated.
iv. k-means is quite sensitive to data with outliers.
v. The k-means clustering technique is computationally intensive since it takes an iterative approach to tackle the problem, which can increase computation time for huge datasets. This makes it inappropriate when there is a large amount of data.

The k-means method has been modified in a number of ways to solve some of its drawbacks. These several variations of k-means, includes ones that limit the centroids to observations of the dataset (k-medoids), select medians (k-medians), and select initial centers less arbitrarily (k-means++), amongst others [9]. These variants are discussed as follows.

7.3.1.1 k-means++

This is an improved k-means clustering algorithm, which develops an initial center at random with a focus on a greater degree of dissimilarity. A consecutive center is chosen from the samples with a uniform probability to the distance between a sample and its closest current center, commencing with a random sample as the initial center as illustrated in Equation 7.1 [61]:

$$P(x) = \frac{d(x)^2}{\sum_{x \in \chi} d(x)^2} \tag{7.1}$$

where $d(x)$ is the separation between x and its closest center, x is a sample, while χ is a set comprising samples apart from the selected centers. This version of k-means reduces the squared distance between the centroids toward enhancing the centroids initialization. Thus, the challenge of subpar clustering outcome consequent upon very close initialization of centroids is abated [62].

7.3.1.2 *k*-Medoids

The k-mediods clustering algorithm represents a cluster's center (or medoid) as a specific point within the cluster [63]. Instead of choosing the mean point of a cluster as the centroid, K-medoids algorithm chooses the most centrally located point in the cluster as the medoids [64]. Due to its resistance to outliers and noise, a medoid can serve as a representative of a valid cluster center [65]. Since k-medoids algorithms employs medoids (actual data points), instead of mean points which might be impacted by outliers, it is more resilient to outliers than K-means. Common example is the PAM (Partitioning Around Medoids) or MCA (Minimum Cluster Analysis). It minimizes an objective function (absolute error) as shown in Equation 7.2 [63]:

$$\text{Absolute error } (E) = \sum_{j=0}^{k} \sum_{p \in c_j}^{n} |p - ob_j| \tag{7.2}$$

where E stands for the total absolute error, p is a data point that represents an object in the cluster (c_j), and ob_j is the object that best reflects c_j. The method repeats until the typical object becomes the medoid or the cluster item with the most centralized location.

7.3.1.3 K-Median

A dataset is divided into k clusters using the K-median clustering algorithm, with each cluster being represented by its median point [66]. The function of the method is to minimize the sum of the distances between each point and its assigned median. The k-median algorithm selects k starting medians at random, and then assigns each data point to the median that is closest to it. The medians are recomputed as the center of all the points assigned to each cluster after the original assignment. Until the medians stop fluctuating or a set number of iterations has been reached, this procedure is repeated [66,67].

7.3.1.4   Fuzzy *k*-Means

Each data point in a conventional *k*-means model only belongs to one cluster; in contrast, each point in fuzzy *k*-means might belong to numerous clusters, each with varying degrees of membership [68]. When there is uncertainty or overlap between groups, this variance enables additional flexibility in grouping. The fuzzy method of clustering is based on a soft grouping of units. The majority of the suggestions take into account object data, which are common unit-variable data matrixes containing numerical variables [69,70].

7.3.2   GAUSSIAN MIXTURE METHOD

As previously revealed, k-means clustering algorithm exhibits a critical drawback which is the use of hard assignment when categorizing data points. Poor performance for several real-life scenarios in k-means application is often attributed to its non-probabilistic structure and the way it employs the simple distance-from-cluster-center to allocate cluster membership. However, to overcome this limitation, the Gaussian Mixture Model (GMM) is an extension and also a viable alternative approach to k-means owing to their soft assignment strategy [71].

When we are unsure of which cluster a specific data point should belong to, instead of employing hard assignment, we can utilize probability to establish the correct cluster assignment for our data point. Being an extension of the K-means methodology, the GMM approach uses the clusters created by k-means as a starting point for data point customization [54]. GMM is a probabilistic model used for clustering or density estimation tasks. A GMM models the data distribution as the weighted sum of many Gaussian distributions (also known as components or clusters). A different subset of the data is represented by each Gaussian distribution. The percentage of each subpopulation in the total distribution is determined by the weights. It is a statistical technique used to represent complicated datasets as a combination of simpler Gaussian-shaped distributions [26].

The mixture component (cluster/distribution) weight and the component mean and variances are the two reconfigurable parameters in a Gaussian mixture model. Each data point is assigned a probability for each cluster, defining its membership, with the number of unique probability values corresponding to the number of clusters selected by the user [54,71]. Equation 7.3 mathematically represents GMM as a sum of M component Gaussian densities [71]:

$$p(\theta|x) = \sum_{i=1}^{k} \phi_i g(\theta \mid \mu_i, \Sigma_i) \qquad (7.3)$$

where θ is a D-dimensional continuous-valued data vector, ϕ_i, $i = 1....k$ are mixture weights, while $g(\theta \mid \mu_i, \Sigma_i)$, $i = 1....k$ are component Gaussian densities and $p(\theta|x)$ is the probability density function. Every element density is depicted by the probability distribution function in Equation 7.4 for representing its D-variate Gaussian function:

$$g(\theta|\mu_i, \Sigma_i) = \frac{1}{(2\pi)^{\frac{D}{2}}|\Sigma_i|} \exp\left\{-\frac{1}{2}(\theta - \mu_i)'\Sigma_i^{-1}(\theta - \mu_i)\right\} \qquad (7.4)$$

Given the covariance matrix Σ_i and the mean vector μ_i, the condition $\sum_{i=1}^{k} \phi_i = 1$ is satisfied by the mixed weights. Finding these parameters that best match the observed data is the aim of a GMM. The Expectation–Maximization (EM) algorithm is commonly used for this. This iterative technique alternates between calculating each data point's odds of belonging to each cluster (the "E-step") and updating each cluster's parameters based on these probabilities (the "M-step"). The cluster parameters are updated repeatedly by the EM algorithm until convergence is attained or a maximum number of repetitions are reached. Each data point can be assigned to the cluster with the highest probability to determine the final cluster allocations.

7.3.3 Density-Based Spatial Clustering of Applications with Noise

DBSCAN is a dynamic clustering algorithm used for clustering datasets with various shapes and sizes [72]. It is especially helpful when the data are noisy or have outliers and can't be effectively clustered using more conventional techniques like k-means or hierarchical clustering. The basic working principle of this algorithm is based on the concept of density connectivity, which entails connecting points that are adjacent to one another in a high-density area of the data space [73]. DBSCAN detects correlations and patterns in data that are challenging to find manually but may be pertinent and valuable to identify patterns and forecast trends. Consider a real-world scenario using DBSCAN. Let's say we operate an online store and want to increase sales by directing customers to pertinent products. We can forecast and recommend a relevant product to a particular consumer based on data collection even though we don't know exactly what they're looking for. By using the DBSCAN on our dataset (which is based on an online store database), we may identify clusters based on the items that customers have purchased. If customer A bought a pen, book, and scissors, and customer B bought a book and scissors, we may recommend a pen to customer B using this clustering technique [74].

DBSCAN can find clusters of any size or density, including nonconvex ones. In addition, the system can handle noisy data and recognize points that don't fit into any cluster. The two parameters that DBSCAN uses to operate are ε and $minPts$ [72]. ε is the minimal number of points needed to establish a dense region, whereas $minPts$ is the minimum number of points needed for two points to be considered to be a part of the same cluster [72]. A pair of data points are neighbors if their distances are equal or smaller than ε [54]. The technique begins by choosing a random location in the dataset and locating every other point that is nearby. A new cluster is created if there are more than $minPts$ within the distance of the starting point [73]. The method then goes through each point in the cluster one more time until no more points can be added. Figure 7.5 illustrates the DBSCAN algorithms and its generated clusters. DBSCAN classifies data points as core, border, or noise based on their proximity to other points in the dataset. These points are defined as follows:

  i. **Core point (C):** A point is a core point if it is surrounded by at least as many other points as $minPts$, including itself. Be aware that these points must also be located inside radius ε. In other words, we say that a point in DBSCAN is

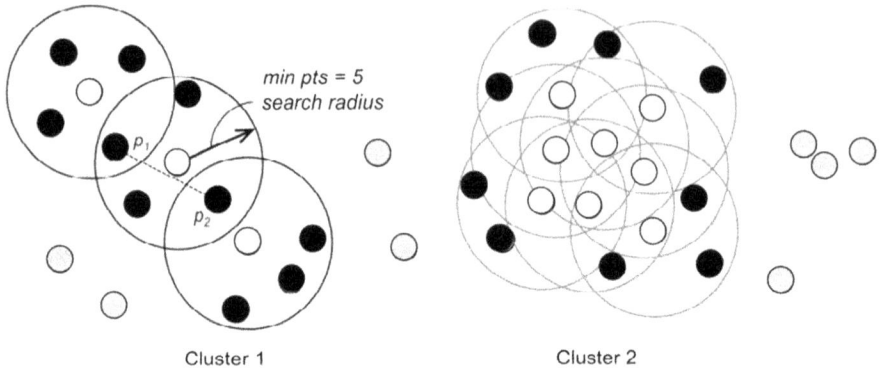

FIGURE 7.5 Density-Based Spatial Clustering of Applications with Noise (DBSCAN) algorithm and generated clusters [75].

considered a core point if it has at least a minimum number of other points (*minPts*) within a specified radius ε around it.

ii. **Border point (B):** A border point is a data point that is located within the radius of a core point but is not a core point. Border points are nonetheless a part of a cluster despite being on the periphery of dense areas and having fewer nearby points than core points. If a point can be reached from a core point and there are fewer than *minPts* surrounding it, it is regarded as a border point.

iii. **Noise point (N):** Noise point is a data point that does not belong to any cluster. Noise or outliers are points that cannot be reached from other core points. Noise points are outliers or low-density data points that do not fit clustering criteria.

The user initially specifies the values for both *minPts* and ε. The technique starts by picking a random data point and utilizing its ε-value to determine its neighborhood. A data point is designated as a core point and cluster formation occurs if there are *minPts* or more nearby data points. In the absence of this, the point is labeled as noise. All of the data points nearby the randomly chosen data point become a component of cluster 1 when cluster formation (let's call it cluster 1) begins. If the recently added data points are also core points, then cluster 1 will also include all of the nearby points [54,73].

Steps in DBSCAN algorithm are summarized as follows:

i. Choose a random location in the dataset that has not yet been explored.
ii. Discover every location that is within ε distance of the starting point.
iii. If there are more than *minPts* within ε distance, create a new cluster and include all of the nearby points. If not, label the data point as noise or an outlier.
iv. For each point in the newly formed cluster, repeat steps i–iii.
v. Repeat steps until all the points have been visited.

The literature is replete with a wide range of applications of the DBSCAN algorithm such as but not limited to automated systems for bridge monitoring in structural engineering [76], detection and sensitivity analysis of river heat in summer [77], field road categorization for farm machinery [78], natural disaster discovery [79], identifying the source of acoustic emissions [80], traffic analysis [81], and particulate matter (PM2.5) concentration monitoring for environmental safety [82]. Because ε and *minPts* might have an impact on DBSCAN, choosing the right settings for these parameters can be difficult [73]. Finding the ideal values for these parameters may need some trial-and-error, while the choice of these parameters might have a considerable impact on the clustering results. DBSCAN might be difficult to utilize on datasets with variable densities or when the actual number of clusters is unclear due to its sensitivity to parameter changes.

7.3.4 PRINCIPAL COMPONENT ANALYSIS

Consider finding trends and correlations in a dataset with multiple variables. Due to many factors, data analysis can be difficult. PCA helps here. As earlier noted, high-dimensional data can pose some challenges for some algorithms, especially when there are more attributes than observations [54]. PCA is an unsupervised machine learning algorithm used for dimensionality reduction. It transforms a series of possibly associated variables into a set of new vectors termed as principal components [83]. In order to extract features, PCA projects a high-dimensional space into a lower-dimensional subspace. PCA attempts to preserve the data section that gives the greatest variance while attempting to eliminate attributes that provide less variance [54,83]. It is a potent statistical method that breaks down complex datasets into smaller sets of variables called principal components by selecting the most crucial variables. These elements indicate the most important variance in the original data and can be used to better describe and evaluate it. PCA measures the link between two variables using covariance. If two variables have high covariance values, they are strongly connected. PCA seeks new, uncorrelated variables that explain the most data variance. These new variables are "principal components". Dimensionality reduction technique with PCA is shown in Figure 7.6. The PCA technique finds principal components in these phases [84]:

i. The first step involves standardizing (normalizing) the variables by dividing by the standard deviation and subtracting the mean. This makes sure that all variables are identical and operate on the identical scale.
ii. Calculate the standardized variable covariance matrix. The covariance matrix, a square matrix with the same number of rows and columns as variables, measures all pairwise connections.
iii. Decompose the covariance matrix's eigenvalues and eigenvectors. The eigenvalues show how much variation each eigenvector (new variables) explains.
iv. Using the eigenvalues as a guide, choose the principal components. As the most important source of variation in the data, the components with the highest eigenvalues are chosen as the principal components.

(a)   (b)

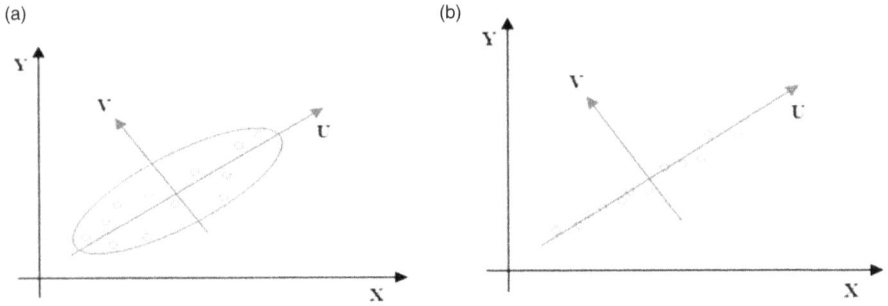

FIGURE 7.6   Dimensionality reduction with Principal Component Analysis (PCA): (a) Adjusted axis system and (b) Variable discarded [85].

PCA benefits include the following:

i. Reduces the number of variables present in a data set, which makes it simpler to interpret and visualize.
ii. PCA enhances data quality. PCA can boost the accuracy of the analysis by removing noise and redundant data from the data.
iii. Improves interpretability: PCA can identify the major causes of data variation, easier for users to understand the conclusions.

PCA, however, also has several drawbacks, such as:

i. **Information loss:** If the most important causes of variation in the data are not taken into account by the principal components, PCA may lead to information loss.
ii. **Complexity:** PCA can be difficult and time-consuming, particularly for large datasets.

7.4   TECHNIQUES IN OPTIMAL UNSUPERVISED LEARNING PROCESSES

Given the variety of methods and strategies available, choosing the optimal unsupervised learning technique and algorithm can be difficult. Using the incorrect approach or algorithm can result in unreliable results or lost opportunities for knowledge and insight. Each algorithm has pros and cons. We will go through the crucial elements to take into account when choosing an algorithm for optimal performance. Some of the key steps in achieving optimal unsupervised learning processes are discussed as follows.

7.4.1   PROBLEM DEFINITION

The problem that has to be solved must be identified as the first and most important stage in selecting an unsupervised learning algorithm. Understanding the information at hand, the desired result, and the kinds of insights being sought are all necessary for problem definition. The problem specification step lays the ground-work for choosing the best unsupervised learning algorithm to help find solutions to the challenge. The problem must be defined along with the criteria for the unsupervised algorithm. For instance, a challenge would be to find patterns in consumer data so that clients can be divided based on their purchase habits. Alternately, it can be to make the data less dimensional in order to perform analysis more quickly. To choose the optimal algorithm that can produce the desired outcomes, the problem must be defined explicitly. Several unsupervised learning methods are created to address various issues. To choose the most suitable unsupervised learning algorithm, it is crucial to comprehend the issue and intended result. The problem statement should contain data type, dimensions, noise or outliers, and distribution. This information helps identify if data preparation is needed before unsupervised learning.

7.4.2 Data Understanding and Pre-Processing

The optimal selection of an unsupervised learning method requires careful under-standing of the data and their pre-processing. Data structure, format, and character-istics must be understood. You must recognize the finest preprocessing methods to choose the best unsupervised learning algorithm. In this step, it is critical to consider cleaning the data to eliminate any discrepancies, mistakes, or missing values from the data. The accuracy and dependability of the data for analysis depend heavily on this stage. Further to this, the data must be transformed to change, scale, modify or categorically encode it. Finding the features that are most pertinent to the analysis is as well crucial in this step as it allows for lowering the data's dimensionality and enhancing the effectiveness of the unsupervised learning algorithm. Exploring the data and finding any patterns or trends that may be there need the use of data visu-alization. This phase might assist you in choosing the best unsupervised learning algorithm to employ.

7.4.3 Hyper-Parameter Tuning

While choosing an unsupervised learning method, hyper-parameter tweaking is vital. Hyper-parameters are data-independent variables that must be provided before the method is run. Hyper-parameters might be the number of clusters in a clustering algorithm, the learning rate in a neural network, or the number of principal components in a PCA in unsupervised learning methods. Tuning these hyper-parameters to their best settings optimizes algorithm performance. Hyper-parameter tuning begins with specifying the range of values for each hyper-parameter that will be considered throughout the search. This range must be big enough to accommodate a variety of sensible values without making the search computa-tionally prohibitive. Grid search, random search, and Bayesian optimization can alter hyper-parameters [86,87]. Previous chapters covered some of these methods. Algorithm performance evaluation metrics must also be chosen. It is also impor-tant to select the kind of evaluation metrics to be used in assessing the performance of the algorithms. Some of the significant evaluation metrics for unsupervised learning has been discussed earlier. This evaluation metric needs to be suitable for the particular activity and issue being dealt with. Run the hyper-parameter search to determine the best values for the hyper-parameters using the search strategy and evaluation metric you've chosen.

7.5 PERFORMANCE EVALUATION OF UNSUPERVISED LEARNING

Labeled data are needed to test unsupervised learning models. Unsupervised learning evaluation criteria assess these models' efficacy. Internal validation helps assess unsupervised learning tasks like grouping and dimensionality reduction. Unsupervised learning restricts cluster evaluation [88]. The clustering approach determines the measuring method, a well-known quirk of the evaluation process [89]. For the validation of the algorithm results, various factors must be considered while assessing the clustering results [88,90]:

i. Assessing propensity of data for clustering (i.e., confirming the existence of nonrandom structure).
ii. Choosing the appropriate number of clusters.
iii. Evaluating the clustering results' quality without using outside data.
iv. Comparing the outcomes with data from outside sources.
v. Comparing two different cluster sets to see which is more effective.

As there is no usage of outside data, the first three problems are resolved through internal or unsupervised validation. The fourth problem is fixed through supervised or external validation, while both supervised and unsupervised validation methods can address the last problem [90]. Evaluation procedures and metrics are presented by Gan et al. [91], and they include methods for both internal and external validation as shown in Figure 7.7.

A supervised learning issue may be related to external validation techniques. By including extra data, such as external class labels for the training instances, in the clustering validation process, external validation moves forward. External validation approaches are not typically employed on clustering problems because unsupervised learning techniques are typically used when such information is unavailable. Nonetheless, they are still applicable when external data are available and when

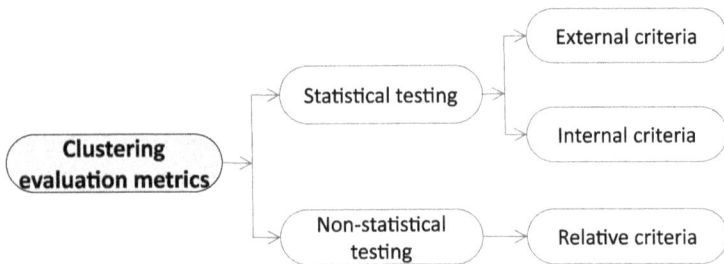

FIGURE 7.7 Clustering evaluation metrics [91].

creating synthetic data from actual data. By relying on the information provided by the data used as the clustering algorithm's input, internal validation methods enable the determination of the clustering structure's quality without requiring access to external data

The following are a few often used internal validation metrics commonly used for assessing the performance of clustering:

1. **Silhouette coefficient:** This metric assesses a data point's degree of fit inside its own cluster in relation to other clusters. A score of +1 denotes a well-clustered data point, a score of 0 denotes a point that is on the cluster boundary, and a score of −1 denotes a misclassified data point. The Silhouette co-efficient (s_i) is represented in Equation 7.5 [88,92]:

$$s(i) = \frac{b(i) - a(i)}{max\{a(i), \, b(i)\}} \tag{7.5}$$

   where $a(i)$ is the distance between each example in a cluster, while $b(i)$ is the average distance between each example and the examples in each cluster that don't contain the analyzed example and are estimated as Equations 7.6 and 7.7, respectively:

$$a(i) = \frac{1}{|C_a|} \sum_{j \in C_a, \, i \neq j} d(i, j) \tag{7.6}$$

$$b(i) = \min_{C_b \neq C_a} \frac{1}{|C_b|} \sum_{j \in C_b} d(i, j) \tag{7.7}$$

2. **Calinski-Harabasz (CH) index:** This metric measures the ratio of the between-cluster variance to the within-cluster variance. Higher values of this metric indicate better-defined clusters. CH is a metric that takes into account the dispersion both within and across clusters. The number of clusters that optimizes the CH value for x clusters would be our choice [93]:

$$CH = \left(\frac{SSB_x}{x - 1} \right) \div \left(\frac{SSE_x}{x} \right) \tag{7.8}$$

   where SSE is the sum of squared error within clusters while SSB is the sum of error between clusters.

3. **Dunn index:** This metric measures the ratio of the minimum distance between clusters to the maximum intra-cluster distance. In other words, Dunn index is the fraction of the maximum distance between clusters and the minimum distance between data from those clusters [88]. Higher values of this metric indicate better-defined clusters [94]. This ratio (D) must be maximized as shown in Equation 7.9:

$$D = \min_{1 < i < k} \left\{ \min_{1 < i < k, \, i \neq j} \left\{ \frac{\delta(C_i, \, C_j)}{\max_{1 < i < k} \{\Delta(C_i)\}} \right\} \right\} \tag{7.9}$$

$$\delta(C_i) = \max_{x, \, y \, \in c_i} \{d(x,y)\} \tag{7.10}$$

$$\delta(C_i, \, C_j) = \min_{x \, \in c_i, \, y \, \in c_j} \{d(x,y)\} \tag{7.11}$$

4. **Ball-Hall index:** The Ball-Hall index is a measure of cluster validity used in cluster analysis to evaluate the quality of clustering results [89]. It is calculated as the square root of the sum of the squared distances between each point in a cluster and its centroid, divided by the cluster's total number of points. It is mathematically expressed as in Equation 7.12:

$$BH = \sum \frac{d^2(i, \, c)}{n} \tag{7.12}$$

where n is the number of points in the cluster and $d^2(i, c)$ is the squared distance between the ith point in the cluster and that cluster's centroid.

5. **Xie-Beni score:** It is a cluster validity score that assesses the performance and quality of clustering outcomes. The ratio between the total distances between each data point and its nearest cluster center and the total distances between each data point and all cluster centers is known as the Xie-Beni score. It can be mathematically expressed as follows [88]:

$$XD = \frac{W_k}{K * D} \tag{7.13}$$

where K is the number of clusters, D is the average distance between all data points and all cluster centers, and W_k is the sum of the squared distances between each data point and its nearest cluster center.

6. **Hartigan index:** By comparing the within-cluster sum of squares (WCSS) of the present solution to the WCSS of a random clustering solution, the Hartigan index evaluates the quality of a clustering solution. The sum of the squared distances between each data point and the centroid of the cluster to which it belongs is the WCSS:

$$H = \frac{RSS_r - RSS}{d - k} \tag{7.14}$$

where d is the number of dimensions in the data, k is the number of clusters in the current solution, RSS_r is the WCSS of a random clustering solution, and RSS is the WCSS of the current solution. Clustering performance improves with higher Hartigan index values. A number of 1 indicates a perfect clustering solution, whereas 0 indicates a random one. When analyzing the effectiveness of hard clustering techniques, where each data point is assigned to a single cluster, the Hartigan index is very helpful. For assessing the performance of soft clustering techniques, where data points can belong to numerous clusters with variable degrees of membership, it might not be as useful.

Common evaluation metrics that are peculiar to dimensionality reductions are as follows:

1. **Explained variance ratio:** To determine how much of the overall variance in the data is explained by each principal component, the explained variance ratio is an evaluation statistic used in dimensionality reduction approaches like PCA. The percentage of variance in the data that is explained by each principal component is measured by the explained variance ratio. Better dimensionality reduction outcomes are indicated by a greater explained variance ratio. PCA projects the data onto a new set of orthogonal axes that captures the most variance in the data. Each primary component's explained variance ratio represents its share of data variation. The first main component accounts for 60% of the data's variation if the explained variance ratio is 0.6. The second primary component may explain 30% of variance with an explained variance ratio of 0.3.

2. **Reconstruction error:** Reconstruction error is a measurement of the difference between the original data and the reconstructed data following dimensionality reduction and is used in dimensionality reduction techniques like autoencoders and matrix factorization [95,96]. After dimensionality reduction, this metric compares the original data to the reconstructed data [95]. Better dimensionality reduction outcomes are indicated by a smaller reconstruction error. These methods use a set of learned parameters to map high-dimensional data onto a lower-dimensional space. The low-dimensional representation of the data is then transformed in an inverse manner to produce the reconstructed data. The reconstruction error is the difference between the original and reconstructed data, defined by a distance metric like MSE or RMSE.

3. **t-SNE visualization:** Visualizing high-dimensional data in two or three dimensions is possible with the t-SNE visualization approach. The t-SNE plot of a competent dimensionality reduction technique should show well-separated clusters. Unsupervised learning uses the evaluation metric t-SNE (t-Distributed Stochastic Neighbor Embedding) to view high-dimensional data in a low-dimensional environment. It is very helpful for displaying intricate and erratic relationships in data. Each data point in the high-dimensional space is represented by a probability distribution over its neighbors in the t-SNE algorithm. Repetition in low-dimensional space reduces the probability distribution discrepancy [97]. Thus, t-SNE can detect local and global data correlations. t-SNE creates 2D or 3D scatter plots for each data point in low-dimensional space. Comparable spots in high-dimensional space are close in low-dimensional space. This simplifies data pattern and cluster visualization.

7.6   CONCLUSION

This chapter has revealed the unsupervised learning as a viable machine learning approach which requires no labeled features, with more focus on its ability to identify hidden patterns and features in data. This chapter also examined some key tasks in unsupervised learning such as problem identification, data preprocessing, hyperparameters settings and performance evaluations. The chapter has demonstrated unique features of unsupervised machine learning as it is capable of creating a flexible and versatile way to unveil hidden features in data owing to its labeled features independency. This significantly contributes to novel knowledge discovery across many fields including thin film deposition. In the context of atomic layer deposition, this approach immensely benefits the discovery of novel materials for enhancing the quality of deposition. The rising potential applications of unsupervised learning are a direct result of the ongoing evolution of technology and computational resources. This advancement continues to drive the trajectories of machine learning space while deciphering intricate patterns and enhancing intelligent decision-making processes in the space of thin film deposition.

REFERENCES

[1] Javapoints, "Unsupervised machine learning," https://www.javatpoint.com/unsupervised-machine-learning, 2021.
[2] P. A. Adedeji, S. Akinlabi, N. Madushele, and O. O. Olatunji, "Wind turbine power output very short-term forecast: A comparative study of data clustering techniques in a PSO-ANFIS model," *Journal of Cleaner Production*, vol. 254, 2020, doi: 10.1016/j.jclepro.2020.120135.
[3] O. Adeleke, S. Akinlabi, T.-C. Jen, and I. Dunmade, "Prediction of municipal solid waste generation: An investigation of the effect of clustering techniques and parameters on ANFIS model Performance," *Environmental Technology*, vol. 43, no. 11, 2022, doi: 10.1080/09593330.2020.1845819.
[4] O. Adeleke, S. Akinlabi, T. C. Jen, P. A. Adedeji, and I. Dunmade, "Evolutionary-based neuro-fuzzy modelling of combustion enthalpy of municipal solid waste," *Neural Computing and Applications*, vol. 2, 2022, doi: 10.1007/s00521-021-06870-2.
[5] M. Pietrzykowski, "Local regression algorithms based on centroid clustering methods," in *Procedia Computer Science*, Elsevier B.V., 2017, pp. 2363–2371. doi: 10.1016/j.procs.2017.08.210.
[6] J. Liu, F. Cao, and J. Liang, "Centroids-guided deep multi-view K-means clustering," *Information Sciences*, vol. 609, pp. 876–896, 2022, doi: 10.1016/j.ins.2022.07.093.
[7] G. Pang and S. Jiang, "A generalized cluster centroid based classifier for text categorization," *Information Processing & Management*, vol. 49, no. 2, pp. 576–586, 2013, doi: 10.1016/j.ipm.2012.10.003.
[8] S. Zahra, M. A. Ghazanfar, A. Khalid, M. A. Azam, U. Naeem, and A. Prugel-Bennett, "Novel centroid selection approaches for KMeans-clustering based recommender systems," *Information Sciences*, vol. 320, pp. 156–189, Nov. 2015, doi: 10.1016/j.ins.2015.03.062.
[9] Shubham Bindal, "Clustering in ML - Part 2: Ccentroids based clustering," 2021, https://appliedsingularity.com/2021/07/13/clustering-in-ml-part-2-centroids-based-clustering/.
[10] J. Wang, H. Wang, and G. Zhao, "A GA-based solution to an NP-hard problem of clustering security events," in *2006 International Conference on Communications, Circuits and Systems, ICCCAS, Proceedings*, 2006, pp. 2093–2097. doi: 10.1109/ICCCAS.2006.284911.

[11] S. K. Shibu and P. Samuel, "Centroid based celestial clustering algorithm: A novel unsupervised learning method for haemogram data clustering," *IEEE Transactions on Emerging Topics in Computational Intelligence*, 2022, doi: 10.1109/TETCI.2022.3211004.

[12] Q. F. Yang *et al.*, "HCDC: A novel hierarchical clustering algorithm based on density-distance cores for data sets with varying density," *Information Sciences*, vol. 114, 2023, doi: 10.1016/j.is.2022.102159.

[13] A. Darányi, T. Czvetkó, A. Kummer, T. Ruppert, and J. Abonyi, "Multi-objective hierarchical clustering for tool assignment," *CIRP Journal of Manufacturing Science and Technology*, vol. 42, pp. 47–54, 2023, doi: 10.1016/j.cirpj.2023.02.002.

[14] E. Burghardt, D. Sewell, and J. Cavanaugh, "Agglomerative and divisive hierarchical Bayesian clustering," *Computational Statistics & Data Analysis*, vol. 176, 2022, doi: 10.1016/j.csda.2022.107566.

[15] A. Subasi, "Clustering examples," in *Practical Machine Learning for Data Analysis Using Python*, Elsevier, 2020, pp. 465–511. doi: 10.1016/b978-0-12-821379-7.00007-2.

[16] P. J. R. L. Kaufman, *Finding Groups in Data: An Introduction To Cluster Analysis*, vol. 344. United States: John Wiley & Sons, 2009.

[17] S. Rastogi, R. Shim, and K. Guha, "CURE: An efficient clustering algorithm for large databases," *ACM Sigmod Record*, vol. 27, no. 2, pp. 73–84, 1998.

[18] I. Oladeji, P. Makolo, R. Zamora, and T. T. Lie, "Density-based clustering and probabilistic classification for integrated transmission-distribution network security state prediction," *Electric Power Systems Research*, vol. 211, 2022, doi: 10.1016/j.epsr.2022.108164.

[19] X. Xu, S. Ding, Y. Wang, L. Wang, and W. Jia, "A fast density peaks clustering algorithm with sparse search," *Information Sciences*, vol. 554, pp. 61–83, 2021, doi: https://doi.org/10.1016/j.ins.2020.11.050.

[20] J. Xie, Z.-Y. Xiong, Y.-F. Zhang, Y. Feng, and J. Ma, "Density core-based clustering algorithm with dynamic scanning radius," *Knowledge-Based Systems*, vol. 142, pp. 58–70, 2018, doi: https://doi.org/10.1016/j.knosys.2017.11.025.

[21] R. Maheshwari, S. K. Mohanty, and A. C. Mishra, "DCSNE: Density-based clustering using graph shared neighbors and entropy," *Pattern Recognition*, vol. 137, 2023, doi: 10.1016/j.patcog.2023.109341.

[22] D. Cheng, R. Xu, B. Zhang, and R. Jin, "Fast density estimation for density-based clustering methods," *Neurocomputing*, 2023, doi: 10.1016/j.neucom.2023.02.035.

[23] A. Fahim, "A varied density-based clustering algorithm," *Journal of Computational Science*, vol. 66, 2023, doi: 10.1016/j.jocs.2022.101925.

[24] K. You and C. Suh, "Parameter estimation and model-based clustering with spherical normal distribution on the unit hypersphere," *Computational Statistics & Data Analysis*, vol. 171, 2022, doi: 10.1016/j.csda.2022.107457.

[25] J. D. Banfield and A. E. Raftery, "Model-based gaussian and non-gaussian clustering," 1993. [Online]. Available: https://about.jstor.org/terms.

[26] H. K. Kim *et al.*, "Probabilistic assessment of potential leachate leakage from livestock mortality burial pits: A supervised classification approach using a Gaussian mixture model (GMM) fitted to a groundwater quality monitoring dataset," *Process Safety and Environmental Protection*, vol. 129, pp. 326–338, 2019, doi: 10.1016/j.psep.2019.07.015.

[27] J. Fan *et al.*, "Convex hull indexed Gaussian mixture model (CH-GMM) for 3D point set registration," *Pattern Recognition*, vol. 59, pp. 126–141, 2016, doi: 10.1016/j.patcog.2016.02.023.

[28] D. H. Yi, D. W. Kim, and C. S. Park, "Prior selection method using likelihood confidence region and Dirichlet process Gaussian mixture model for Bayesian inference of building energy models," *Energy and Buildings*, vol. 224, 2020, doi: 10.1016/j.enbuild.2020.110293.

[29] T. Li and J. Ma, "Dirichlet process mixture of Gaussian process functional regressions and its variational EM algorithm," *Pattern Recognition*, vol. 134, 2023, doi: 10.1016/j.patcog.2022.109129.

[30] V. Melnykov and Y. Wang, "Conditional mixture modeling and model-based cluster-ing," *Pattern Recognition*, vol. 133, 2023, doi: 10.1016/j.patcog.2022.108994.

[31] Rukshan Pramoditha, "11 Dimensionality reduction techniques you should know in 2021," 2021, https://towardsdatascience.com/11-dimensionality-reduction-techniques-you-should-know-in-2021-dcb9500d388b.

[32] N. Trendafilov and M. Gallo, "PCA and other dimensionality-reduction techniques," in *International Encyclopedia of Education*, 4thFourth Eedition, R. J. Tierney, F. Rizvi, and K. Ercikan, Eds., Oxford: Elsevier, 2023, pp. 590–599. doi: https://doi.org/10.1016/B978-0-12-818630-5.10014-4.

[33] F. Anowar, S. Sadaoui, and B. Selim, "Conceptual and empirical comparison of dimen-sionality reduction algorithms (PCA, KPCA, LDA, MDS, SVD, LLE, ISOMAP, LE, ICA, t-SNE)," *Computer Science Review*, vol. 40, 2021. doi: 10.1016/j.cosrev.2021.100378.

[34] U. W. Lok *et al.*, "Real time SVD-based clutter filtering using randomized singular value decomposition and spatial downsampling for micro-vessel imaging on a Verasonics ultrasound system," *Ultrasonics*, vol. 107, 2020, doi: 10.1016/j.ultras.2020.106163.

[35] Y. Chen, L. Zhang, and B. Zhao, "Application of singular value decomposition (SVD) in extraction of gravity components indicating the deeply and shallowly buried gra-nitic complex associated with tin polymetallic mineralization in the Gejiu tin ore field, Southwestern China," *Journal of Applied Geophysics*, vol. 123, pp. 63–70, 2015, doi: 10.1016/j.jappgeo.2015.09.022.

[36] M. Vandecappelle and L. de Lathauwer, "From multilinear SVD to multilinear UTV decomposition," *Signal Processing*, vol. 198, 2022, doi: 10.1016/j.sigpro.2022.108575.

[37] H. Xu, "Unsupervised manifold learning with polynomial mapping on symmetric positive definite matrices," *Information Sciences*, vol. 609, pp. 215–227, 2022, doi: 10.1016/j.ins.2022.07.077.

[38] B. Yang, M. Xiang, and Y. Zhang, "Multi-manifold discriminant isomap for visualiza-tion and classification," *Pattern Recognition*, vol. 55, pp. 215–230, 2016, doi: 10.1016/j.patcog.2016.02.001.

[39] Y. Duan, C. Liu, S. Li, X. Guo, and C. Yang, "An automatic affinity propagation cluster-ing based on improved equilibrium optimizer and t-SNE for high-dimensional data," *Information Sciences*, vol. 623, pp. 434–454, 2023, doi: 10.1016/j.ins.2022.12.057.

[40] A. Bibal, V. Delchevalerie, and B. Frénay, "DT-SNE: t-SNE discrete visualizations as decision tree structures," *Neurocomputing*, vol. 529, pp. 101–112, 2023, doi: 10.1016/j.neucom.2023.01.073.

[41] H. Qu, L. Li, Z. Li, and J. Zheng, "Supervised discriminant isomap with maximum margin graph regularization for dimensionality reduction," *Expert Systems with Applications*, vol. 180, 2021, doi: 10.1016/j.eswa.2021.115055.

[42] X. Liu, D. Tosun, M. W. Weiner, and N. Schuff, "Locally linear embedding (LLE) for MRI based Alzheimer's disease classification," *Neuroimage*, vol. 83, pp. 148–157, 2013, doi: 10.1016/j.neuroimage.2013.06.033.

[43] I. Cetin, M. Stephens, O. Camara, and M. A. González Ballester, "Attri-VAE: Attribute-based interpretable representations of medical images with variational autoencod-ers," *Computerized Medical Imaging and Graphics*, vol. 104, 2023, doi: 10.1016/j.compmedimag.2022.102158.

[44] A. Caciularu and J. Goldberger, "An entangled mixture of variational autoencod-ers approach to deep clustering," *Neurocomputing*, vol. 529, pp. 182–189, 2023, doi: 10.1016/j.neucom.2023.01.069.

[45] T. Toe and T. Z. Rong, "Artificial intelligence with python." [Online]. Available: https://link.springer.com/bookseries/16715.

[46] Analyn Ng, "Association rules and the apriori algorithm: A tutorial," 2016, https://www.kdnuggets.com/2016/04/association-rules-apriori-algorithm-tutorial.html.

[47] L. Linwei, W. Yiping, H. Yepiao, L. Bo, M. Fasheng, and D. Ziqiang, "Optimized apri-ori algorithm for deformation response analysis of landslide hazards," *Computers & Geosciences*, vol. 170, 2023, doi: 10.1016/j.cageo.2022.105261.

[48] R. Papi, S. Attarchi, A. Darvishi Booorani, and N. Neysani Samany, "Knowledge discovery of middle east dust sources using apriori spatial data mining algorithm," *Ecological Informatics*, vol. 72, 2022, doi: 10.1016/j.ecoinf.2022.101867.

[49] S. Bagui, K. Devulapalli, and J. Coffey, "A heuristic approach for load balancing the FP-growth algorithm on MapReduce," *Array*, vol. 7, p. 100035, 2020, doi: 10.1016/j.array.2020.100035.

[50] S. P. Lloyd, "Least squares quantization in PCM," *IEEE Transactions on Information Theory*, vol. 28, no. 2, pp. 129–137, 1982, doi: 10.1109/TIT.1982.1056489.

[51] M. Haonan, Y. C. He, M. Huang, Y. Wen, Y. Cheng, and Y. Jin, Application of K-means clustering algorithms in optimizing logistics distribution routes, in *2019 6th International Conference on Systems and Informatics (ICSAI)*, New York: IEEE, pp. 1466–1470, 2019.

[52] S. Javadi, S. M. Hashemy, K. Mohammadi, K. W. F. Howard, and A. Neshat, "Classification of aquifer vulnerability using K-means cluster analysis," *Journal of Hydrology*, vol. 549, pp. 27–37, 2017, doi: 10.1016/j.jhydrol.2017.03.060.

[53] M. A. Syakur, B. K. Khotimah, E. M. S. Rochman, and B. D. Satoto, "Integration K-means clustering method and elbow method for identification of the best customer profile cluster," in *IOP Conference Series: Materials Science and Engineering*, Institute of Physics Publishing, 2018. doi: 10.1088/1757-899X/336/1/012017.

[54] C. C. Paul Fergus, *Applied Deep Learning: Tools, Techniques and Implementation*, 1stFrist edition. France: Springers, 2022.

[55] A. Ghorbani and S. Farzai, "Fraud detection in automobile insurance using a data min-ing based approach," 2018. [Online]. Available: www.aeuso.org.

[56] V. Jain, Y. Sharma, A. Bhatia, and V. Arora, "Crime prediction using K-means algorithm," *GRD Journal for Engineering*, vol. 2, 2017, [Online]. Available: www.grdjournals.com.

[57] M. Zulfadhilah, Y. Prayudi, and I. Riadi, "Cyber profiling using log analysis and K-means clustering a case study higher education in indonesia," 2016. [Online]. Available: www.ijacsa.thesai.org.

[58] N. Kumar, S. K. Yadav, and D. S. Yadav, "An Approach for Documents Clustering Using K-Means Algorithm," in *Innovations in Information and Communication Technologies (IICT-2020)*, P. K. Singh, Z. Polkowski, S. Tanwar, S. K. Pandey, G. Matei, and D. Pirvu, Eds., Cham: Springer International Publishing, pp. 453–460, 2021.

[59] S. M. Ferrandez, T. Harbison, T. Weber, R. Sturges, and R. Rich, "Optimization of a truck-drone in tandem delivery network using k-means and genetic algorithm," *Journal of Industrial Engineering and Management*, vol. 9, no. 2, pp. 374–388, 2016, doi: 10.3926/jiem.1929.

[60] S. G. and A. M. S. E. Y. Nayini, "A novel threshold-based clustering method to solve K-means weaknesses," in *International Conference on Energy, Communication, Data Analytics and Soft Computing (ICECDS-2017)*, pp. 47–52, Chennai, India: IEEExplore, 2017.

[61] H. Li and J. Wang, "Collaborative annealing power k-means++ clustering," *Knowledge-Based Systems*, vol. 255, 2022, doi: 10.1016/j.knosys.2022.109593.

[62] H. Li and J. Wang, "CAPKM++2.0: An upgraded version of the collaborative anneal-ing power k-means++ clustering algorithm," *Knowledge-Based Systems*, vol. 262, 2023, doi: 10.1016/j.knosys.2022.110241.

[63] N. Sureja, B. Chawda, and A. Vasant, "An improved K-medoids clustering approach based on the crow search algorithm," *Journal of Computational Mathematics and Data Science*, vol. 3, p. 100034, 2022, doi: 10.1016/j.jcmds.2022.100034.

[64] M. J. van der Laan, K. S. Pollard, and J. Bryan, "A new partitioning around medoids algorithm," *Journal of Statistical Computation and Simulation*, vol. 73, no. 8, pp. 575–584, 2003, doi: 10.1080/0094965031000136012.

[65] Q. Couloigner and I. Zhang, "A new and efficient K-medoid algorithm for spatial clustering," in *Lecture Notes in Computer Science*, 3rd edition, O. Gervasi, M. L. Gavrilova, V. Kumar, A. Laganà, H. P. Lee, Y. Mun, D. Taniar, C. J. K. Tan, Eds., pp. 181–189, 2005.

[66] V. Braverman, H. Lang, K. Levin, and Y. Rudoy, "Metric k-median clustering in insertion-only streams," *Discrete Applied Mathematics*, vol. 304, pp. 164–180, 2021, doi: 10.1016/j.dam.2021.07.025.

[67] Z. Zhang, Y. Zhou, and S. Yu, "Better guarantees for k-median with service installation costs," *Theoretical Computer Science*, vol. 923, pp. 292–303, 2022, doi: 10.1016/j.tcs.2022.05.014.

[68] Y. Wang, J. Ma, N. Gao, Q. Wen, L. Sun, and H. Guo, "Federated fuzzy k-means for privacy-preserving behavior analysis in smart grids," *Applied Energy*, vol. 331, 2023, doi: 10.1016/j.apenergy.2022.120396.

[69] X. Zhao, F. Nie, R. Wang, and X. Li, "Improving projected fuzzy K-means clustering via robust learning," *Neurocomputing*, vol. 491, pp. 34–43, 2022, doi: 10.1016/j.neucom.2022.03.043.

[70] M. B. Ferraro, "Fuzzy k-means: History and applications," *Econometrics and Statistics*, 2021, doi: 10.1016/j.ecosta.2021.11.008.

[71] D. Reynolds, "Gaussian mixture models," in *Encyclopedia of Biometrics*, S. Z. Li and A. Jain, Eds., Boston, MA: Springer US, 2009, pp. 659–663. doi: 10.1007/978-0-387-73003-5_196.

[72] KDD-96.final.frame.

[73] A. Latifi-Pakdehi and N. Daneshpour, "DBHC: A DBSCAN-based hierarchical clustering algorithm," *Data & Knowledge Engineering*, vol. 135, 2021, doi: 10.1016/j.datak.2021.101922.

[74] Kelvin Salton do Prado, "How DBSCAN works and why should we use it?" 2017, https://towardsdatascience.com/how-dbscan-works-and-why-should-i-use-it-443b4a191c80.

[75] P. M. DiFrancesco, D. Bonneau, and D. J. Hutchinson, "The implications of M3C2 projection diameter on 3D semi-automated rockfall extraction from sequential terrestrial laser scanning point clouds," *Remote Sensing*, vol. 12, no. 11, 2020, doi: 10.3390/rs12111885.

[76] M. Civera, L. Sibille, L. Zanotti Fragonara, and R. Ceravolo, "A DBSCAN-based automated operational modal analysis algorithm for bridge monitoring," *Measurement (Lond)*, vol. 208, 2023, doi: 10.1016/j.measurement.2023.112451.

[77] Z. Liu, W. Zhou, and Y. Yuan, "3D DBSCAN detection and parameter sensitivity of the 2022 Yangtze river summertime heatwave and drought," *Atmospheric and Oceanic Science Letters*, 2023, doi: 10.1016/j.aosl.2022.100324.

[78] X. Zhang, Y. Chen, J. Jia, K. Kuang, Y. Lan, and C. Wu, "Multi-view density-based field-road classification for agricultural machinery: DBSCAN and object detection," *Computers and Electronics in Agriculture*, vol. 200, 2022, doi: 10.1016/j.compag.2022.107263.

[79] F. Ros, S. Guillaume, R. Riad, and M. el Hajji, "Detection of natural clusters via S-DBSCAN a self-tuning version of DBSCAN," *Knowledge-Based Systems*, vol. 241, 2022, doi: 10.1016/j.knosys.2022.108288.

[80] Y. Rui, Z. Zhou, X. Cai, and L. Dong, "A novel robust method for acoustic emission source location using DBSCAN principle," *Measurement (Lond)*, vol. 191, 2022, doi: 10.1016/j.measurement.2022.110812.

[81] J. Holmgren, L. Knapen, V. Olsson, and A. P. Masud, "On the use of clustering analysis for identification of unsafe places in an urban traffic network," in *Procedia Computer Science*, Elsevier B.V., 2020, pp. 187–194. doi: 10.1016/j.procs.2020.03.024.

[82] X. Lu, J. Wang, Y. Yan, L. Zhou, and W. Ma, "Estimating hourly PM2.5 concentrations using Himawari-8 AOD and a DBSCAN-modified deep learning model over the YRDUA, China," *Atmospheric Pollution Research*, vol. 12, no. 2, pp. 183–192, 2021, doi: 10.1016/j.apr.2020.10.020.

[83] M. Schäfer et al., "Aortic shape variation after frozen elephant trunk procedure predicts aortic events: Principal component analysis study," *JTCVS Open*, 2023, doi: 10.1016/j.xjon.2023.01.015.

[84] Ł. Witanowski, P. Ziółkowski, P. Klonowicz, and P. Lampart, "A hybrid approach to optimization of radial inflow turbine with principal component analysis," *Energy*, 2023, doi: 10.1016/j.energy.2023.127064.

[85] S. C. Ng, "Principal component analysis to reduce dimension on digital image," *Procedia Computer Science*, pp. 113–119, 2017. doi: 10.1016/j.procs.2017.06.017.

[86] W. Gao, R. Fan, R. Huang, Q. Huang, W. Gao, and L. Du, "Augmented random search based inter-area oscillation damping using high voltage DC transmission," *Electric Power Systems Research*, vol. 216, 2023, doi: 10.1016/j.epsr.2022.109063.

[87] A. A. Ewees, F. H. Ismail, and A. T. Sahlol, "Gradient-based optimizer improved by slime mould algorithm for global optimization and feature selection for diverse computation problems," *Expert Systems with Applications*, vol. 213, 2023, doi: 10.1016/j.eswa.2022.118872.

[88] J.-O. Palacio-Niño and F. Berzal, "Evaluation metrics for unsupervised learning algorithms," 2019, [Online]. Available: https://arxiv.org/abs/1905.05667.

[89] A. C. Benabdellah, A. Benghabrit, and I. Bouhaddou, "A survey of clustering algorithms for an industrial context," *Procedia Computer Science*, pp. 291–302, 2019. doi: 10.1016/j.procs.2019.01.022.

[90] T. Michael S. Vipin, and K. Pang-Ning, *Introduction to Data Mining*, New York: Pearson Education, 2005.

[91] G. Gan, C. Ma, and J. Wu, *"Data Clustering: Theory, Algorithms, and Applications,"* Society for Industrial and Applied Mathematics, 2007. doi: 10.1137/1.9780898718348.

[92] A. M. Bagirov, R. M. Aliguliyev, and N. Sultanova, "Finding compact and well-separated clusters: Clustering using silhouette coefficients," *Pattern Recognition*, vol. 135, p. 109144, 2023, doi: https://doi.org/10.1016/j.patcog.2022.109144.

[93] MathWorks, "Calinski Harabasz Eevaluation," 2022, https://www.mathworks.com/help/stats/clustering.evaluation.calinskiharabaszevaluation.html.

[94] C.-E. ben Ncir, A. Hamza, and W. Bouaguel, "Parallel and scalable dunn index for the validation of big data clusters," *Parallel Computing*, vol. 102, p. 102751, 2021, doi: https://doi.org/10.1016/j.parco.2021.102751.

[95] W. Song, W. Li, Z. Hua, and F. Zhu, "A new deep auto-encoder using multiscale reconstruction errors and weight update correlation," *Information Sciences*, vol. 559, pp. 130–152, 2021, doi: https://doi.org/10.1016/j.ins.2021.01.064.

[96] C. Milbradt and M. Wahl, "High-probability bounds for the reconstruction error of PCA," *Statistics & Probability Letters*, vol. 161, p. 108741, 2020, doi: https://doi.org/10.1016/j.spl.2020.108741.

[97] A. Bibal, V. Delchevalerie, and B. Frénay, "DT-SNE: t-SNE discrete visualizations as decision tree structures," *Neurocomputing*, vol. 529, pp. 101–112, 2023, doi: https://doi.org/10.1016/j.neucom.2023.01.073.

8 Deep Learning

8.1 INTRODUCTION

This chapter extends the previous discussions on machine learning's principles by examining another vital subset of the machine learning i.e., deep learning which is based on deep neural network framework. To better comprehend the complexities of real-world scenarios and derive useful insights, deep learning has emerged as a revolutionary algorithm that has transformed the way we understand and process real-life data. Hence, to fully comprehend the realms of deep learning applications in thin films deposition, this chapter provides a strong preliminary background to deep learning methods, types, algorithms, and applications. Deep learning's capacity to unveil and identify hidden features and patterns automatically in data without needing manual feature engineering is what makes it so exciting. Additionally, other traditional machine learning techniques are incapable of handling and deriving insights from unstructured data such as image and video data. With a working principle similar to the human brain, deep learning algorithms handles a vast amount of data to inform an intelligent decision making. However, this approach grapples with some challenges in its applications. This chapter concludes by highlighting some of the limitation of the deep learning techniques which impedes its exploration while shaping the trajectories of industries, businesses, and future applications.

8.2 WHAT IS DEEP LEARNING

Deep learning, a kind of machine learning, employs deep neural networks to carry out challenging tasks. Instead of depending on manually created features or subject-matter knowledge, deep learning algorithms can discover features directly from the raw data [1]. A deep neural network's basic design is made up of various layers, each of which applies a particular operation to the incoming data. The first hidden layer is given the raw data by the input layer. The input data are transformed nonlinearly by each hidden layer before being passed on to the following layer, and so on, until the desired result is achieved [2,3]. The model's ultimate output, which can be a classification, regression, or prediction, is provided by the output layer. Deep learning's primary benefit is its capacity to learn from enormous volumes of unstructured data [4,5]. For instance, by studying millions of photos with identified objects, a deep learning model can be trained to identify things in an image [1]. In applications like image identification, speech recognition, and natural language processing (NLP), this strategy has demonstrated to be incredibly effective [6]. Figure 8.1 represents the workflow of deep learning in solving real-life problems.

Backpropagation, a method used to adjust the neural network weights during training, is another feature of deep learning. In backpropagation, the weights of the network are updated in the direction of the negative gradient after computing the

DOI: 10.1201/9781003346234-10

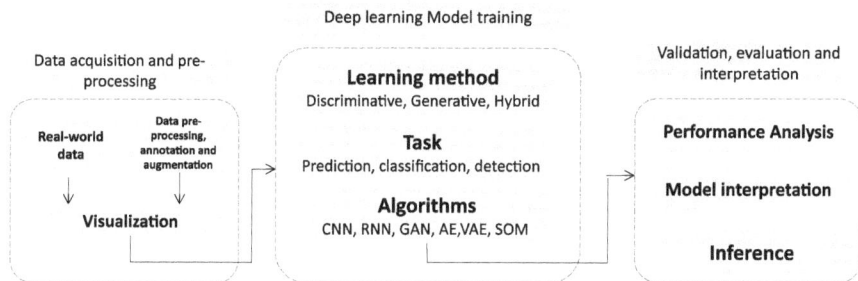

FIGURE 8.1   A common deep learning flowchart for solving real-world problems.

gradient of the loss function with respect to the weights [7]. Iteratively repeating this procedure allows the model to reach a minimum of the loss function. The issue of overfitting, when the model gets overly specialized to the training data and performs badly on unseen data, is one of the most important difficulties in deep learning. Many regularization methods, including dropout, early halting, and weight decay, can be used to prevent overfitting. These methods aid the model's improved generalization to fresh data. Some prominent deep learning architecture has been extensively applied in numerous applications. The convolutional neural network (CNN), which is largely employed for image and video recognition applications, is one of the most widely utilized designs [8]. Convolutional layers are used repeatedly in CNNs to extract features from the input image, which are then down sampled by pooling layers. The recurrent neural network (RNN), which is extensively utilized for sequential data such as speech and text, is another well-liked architecture. RNNs are particularly good at processing input sequences with varied lengths because they use recurrent connections to model temporal dependencies in the data [9].

8.3   DEEP LEARNING ALGORITHMS

Deep learning has immense potential for addressing some of the most difficult issues facing society today and is a topic that is quickly developing. We'll examine the fundamentals of deep learning algorithms and their specific applications. Figure 8.2 represents different categories of deep learning techniques and their respective algorithms.

8.3.1   CONVOLUTIONAL NEURAL NETWORK

CNN is a prominent discriminative deep learning architecture [10]. Thus, the CNN improves regularized multi-layer perceptron (MLP) network architecture. CNN layers optimize parameters for meaningful output and reduce model complexity while utilizing a "dropout" to prevent overfitting in traditional networks [1]. CNNs are used in visual recognition, medical image analysis, image segmentation, NLP, and more [11]. It is more powerful than a regular network since it automatically discovers key traits from input. CNNs like visual geometry group (VGG) [12], AlexNet [13],

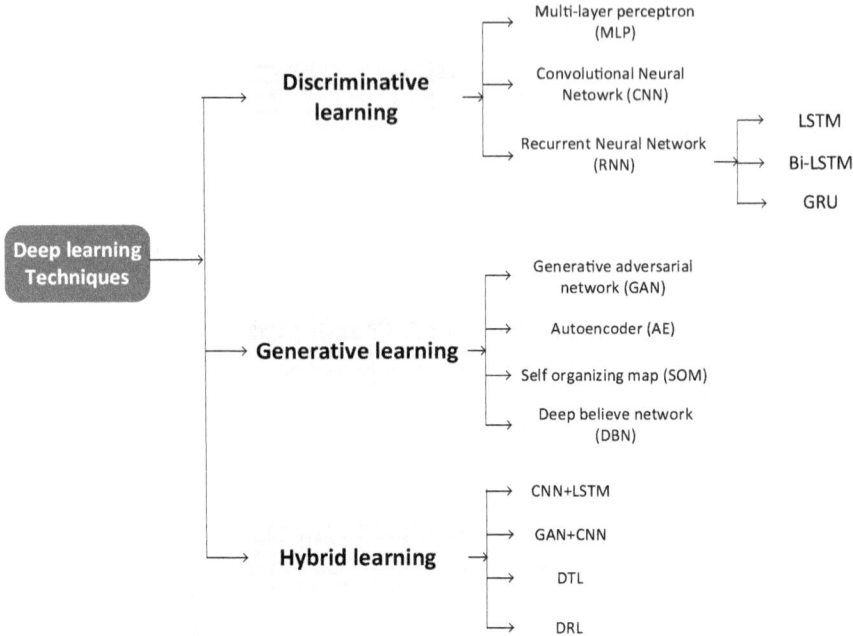

FIGURE 8.2   Deep learning techniques.

Xception [14], Inception [15], ResNet [16], and others can be employed in different application domains depending on their learning capabilities. Consider a sample set depicted as X_{ij}, where $i = 1,...p$ and $j = 1,....q$ (p is the input length while q is the feature size). The convolution kernel (m) is denoted by the amount of extracted features, where n is the stride size. The output f_{lh} of CNN in this scenario is expressed as shown as equation in Equation 8.1:

$$f_{lh} = \sum_{i=1}^{p}\sum_{i=1}^{q} w_{hij} \times X_{ij} \qquad (8.1)$$

where h designates the hth convolution kernel and spans from 1 to k, while l denotes the lth sample in the output vector and ranges from 1 to r, depicting the length of the output vector, which is determined by kernel size m and stride size n as indicated in Equation 8.2 [8]:

$$r = \frac{q - m}{n} + 1 \qquad (8.2)$$

Moreover, w_{hij} is used to denote the weight that each convolution kernel possesses because each one's weight is distinct from the other. This model will enable you to obtain the parameter. The final step involves combining the numerous outputs

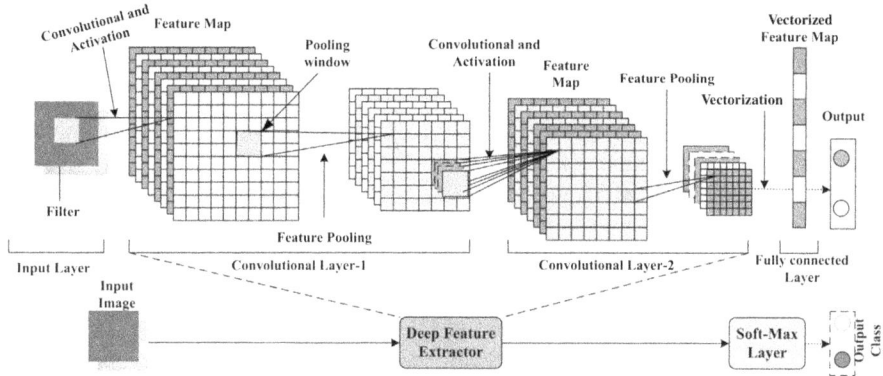

FIGURE 8.3   Convolutional neural network framework [17].

produced by each of the k convolution kernels into a single result. The CNN model's simplified training process, which is made possible by using lesser weights than the fully linked design, is a crucial feature [8]. Figure 8.3 displays a CNN with numerous convolutions and pooling layers

8.3.2   Recurrent Neural Network

Another well-known neural network is the RNN, which uses sequential or time-series data and inputs the results of the previous stage into the current stage [1]. RNNs are able to process sequences of inputs and keep a "memory" of previous inputs, in contrast to typical feedforward neural networks, which only analyze data in a single pass [18]. Recurrent networks, like feedforward and CNN, also learn from training input, but they set themselves apart by having a "memory" that lets them use data from earlier inputs to influence current input and output [19]. RNNs have a "hidden state" that updates with each input. This concealed state represents the network's "memory" of prior inputs. Training-learned weights and biases update the hidden state. RNN output depends on sequence elements, unlike standard DNN, which believes inputs and outputs are independent. Standard recurrent networks have vanishing gradients, making learning long data sequences difficult [1]. The Long Short-Term Memory (LSTM) network is a prominent RNN variation that handles input sequence dependencies well [20]. Three gates control data flow into and out of LSTM memory cells, which may store data for lengthy durations [20]. The "Forget Gate" determines what information from the previous state cell will be memorized and what will be removed, while the "Input Gate" determines which information enters the cell state and the "Output Gate" controls outputs [1,20]. LSTM networks are successful RNNs because they handle training problems. Bidirectional RNNs is another variant of RNN which accept past and future data by connecting two hidden layers that run in opposite directions to a single output [21]. Bidirectional RNNs can predict both positive and negative temporal directions. Bidirectional LSTMs (BiLSTMs) can improve sequence classification model performance.

8.3.3 Generative Adversarial Network

The Generative Adversarial Network (GAN), a generative deep learning model, proposed by Goodfellow in 2014, has been called the most revolutionary machine learning idea of the last decade [22]. The GAN has a generator and a discriminator to generate synthetic data that look like genuine data. The generator and discriminator models are antagonistic and increase each other's performance during training because their goals are opposing [23]. The generator and discriminator train together to improve their ability to distinguish real data from bogus data. The generator is trained until it produces data that are indistinguishable from real data. By replicating x distribution, the generator G generates $G(z)$ that is indistinguishable from x. The initial GAN model used random variables and input noise, z. After that, controlled z or extra labels were used to control $G(z)$[24,25]. However, the discriminator D calculates the chance for each case to discriminate between input data from the training data x or the created fake data $G(z)$. As with $D(z)=1$ and $DG(z)=0$, the purpose is to recognize actual data as "1" and artificially generated data as "0". GAN's loss function as in equation 3, indicates that the generator and discriminator's goal functions oppose each other [23]:

$$\min_{G} \max_{D} V(D,G) = E_{x \sim P_{data}(x)}\left[\log D(x)\right] + E_{z \sim P_x(z)}\left[\log(1 - D(G(z)))\right] \quad (8.3)$$

Although GAN may train a generator that creates high-quality data quickly, model collapse that cannot dependably handle the balance between generator and discriminator has been raised. Accordingly, models like DCGAN, WAGAN, BEGAN, and StarGAN have been developed to generate high-quality image data by compensating for model instability and generating the required image [23].

8.3.4 Autoencoder

Autoencoder neural networks use significant properties to compress and decompress data. Data compression, anomaly detection, picture and speech recognition, and autoencoders are common autoencoder applications [26,27]. Autoencoders contain input, hidden, and output layers. Hidden layers encode input data into a lower-dimensional latent space while the encoder maps input data to latent space, whereas the decoder maps latent space to output space. Training an autoencoder reduces reconstruction error, the discrepancy between input and output data [28]. As shown in Figure 8.4, we can structure an unlabeled dataset as a supervised learning problem to rebuild. By minimizing the reconstruction error, the difference between the original input and the reconstruction trains this network. Our network design relies on the bottleneck because without it, our network could readily memorize input data by passing them through the network [29]. Autoencoders are neural networks that copy input to output. An autoencoder encodes and decodes a handwritten digit image. Autoencoders compress data while decreasing reconstruction error.

Input Layer   Output Layer

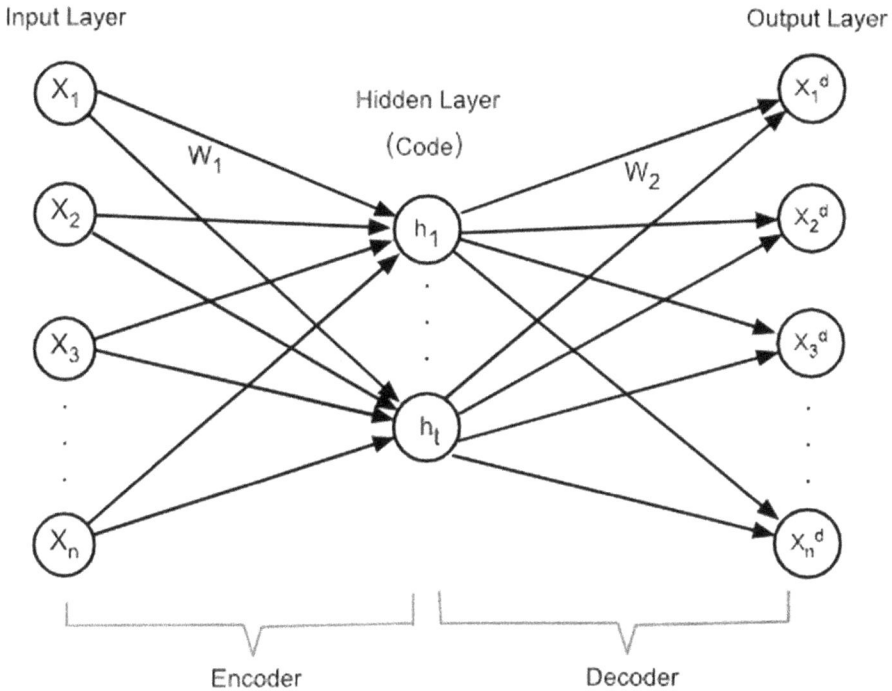

FIGURE 8.4 Framework of autoencoder [28].

8.3.5 Self Organizing Map

The Self-Organizing Map (SOM), also referred to as the Kohonen map, is an unsupervised neural network technique that represents high-dimensional data in a lower-dimensional space through a competitive learning process [30,31]. It is frequently employed for pattern detection, grouping, and data visualization [32]. A competitive learning process that organizes high-dimensional data in a lower-dimensional space is the basis of the self-organizing map. A prototype or cluster in the data space is represented by each node or neuron in the map layer of the SOM, which is made up of an input layer and a map layer [30]. In order to reduce the distance between the input data and the nearest neuron, the SOM modifies the weights of the neurons in the map layer during training. A dataset with p variables observed in n observations could be clustered by variable values. These clusters could be viewed as a two-dimensional "map" with proximal clusters having more comparable values than distal clusters. This helps view and analyze high-dimensional data. Like most artificial neural networks, SOM's core is training and mapping [33]. Training first generates a lower-dimensional representation of an input dataset (the "map space") from the "input space." Second, the map classifies more input data. Training typically represents a p-dimensional input space as a two-dimensional map space. A p-variable input space has p dimensions. Map spaces have hexagonal or rectangular grids of "nodes" or "neurons" in two dimensions. Data analysis and exploration goals determine the number and placement of nodes [31].

8.4 A BRIEF OVERVIEW OF SOME REAL-LIFE APPLICATIONS OF DEEP LEARNING

Deep learning's ability to absorb massive amounts of data and discover intricate patterns has opened up an endless number of potentials for a range of real-life applications. The notable deep learning real-world applications are discussed as follows.

8.4.1 CHATBOTS

A chatbot is a text- or text-to-speech-based Artificial Intelligence (AI) application for online communication. Chatbots can quickly fix consumer issues. The basic concept of Chatbots operation lies in the NLP methods to comprehend and respond to inputs from user inputs, and learn from prior interactions. It has the ability to interact with people and carry out human-like tasks. Chatbots are frequently employed in customer service, social media marketing, and instant messaging clients [34]. Inputs from users are met with automated responses. To produce various forms of reactions, it employs machine learning and deep learning algorithms. A chatbot may very well engage in informal conversations while also being aware of users' requirements and attempting to meet those needs [35]. People converse, whereas chatbots speak to consumers in a very similar way. It's an ideal tool for e-learning as well as for sales, customer support, and searching [35]. Microsoft's bots can also fulfill private users' business wishes. They schedule conferences and seminars, making it simple to keep track of tasks and maintain order. Figure 8.5 shows typical chatbot training using ensemble-based training technique proposed by Cuayáhuitl et al. [36] This process in the framework presented in Figure 8.5 is repeated until the end of a discussion for the required dialogue, or until there is no longer any improvement in the performance of the agents.

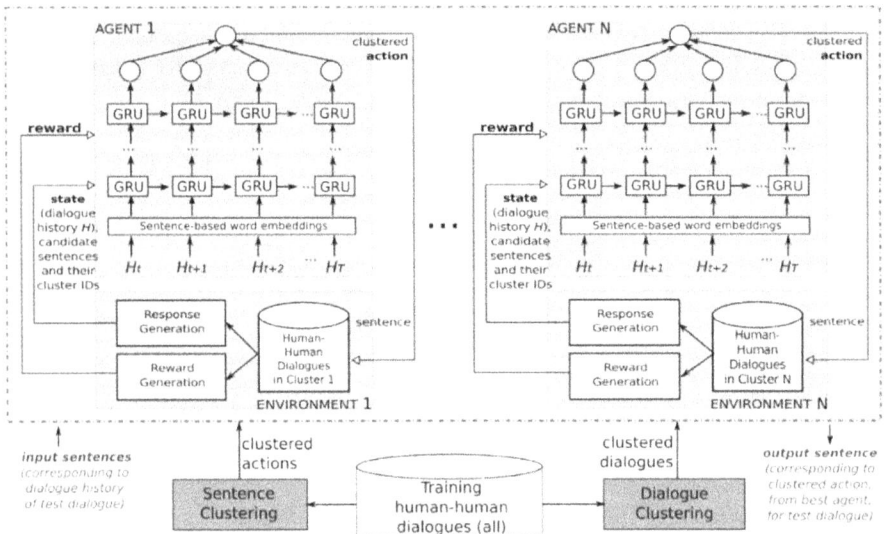

FIGURE 8.5 A typical chatbot training framework using an ensemble-based training approach [36].

8.4.2 Autonomous Vehicle

The concept of autonomous self-driving vehicles is propelled by deep learning [37,38]. To accurately recognize and track items in the car's environment, such as pedestrians, and obstructions, and other vehicles, deep learning models can be utilized [39]. Training these algorithms to distinguish these objects and their motions involves massive datasets of tagged photos. Deep learning can be used to design a self-driving car's secure and efficient route as well as to predict the future actions of other cars and pedestrians [39]. In order to maintain a safe and pleasurable ride, the speed and steering of the car can also be managed by deep learning models. A newly developed technique for learning driving rules is deep reinforcement learning (DRL). Without interference from humans, the autonomous learner may use a DRL algorithm to learn to drive by trial-and-error method [40]. To recognize road signs, lane lines, and other objects, semantic segmentation involves splitting a picture into distinct parts based on their semantic significance. To help the car comprehend its surroundings and make judgments, deep learning models can conduct semantic segmentation on camera images [41]. Uber Artificial Intelligence Laboratories are empowering new autonomous vehicles and creating autonomous vehicle for on-demand food delivery in an effort to diversify their business infrastructure while Amazon has used drones to transport its goods in a few locations throughout the world [42].

8.4.3 Robotics

Deep learning is employed while creating robots that can carry out activities that humans do [43]. Robotics is increasingly using deep learning to enhance robot performance in a variety of tasks, including object detection, navigation, manipulation, and control [44–46]. Robots with deep learning capabilities employ real-time updates to detect obstructions in their way and instantaneously organize their route [46]. It can be used to transport things in medical facilities, manufacturing facilities, warehouses, inventory control, etc. [42]. Further to this, Robotic gripping and manipulation capabilities can also be enhanced via deep learning. The advancements in computer vision aspects of deep learning have caused a paradigm in potential robot applications. A significant limitation of robots is the restricted movements. However, robot's capabilities in handling vast data sets upon training, often gives them more grasping possibilities while confirming to diverse shapes and sizes [45]. As a vital preprocessing element, deep learning can process sensor data into feature space with lesser dimensions. This is utilized in control process. Furthermore, the sensor data analysis can assist informed decision regarding maintenance planning as it can predict when parts are likely to break [47]. Thus, the restricted limitation challenge is overcome through deep learning, as more control is achieved effectively and precisely to carry out difficult activities like walking, sprinting, and jumping. One crucial element pertaining to robots is their capacity to alter their surroundings, a skill that has proven to be challenging to acquire. Additionally, deep reinforcement learning, for instance, can be used to teach robots to master difficult locomotive tasks through trial and error [48].

8.4.4 HEALTHCARE

Beyond technological and industrial benefits, deep learning technique has extended their reach in healthcare services by facilitating rapid, effective and precise operations across the entire spectrum of healthcare delivery and clinical practise. Computer vision, natural language processing, and reinforcement learning are widely employed deep learning methodologies within the healthcare domain. The ability of healthcare practitioners to swiftly and effectively evaluate massive amounts of data has changed the healthcare sector [49]. Image processing and microstructural analysis functionality of the deep learning has been a game changer in medical image analysis. Other prominent applications in this domain include disease diagnosis, prediction and monitoring of patient outcomes, and drug development [50–54]. Through the use of medical imaging, it is frequently employed for medical research, medication discovery, and the identification of serious illnesses like cancer and diabetic retinopathy [48,55]. Deep learning algorithms have the ability to analyze medical images such as MRI, CT scans, and X-rays to diagnose anomalies and identify infections [49,56]. Deep learning models can analyze electronic health records (EHR) more precisely and effectively such as laboratory test results, diagnoses, and prescription information. A deep learning algorithm, for instance, can precisely identify malignant cells in mammograms, assisting doctors in early breast cancer detection. Deep learning has the ability to forecast patient outcomes [50], including the possibility that a patient will contract a disease, experience complications, and respond to therapy. Deep learning algorithms can evaluate enormous volumes of data to find risk factors and forecast results, enabling individualized therapy and care. Deep learning has significantly contributed to drug delivery and interaction prediction. They are able to recognize effective therapy by integrating genomics, clinical and population datasets. This is a critical advancements in the pharmaceutical industry. Deep learning algorithms have the ability to spot anomalies and notify medical staff of potential health issues, allowing for early intervention and prevention [51].

8.4.5 RECOMMENDER SYSTEM

The subject of recommender systems has undergone a revolution because of deep learning, which has made it possible to model intricate connections and patterns in user–item interaction data [57–63]. Deep learning in advertising enables user experience optimization [63]. Deep Learning assists publishers and marketers in boosting advertising campaigns and increasing the significance of the advertisements. Targeted display advertising, real-time ad bidding, and data-driven predictive advertising are all options [60]. A popular method in recommender systems that bases recommendations on previous interactions between users is collaborative filtering [64]. Learning embeddings that describe users and objects in a low-dimensional space can be accomplished using deep learning models like autoencoders, matrix factorization, and neural networks [57]. The user–item interactions can then be predicted using these embeddings. An approach called content-based recommendation suggests products based on their characteristics, such as their genre, director, or actor [65]. Convolutional neural networks and RNNs are examples of deep learning models that can be used to learn representations of items based on their qualities and then suggest related items based on the learned representations [66,67]. Deep learning has shown to be an effective method in recommender systems, giving users precise and customized recommendations. To attain ideal performance, it also needs a lot of data and computer power, as well as thorough model selection and optimization [61].

8.4.6 Voice Assistance

Virtual assistants are web-based programs that take user commands in natural language and carry them out. Virtual assistants like Amazon Alexa, Cortana, Siri, and Google Assistant are common examples [42]. To operate to their greatest potential, they require computers with an internet connection. Based on prior interactions and the usage of Deep learning algorithms, the assistant generally offers a better user experience each time a command is given to it [68]. Neural networks are frequently used in deep learning voice assistance models [68–70]. Large volumes of data are used to train these algorithms to identify patterns and comprehend the subtleties of human speech. Utilizing NLP techniques to decipher the intent behind user voice commands is a crucial aspect of deep learning models for voice support [71]. This entails analyzing the user's words to determine whether they are an inquiry, a request for information, or a command to carry out an action.

8.5 DEEP LEARNING PLATFORMS AND FRAMEWORK

Deep learning model creation and training require a lot of processing power and expertise. Fortunately, some platforms offer entire deep learning model creation and deployment packages. These platforms offer cloud-based training and deployment, neural network libraries, and model optimization. Some of the most popular deep learning platforms are discussed as follows.

8.5.1 TensorFlow

TensorFlow is a complete open-source machine learning framework and library for deep neural network training and decision-making [72]. TensorFlow allows programmers to design and train deep learning models with a number of tools and modules [73]. TensorFlow's computational power with simplicity is one of its key merits. Data flow graphs are used by TensorFlow to depict computing in a flexible and effective manner [73,74]. This enables developers to easily design complicated models and fully utilize the available computational capacity. With TensorFlow, all computation and states in a machine learning algorithm are represented by a single dataflow graph, including each individual mathematical operation, each parameter's update rules, and the input preprocessing as Figure 8.6 depicts [75]. Users have the option of creating new functions from scratch or describing custom functions based on already existing procedures [73]. A considerable advantage over CPU-based training applications exists for TensorFlow operators with GPU (Graphics Processing Unit) and TPU (Tensor Processing Unit) capabilities. Developers can create their own layers by employing custom layer definitions in addition to the existing layers

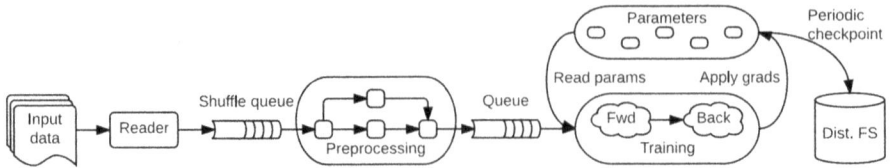

FIGURE 8.6   An illustration of a TensorFlow dataflow graph [75].

of TensorFlow, such as convolutions, pooling, and dense layers [74]. Hardware acceleration is also used in layers created with TensorFlow operators. Researchers may mix these operators to create new layers in TensorFlow because it contains a number of fundamental and complex operators in various categories, including mathematics, image processing, and neural networks. TensorFlow's own layer classes make it simpler to include built layers in deep learning models [72–75]. The majority of the layers and functions for creating and refining deep learning models are included in the TensorFlow library [73]. Furthermore, it has the adaptability of custom layers, allowing programmers to create their own layers [74].

TensorFlow is a powerful machine learning framework with simple tools for developing and testing deep learning architectures [73]. A low-level Application Programming Interface (API) for TensorFlow communicates with hardware like the CPU, GPU, and TPU through a tiered structure, as depicted in Figure 8.7 [73]. The use of TensorFlow operators on GPU and TPU resources is made possible by low-level API, which also speeds up the training and application of deep learning models. A large selection of TensorFlow operators offer effective machine learning methods on top of the low-level API. The unique class definitions established in TensorFlow can be used by developers to construct their own custom layers. The custom layer's algorithm may be expressed using TensorFlow operators, which makes it simple to include the new layer into the desired model architecture [73].

8.5.2   PyTorch

The deep learning and machine learning tool, PyTorch, was developed by Facebook's (Now Meta) artificial intelligence group to perform large-scale image analysis; as a result, its features include object discovery, classification, and division. It is not limited to these tasks, though, and can be used to evaluate sophisticated algorithms in conjunction with other tools. It is based on Torch library while performing tasks such as natural language processing and computer vision [76]. It requires a few modifications to handle large-scale calculations in a GPU environment because it is built in Python and C++. However, PyTorch offers a great platform for creating automated functions in these GPU contexts [77]. PyTorch is well renowned for its adaptability and simplicity. Python's high-level features and modules can create and train deep learning models. PyTorch's dynamic computation graph is built as the program runs. This allows testing out various architectures and debugging the code simple. Hybrid graphs, which blend static and dynamic graphs, are another feature of PyTorch. The powerful GPU acceleration support that PyTorch offers can

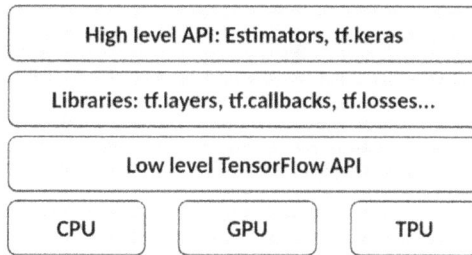

FIGURE 8.7   TensorFlow hierarchical framework for APIs [73].

significantly shorten deep neural network training times. In addition, it has a sizable and vibrant developer community, which results in the availability of numerous pre-built models and libraries for typical tasks. Two high-level features offered by PyTorch are deep neural networks constructed using a tape-based automatic differentiation mechanism and tensor computation with strong GPU acceleration [76].

8.5.3 KERAS

Google created the high-level Keras deep learning API to implement neural networks. It is used to make the implementation of neural networks simple and is developed in Python [78]. In addition, various backend neural network computations are supported [79]. Deep learning models may be created and trained quickly and easily with the help of Keras. With the high-level interface that Keras offers for creating and training neural networks, programmers can quickly prototype and test out various topologies and hyperparameters. Keras builds on low-level machine learning frameworks like TensorFlow and Theano to simplify neural network creation and training [78]. It supports feedforward, convolutional, and recurrent networks and has many built-in layers, activation functions, and optimization techniques [79]. A variety of data pretreatment tools, such as those for text and picture pre-processing, model evaluation tools, and visualization tools are also supported by Keras. Researchers and developers in the field of machine learning frequently employ Keras because it is a strong and adaptable tool for creating and training deep learning models [80,81]. As well as a variety of tools to make dealing with picture and text data easier, Keras includes multiple implementations of widely used neural-network building blocks like layers, objectives, activation functions, and optimizers. This helps to simplify the coding required to create deep neural networks. To make deep learning more approachable for both researchers and developers, Keras was created; even people with modest machine learning expertise may easily develop and train deep learning models thanks to its user-friendly API and modular design [79].

8.5.4 CAFFE

Caffe was developed at the University of California Berkeley Vision and Learning Center (BVLC) [82]. It is an open-source framework with a quick, effective, and scalable design. Caffe uses CUDA for GPU acceleration and is developed in C++ [83].

Caffe was originally developed to help picture classification and other computer vision tasks, but it has now been expanded to serve additional deep learning applications like object detection and segmentation. The structure of Caffe is built on a data flow paradigm, where data passes through a network of layers that carry out different operations like convolution, pooling, and activation. The foundation of Caffe's design is a data flow model, in which data passes through a network of layers that carry out different tasks [84]. Each layer gets input from the layer before and delivers output to the layer after via a directed acyclic network [85]. Caffe layers include convolutional, pooling, normalizing, and activation functions. Windows, Linux, and macOS support Caffe. It supports Python, C++, and MATLAB. APIs connect Caffe to TensorFlow and PyTorch [84].

8.5.5 THEANO

The Montreal Institute for Learning Algorithms (MILA) at the University of Montreal created Theano, an open-source numerical computation package for Python [86–88]. It enables programmers to effectively define, optimize, and test multi-dimensional array-based mathematical equations. Theano can run on both CPUs and GPUs and was created to be used for deep learning, namely, for the implementation of neural networks [87]. It was one of the first libraries to offer automatic differentiation, a feature required for backpropagation-based deep learning training of deep neural networks. The popular Python library for numerical calculation, NumPy, had a significant impact on the creation of Theano [88]. Similar to NumPy in syntax, Theano enables programmers to operate on multi-dimensional arrays in a number of ways, including addition, element-wise multiplication, and matrix multiplication. However, Theano is substantially quicker than NumPy for some sorts of computations because to its optimizations and support for GPU acceleration [89].

8.6 CHALLENGES AND LIMITATIONS OF DEEP LEARNING

Despite deep learning's widespread success, several issues and obstacles impede its practicality. Deep learning's biggest problem is the massive data needed to train neural networks. Unlike deep learning models, supervised and unsupervised learning algorithms can often perform well with small amounts of data. This section discusses challenges of deep learning. Despite these obstacles, deep learning has led to many industry-changing advancements. By understanding and overcoming its limitations, we can maximize deep learning's potential and advance artificial intelligence.

8.6.1 COMPUTING POWER, RESOURCES, AND HARDWARE REQUIREMENT

Deep learning's biggest drawback is its need for powerful computing power and resources. Deep learning techniques require massive data for training due to their complexity. This massive dataset requires great computational and processing power. GPUs and TPUs can only deliver such computing power. Small enterprises may be unable to afford these hardware gadgets. Training may take days or weeks depending on the model's complexity and data volume. If employed on a big scale or in real time, deep learning models may require a lot of computing resources. This may require more hardware and infrastructure to control load and system performance. Due to its data storage requirement, tiny or low-budget organizations may struggle to meet this storage need.

8.6.2 Data Availability, Integrity, and Quality

Data quantity, quality, and integrity determine deep learning model's success. Deep learning algorithms use massive amounts of high-quality data to learn and predict. Data integrity is crucial because the model's accuracy and efficiency depend on its training data. Noisy or inadequate data may lead to inaccurate predictions and poor model performance. When data are scarce, large-scale data collection and curation can take time and money. This need may be problematic for organizations with limited data access. Skewed training data can bias deep learning algorithms. If the data used to train the model does not match real-world scenarios and demographics, predictions may be inaccurate or unfair. Deep learning's use of sensitive or classified personal data might also raise confidentiality concerns. It's crucial to protect deep learning model data to maintain consumer trust.

8.6.3 Explainability and Transparency of Model's Outcome

One of the biggest problems with deep learning is transparency in explaining its outcome. Understanding how and why a deep learning model makes a specific choice or prediction is referred to as interpretability. In some circumstances, deep learning models' outcome with higher accuracy may also have lower interpretability, and vice versa. This trade-off might be difficult since, in some circumstances, accuracy may have to be given up for interpretability. Understanding how deep learning models' function is challenging due to their complexity and extensive number of parameters. When it's critical to understand how the model came to its conclusion, as it is in areas like healthcare or finance, this lack of transparency and interpretability can pose a big issue. It can be difficult to give feedback on each choice that a deep learning model makes because it is trained on vast volumes of data. Because of this, it could be challenging to find and fix model biases or inaccuracies. Deep learning model comparison and validation might be difficult because there is currently no accepted approach for interpreting deep learning models. It can be difficult to understand the model's results in the case of its application in analyzing complex data, such as photos or natural language. It is crucial to create approaches for comprehending how deep learning models function and why they produce particular predictions as they get more complicated and are utilized in more crucial applications. To overcome this obstacle, new interpretive techniques for deep learning models must be created to ensure a compromise between its accuracy and interpretability

8.6.4 Overfitting

Deep learning models are able to learn complex relationships in the data since they can include millions of parameters. Due to the possibility of learning to memorize the training data rather than generalizing to new data, they are rendered vulnerable to overfitting. A number of strategies, including regularization, early halting, and dropout, can be used to combat overfitting. By adding a penalty term to the loss function, approaches like L1 or L2 regularization prevent the model from inferring too many intricate correlations from the data. Early stopping entails keeping track of the model's performance during training on a validation dataset and terminating training when the performance begins to deteriorate. Dropout is a strategy that, while the model is being trained, randomly removes some neurons, assisting in preventing the model from becoming overfit to particular data patterns.

8.7   FUTURE PROSPECT OF DEEP LEARNING

Deep learning has great potential for innovation and growth across many industries. Deep learning has transformed how we solve complex challenges in several industries. New tools and technologies are being developed for deep learning, creating great prospects for innovation and groundbreaking research. Deep learning techniques and technologies promise to transform data-driven difficulties and usher in a new era of creation and discovery. These techniques range from specialized hardware accelerators to automated machine learning and quantum computing.

Tensor Processing Units (TPUs) are specialized hardware accelerators made to expedite the inference and training of deep learning models [90]. These processors are perfect for deep learning workloads because they can multiply matrices significantly quicker than conventional CPUs and GPUs [91]. The deep learning architecture known as "generative adversarial networks" (GANs) is capable of producing new data samples that are comparable to the training data [92]. This technology has many uses, including improving data augmentation methods and producing realistic photos and movies. In a distributed machine learning strategy known as federated learning, only the trained model is shared; the training data are kept on individual computers or servers. By enabling the training set to stay private, this method may help to overcome privacy problems in deep learning [93].

Automated machine learning (AutoML) is machine learning that has been partially or completely automated [94,95]. By democratizing machine learning and making it more approachable for nonexperts, this technology could hasten the development of novel deep learning applications [96]. By providing exponential speedups for some tasks, quantum computing has the potential to transform deep learning [97]. Quantum computing is still in its infancy, but experts are investigating its utility in deep learning. GNNs (Graph Neural Networks) are a deep learning architecture for graph-data-like social networks, chemical structures, and internet structures [98,99]. Many uses for this technology exist, ranging from the development of new drugs to personalized recommendations.

Drones and self-driving cars use deep learning algorithms. As technology advances, robots may be able to perform challenging tasks without human aid. Deep learning can analyze massive medical data to create patient-specific treatment plans. Deep learning models could assist doctors in creating more effective and tailored medicines by examining elements including genetics, lifestyle, and medical history [100]. Deep learning is a rapidly developing discipline, and a number of fascinating research trends are emerging that could fundamentally alter how we approach hard issues in a variety of fields. We can anticipate seeing many more fascinating discoveries and achievements in the years to come as academics continue to investigate these tendencies. The necessity for interpretability and transparency in deep learning models has grown in significance as they get more complicated. To help users better comprehend the logic underlying deep learning models' predictions, researchers are looking into novel approaches to make the models more comprehensible and interpretable.

8.8 CONCLUSION

Deep learning has demonstrated superior characteristics of automatic feature extraction from unprocessed dataset and unravel complex hidden pattern. This has been a game changer in several spectrum of human endeavors spanning from entertainment, healthcare, transportation, and finance amongst others. Despite deep learning's impressive progress, there are still several issues to address, such as improving deep learning models' interpretability and explainability, addressing AI ethics, and making deep learning more accessible and affordable. This chapter provided basic information regarding the types, algorithms, and some real-life applications of deep learning. While this chapter has laid the foundations for understanding the encompassing realms of deep learning applications in ALD and thin film technology, information provided in this chapter are not exhaustive, readers interested in more details are advised to refer to more in-depth text on deep learning. This space continues to emerge as the major driver of innovative research and expanding the capabilities of machines while integrating intelligent systems and human comprehension.

REFERENCES

[1] I. H. Sarker, "Deep learning: A comprehensive overview on techniques, taxonomy, applications and research directions," *SN Computer Science*, vol. 2, no. 6, 2021. doi: 10.1007/s42979-021-00815-1.
[2] Mathworks, "What is deep learning?" https://www.mathworks.com/discovery/deep-learning.html.
[3] B. Y. C. A. Goodfellow, *A Deep Learning*, Cambridge, MA: MIT Press, 2016.
[4] C. C. Paul Fergus, *Applied Deep Learning: Tools, Techniques and Implementation*, 1stFrist edition. France: Springers, 2022.
[5] Y. LeCun, Y. Bengio, and G. Hinton, "Deep learning," *Nature*, vol. 521, no. 7553, pp. 436–444, 2015, doi: 10.1038/nature14539.
[6] D. Kaul, H. Raju, and B. K. Tripathy, "Deep learning in healthcare," in *Studies in Big Data*, Springer Science and Business Media Deutschland GmbH, 2022, pp. 97–115. doi: 10.1007/978-3-030-75855-4_6.
[7] M. Vogt, "An overview of deep learning and its applications," *Proceedings*, no. January, pp. 178–202, 2019, doi: 10.1007/978-3-658-23751-6_17.
[8] Z. Guo, C. Yang, D. Wang, and H. Liu, "A novel deep learning model integrating CNN and GRU to predict particulate matter concentrations," *Process Safety and Environmental Protection*, vol. 173, pp. 604–613, 2023, doi: 10.1016/j.psep.2023.03.052.
[9] R. Khaldi, A. El Afia, R. Chiheb, and S. Tabik, "What is the best RNN-cell structure to forecast each time series behavior?" *Expert Systems with Applications*, vol. 215, p. 119140, 2023, doi: https://doi.org/10.1016/j.eswa.2022.119140.

[10] Y. Lecun, L. Bottou, Y. Bengio, and P. Haffner, "Gradient-based learning applied to document recognition," *Proceedings of the IEEE*, vol. 86, no. 11, pp. 2278–2324, 1998, doi: 10.1109/5.726791.

[11] I. H. Sarker, "Deep cybersecurity: A comprehensive overview from neural network and deep learning perspective," *SN Comput Science*, vol. 2, no. 3, 2021, doi: 10.1007/s42979-021-00535-6.

[12] K. He, X. Zhang, S. Ren, and J. Sun, "Spatial pyramid pooling in deep convolutional networks for visual recognition," *IEEE Transactions on Pattern Analysis and Machine Intelligence*, vol. 37, no. 9, pp. 1904–1916, 2015, doi: 10.1109/TPAMI.2015.2389824.

[13] A. Krizhevsky, I. Sutskever, and G. E. Hinton, "ImageNet classification with deep convolutional neural networks." [Online]. Available: https://code.google.com/p/cuda-convnet/.

[14] F. Chollet, "Xception: Deep learning with depthwise separable convolutions," in *2017 IEEE Conference on Computer Vision and Pattern Recognition (CVPR)*, pp. 1800–1807, 2017. doi: 10.1109/CVPR.2017.195.

[15] C. Szegedy *et al.*, "Going deeper with convolutions," in *2015 IEEE Conference on Computer Vision and Pattern Recognition (CVPR)*, pp. 1–9, 2015. doi: 10.1109/CVPR.2015.7298594.

[16] K. He, X. Zhang, S. Ren, and J. Sun, "Deep residual learning for image recognition," in *2016 IEEE Conference on Computer Vision and Pattern Recognition (CVPR)*, pp. 770–778, 2016. doi: 10.1109/CVPR.2016.90.

[17] I. Kumar, S. P. Singh, and Shivam, "Chapter 26- Machine learning in bioinformatics," in *Bioinformatics*, D. B. Singh and R. K. Pathak, Eds., Academic Press, 2022, pp. 443–456. doi: https://doi.org/10.1016/B978-0-323-89775-4.00020-1.

[18] C. G. S. Capanema, G. S. de Oliveira, F. A. Silva, T. R. M. B. Silva, and A. A. F. Loureiro, "Combining recurrent and Graph Neural Networks to predict the next place's category," *Ad Hoc Networks*, vol. 138, p. 103016, 2023, doi: https://doi.org/10.1016/j.adhoc.2022.103016.

[19] M. Ibrahim and R. Elhafiz, "Modeling an intrusion detection using recurrent neural networks," *Journal of Engineering Research*, vol. 11, no. 1, p. 100013, 2023, doi: https://doi.org/10.1016/j.jer.2023.100013.

[20] G. Van Houdt, C. Mosquera, and G. Nápoles, "A review on the long short-term memory model," *Artificial Intelligence Review*, vol. 53, no. 8, pp. 5929–5955, 2020, doi: 10.1007/s10462-020-09838-1.

[21] N. Singh, R. Nath, and D. B. Singh, "Splice-site identification for exon prediction using bidirectional LSTM-RNN approach," *Biochemistry and Biophysics Reports*, vol. 30, p. 101285, 2022, doi: https://doi.org/10.1016/j.bbrep.2022.101285.

[22] I. Goodfellow *et al.*, "Generative adversarial nets," in *Advances in Neural Information Processing Systems*, Z. Ghahramani, M. Welling, C. Cortes, N. Lawrence, and K. Q. Weinberger, Eds., Curran Associates, Inc., 2014. [Online]. Available: https://proceedings.neurips.cc/paper_files/paper/2014/file/5ca3e9b122f61f8f06494c97b1afccf3-Paper.pdf.

[23] S. Kim, H. Jang, and B. Yoon, "Developing a data-driven technology roadmapping method using generative adversarial network (GAN)," *Computers in Industry*, vol. 145, p. 103835, 2023, doi: https://doi.org/10.1016/j.compind.2022.103835.

[24] J. Bao, D. Chen, H. Li, and G. Hua, "CVAE-GAN: Fine-grained image generation through asymmetric training." in *Proceedings of the IEEE International Conference on Computer Vision*, IEEE, pp. 2745–2754, 2017.

[25] Y. Choi, M. Choi, M. Kim, J.-W. Ha, S. Kim, and J. Choo, "StarGAN: Unified generative adversarial networks for multi-domain image-to-image translation," 2017, [Online]. Available: https://arxiv.org/abs/1711.09020.

[26] M. Schultz and M. Tropmann-Frick, "Autoencoder neural networks versus external auditors: Detecting unusual journal entries in financial statement audits," in *Proceedings of the Annual Hawaii International Conference on System Sciences*, IEEE Computer Society, pp. 5421–5430, 2020. doi: 10.24251/hicss.2020.666.

[27] S. E. Otto and C. W. Rowley, "Linearly recurrent autoencoder networks for learning dynamics," *SIAM Journal on Applied Dynamical Systems*, vol. 18, no. 1, pp. 558–593, 2019, doi: 10.1137/18M1177846.

[28] P. Li, Y. Pei, and J. Li, "A comprehensive survey on design and application of auto-encoder in deep learning," *Applied Soft Computing*, vol. 138, p. 110176, 2023, doi: 10.1016/j.asoc.2023.110176.

[29] J. Zhai, S. Zhang, J. Chen, and Q. He, "Autoencoder and its various variants," in *2018 IEEE International Conference on Systems, Man, and Cybernetics (SMC)*, pp. 415–419, 2018. doi: 10.1109/SMC.2018.00080.

[30] D. Miljkovic, "Brief review of self-organizing maps," in *2017 40th International Convention on Information and Communication Technology, Electronics and Microelectronics, MIPRO 2017 - Proceedings*, Institute of Electrical and Electronics Engineers Inc., pp. 1061–1066, 2017. doi: 10.23919/MIPRO.2017.7973581.

[31] U. Asan and S. Ercan, "An introduction to self-organizing maps," in *Computational Intelligence Systems in Industrial Engineering. Atlantis Computational Intelligence Systems*, C. Kahraman, Eed., Paris: Atlantis Press, pp. 295–315, 2012. doi: 10.2991/978-94-91216-77-0_14.

[32] S. Lakshminarayanan, "Application of self-organizing maps on time series data for identifying interpretable driving manoeuvres," *European Transport Research Review*, vol. 12, no. 1, p. 25, 2020, doi: 10.1186/s12544-020-00421-x.

[33] E.-W. Augustijn and R. Zurita-Milla, "Self-organizing maps as an approach to exploring spatiotemporal diffusion patterns," *International Journal of Health Geographics*, vol. 12, no. 1, p. 60, 2013, doi: 10.1186/1476-072X-12-60.

[34] G. Sperlí, "A cultural heritage framework using a deep learning based chatbot for supporting tourist journey," *Expert Systems with Applications*, vol. 183, p. 115277, 2021, doi: https://doi.org/10.1016/j.eswa.2021.115277.

[35] E. Kasthuri and S. Balaji, "Natural language processing and deep learning chatbot using long short term memory algorithm," *Materials Today: Proceedings*, 2021, doi: https://doi.org/10.1016/j.matpr.2021.04.154.

[36] H. Cuayáhuitl *et al.*, "Ensemble-based deep reinforcement learning for chatbots," *Neurocomputing*, vol. 366, pp. 118–130, 2019, doi: https://doi.org/10.1016/j.neucom.2019.08.007.

[37] S. Grigorescu, B. Trasnea, T. Cocias, and G. Macesanu, "A survey of deep learning techniques for autonomous driving," *Journal of Field Robotics*, vol. 37, no. 3, pp. 362–386, 2020, doi: 10.1002/rob.21918.

[38] J. Ren, H. Gaber, and S. S. Al Jabar, "Applying deep learning to autonomous vehicles: A survey," in *2021 4th International Conference on Artificial Intelligence and Big Data (ICAIBD)*, pp. 247–252, 2021. doi: 10.1109/ICAIBD51990.2021.9458968.

[39] A. Khanum, C. Y. Lee, and C. S. Yang, "Deep-learning-based network for lane following in autonomous vehicles," *Electronics (Switzerland)*, vol. 11, no. 19, 2022, doi: 10.3390/electronics11193084.

[40] S. Alagumuthukrishnan, S. Deepajothi, R. Vani, and S. Velliangiri, "Reliable and efficient lane changing behaviour for connected autonomous vehicle through deep reinforcement learning," *Procedia Computer Science*, vol. 218, pp. 1112–1121, 2023, doi: https://doi.org/10.1016/j.procs.2023.01.090.

[41] Q. Sellat, S. Bisoy, R. Priyadarshini, A. Vidyarthi, S. Kautish, and R. K. Barik, "Intelligent semantic segmentation for self-driving vehicles using deep learning," *Computational Intelligence and Neuroscience*, vol. 2022, p. 6390260, 2022, doi: 10.1155/2022/6390260.

[42] By Avijeet Biswal, "Top 25 deep learning applications used across industries," 2023, https://www.simplilearn.com/tutorials/deep-learning-tutorial/deep-learning-applications.

[43] N. Sünderhauf *et al.*, "The limits and potentials of deep learning for robotics," *The International Journal of Robotics Research*, vol. 37, no. 4–5, pp. 405–420, 2018, doi: 10.1177/0278364918770733.

[44] R. A. Mouha, "Deep learning for robotics," *Journal of Data Analysis and Information Processing*, vol. 9, no. 2, pp. 63–76, 2021, doi: 10.4236/jdaip.2021.92005.

[45] C. C. Johnson, T. Quackenbush, T. Sorensen, D. Wingate, and M. D. Killpack, "Using first principles for deep learning and model-based control of soft robots," *Frontiers in Robotics and AI*, vol. 8, 2021, doi: 10.3389/frobt.2021.654398.

[46] M. O. Macaulay and M. Shafiee, "Machine learning techniques for robotic and autonomous inspection of mechanical systems and civil infrastructure," *Autonomous Intelligent Systems*, vol. 2, no. 1, p. 8, 2022, doi: 10.1007/s43684-022-00025-3.

[47] M. Soori, B. Arezoo, and R. Dastres, "Artificial intelligence, machine learning and deep learning in advanced robotics, a review," *Cognitive Robotics*, vol. 3, pp. 54–70, 2023, doi: https://doi.org/10.1016/j.cogr.2023.04.001.

[48] R. Liu, F. Nageotte, P. Zanne, M. De Mathelin, and B. Dresp-Langley, "Deep reinforcement learning for the control of robotic manipulation: A focussed mini-review." *Robotics*, vol. 10, no. 1, p. 22, 2021.

[49] D.-E.-M. Nisar, R. Amin, N.-U.-H. Shah, M. A. A. Ghamdi, S. H. Almotiri, and M. Alruily, "Healthcare techniques through deep learning: Issues, challenges and opportunities," *IEEE Access*, vol. 9, pp. 98523–98541, 2021, doi: 10.1109/ACCESS.2021.3095312.

[50] A. Esteva *et al.*, "A guide to deep learning in healthcare," *Nature Medicine*, vol. 25, no. 1, pp. 24–29, 2019, doi: 10.1038/s41591-018-0316-z.

[51] S. Yang, F. Zhu, X. Ling, Q. Liu, and P. Zhao, "Intelligent health care: Applications of deep learning in computational medicine," *Frontiers in Genetics*, vol. 12, 2021. doi: 10.3389/fgene.2021.607471.

[52] D. Kaul, H. Raju, and B. K. Tripathy, "Deep learning in healthcare," in *Studies in Big Data*, Springer Science and Business Media Deutschland GmbH, pp. 97–115, 2022. doi: 10.1007/978-3-030-75855-4_6.

[53] T. J. Loftus *et al.*, "Uncertainty-aware deep learning in healthcare: A scoping review," *PLOS Digital Health*, vol. 1, no. 8, p. e0000085, 2022, doi: 10.1371/journal.pdig.0000085.

[54] D. Williams, H. Hornung, A. Nadimpalli, and A. Peery, "Deep learning and its application for healthcare delivery in low and middle income countries," *Frontiers in Artificial Intelligence*, vol. 4, 2021, doi: 10.3389/frai.2021.553987.

[55] R. Miotto, F. Wang, S. Wang, X. Jiang, and J. T. Dudley, "Deep learning for healthcare: Review, opportunities and challenges," *Briefings in Bioinformatics*, vol. 19, no. 6, pp. 1236–1246, 2017, doi: 10.1093/bib/bbx044.

[56] J. Egger *et al.*, "Medical deep learning-A systematic meta-review," *Computer Methods and Programs in Biomedicine*, vol. 221, p. 106874, 2022, doi: https://doi.org/10.1016/j.cmpb.2022.106874.

[57] D. Roy and M. Dutta, "A systematic review and research perspective on recommender systems," *Journal of Big Data*, vol. 9, no. 1, p. 59, 2022, doi: 10.1186/s40537-022-00592-5.

[58] S. Yang, F. Zhu, X. Ling, Q. Liu, and P. Zhao, "Intelligent health care: Applications of deep learning in computational medicine," *Frontiers in Genetics*, vol. 12, 2021. doi: 10.3389/fgene.2021.607471.

[59] H. Steck, L. Baltrunas, E. Elahi, D. Liang, Y. Raimond, and J. Basilico, "Deep learning for recommender systems: A netflix case study," *AI Magazine*, vol. 42, pp. 7–18, 2021, doi: 10.1609/aaai.12013.

[60] T. J. Loftus *et al.*, "Uncertainty-aware deep learning in healthcare: A scoping review," *PLOS Digital Health*, vol. 1, no. 8, p. e0000085, 2022, doi: 10.1371/journal.pdig.0000085.

[61] R. Mu, "A survey of recommender systems based on deep learning," *IEEE Access*, vol. 6, pp. 69009–69022, 2018, doi: 10.1109/ACCESS.2018.2880197.

[62] S. Zhang, L. Yao, A. Sun, and Y. Tay, "Deep learning based recommender system: A survey and new perspectives," *ACM Computing Surveys*, vol. 52, no. 1, 2019. doi: 10.1145/3285029.

[63] B. Liu, Q. Zeng, L. Lu, Y. Li, and F. You, "A survey of recommendation systems based on deep learning," *Journal of Physics: Conference Series*, 2021. doi: 10.1088/1742-6596/1754/1/012148.

[64] M. F. Aljunid and M. Dh, "An efficient deep learning approach for collaborative filtering recommender system," *Procedia Computer Science*, vol. 171, pp. 829–836, 2020, doi: https://doi.org/10.1016/j.procs.2020.04.090.

[65] D. Wang, Y. Liang, D. Xu, X. Feng, and R. Guan, "A content-based recommender system for computer science publications," *Knowledge-Based Systems*, vol. 157, pp. 1–9, 2018, doi: https://doi.org/10.1016/j.knosys.2018.05.001.

[66] K. V. Dudekula *et al.*, "Convolutional neural network-based personalized program recommendation system for smart television users," *Sustainability (Switzerland)*, vol. 15, no. 3, 2023, doi: 10.3390/su15032206.

[67] Y. H. Low, W.-S. Yap, and Y. K. Tee, "Convolutional neural network-based collaborative filtering for recommendation systems," in *Robot Intelligence Technology and Applications*, J.- H. Kim, H. Myung, and S.- M. Lee, Eds., Singapore: Springer Singapore, pp. 117–131, 2019.

[68] G. K. Venayagamoorthy, V. Moonasar, and K. Sandrasegaran, "Voice recognition using neural networks," in *Proceedings of the 1998 South African Symposium on Communications and Signal Processing-COMSIG' '98 (Cat. No. 98EX214)*, IEEE, pp. 29–32. doi: 10.1109/COMSIG.1998.736916.

[69] C. Yang, W. Yang, and S. Wang, "Based on artificial neural networks for voice recognition word segment," in *2011 IEEE 3rd International Conference on Communication Software and Networks*, pp. 394–396, 2011. doi: 10.1109/ICCSN.2011.6014920.

[70] V. Lyashenko, F. Laariedh, S. Sotnik, and M. Ayaz Ahmad, "Recognition of voice commands based on neural network," *TEM Journal*, vol. 10, no. 2, pp. 583–591, 2021, doi: 10.18421/TEM102-13.

[71] Lalit Kumar, "Desktop voice assistant using natural language processing (NLP)," *International Journal for Modern Trends in Science and Technology*, vol. 6, no. 12, pp. 332–335, 2020, doi: 10.46501/ijmtst061262.

[72] M. Ramchandani *et al.*, "Survey: Tensorflow in machine learning," in *Journal of Physics: Conference Series*, Institute of Physics, 2022. doi: 10.1088/1742-6596/2273/1/012008.

[73] D. Akgun, "TensorFlow based deep learning layer for local derivative patterns," *Software Impacts*, vol. 14, 2022, doi: 10.1016/j.simpa.2022.100452.

[74] A. Agrawal *et al.*, "TensorFlow eager: A multi-stage, python-embedded DSL for machine learning," 2019, [Online]. Available: https://arxiv.org/abs/1903.01855.

[75] USENIX Association, "ACM Sigmobile, ACM special interest group in operating systems., and ACM Digital Library," *Papers presented at the Workshop on Wireless Traffic Measurements and Modeling: June 5, 2005*, Seattle, WA: USENIX Association, 2005.

[76] N. Ketkar, "Introduction to PyTorch," in *Deep Learning with Python: A Hands-on Introduction*, N. Ketkar, Ed., Berkeley, CA: Apress, pp. 195–208, 2017. doi: 10.1007/978-1-4842-2766-4_12.

[77] I. Oleksiienko, D. T. Tran, and A. Iosifidis, "Variational neural networks implementation in pytorch and JAX[Formula presented]," *Software Impacts*, vol. 14, 2022, doi: 10.1016/j.simpa.2022.100431.

[78] N. Ketkar, "Introduction to Keras," in *Deep Learning with Python: A Hands-on Introduction*, N. Ketkar, Ed., Berkeley, CA: Apress, pp. 97–111, 2017. doi: 10.1007/978-1-4842-2766-4_7.

[79] F. J. J. Joseph, S. Nonsiri, and A. Monsakul, "Keras and TensorFlow: A hands-on experience," in *Advanced Deep Learning for Engineers and Scientists: A Practical Approach*, K. B. Prakash, R. Kannan, S. A. Alexander, and G. R. Kanagachidambaresan, Eds., Cham: Springer International Publishing, pp. 85–111, 2021. doi: 10.1007/978-3-030-66519-7_4.

[80] B. T. Chicho and A. Bibo Sallow, "A comprehensive survey of deep learning models based on Keras framework," *Journal of Soft Computing and Data Mining*, vol. 2, no. 2, 2021, doi: 10.30880/jscdm.2021.02.02.005.

[81] V.-H. Nhu *et al.*, "Effectiveness assessment of Keras based deep learning with different robust optimization algorithms for shallow landslide susceptibility mapping at tropical area," *Catena (Amst)*, vol. 188, p. 104458, 2020, doi: https://doi.org/10.1016/j.catena.2020.104458.

[82] A. Kishore, S. Jindal, and S. Singh, "Designing deep learning neural networks using caffe," 2015. [Online]. Available: https://github.com/.

[83] F. Guo, H. Huang, Y. Liu, and J. Xu, "Application of neural network based on Caffe framework for object detection in Hilens," in *2019 Chinese Automation Congress (CAC)*, pp. 4355–4359, 2019. doi: 10.1109/CAC48633.2019.8996226.

[84] Y. Jia *et al.*, "Caffe: Convolutional architecture for fast feature embedding," in *Proceedings of the 22nd ACM International Conference on Multimedia*, in MM' '14. New York, NY: Association for Computing Machinery, pp. 675–678, 2014. doi: 10.1145/2647868.2654889.

[85] Y. Jia *et al.*, "Caffe: Convolutional architecture for fast feature embedding," in *Proceedings of the 22nd ACM International Conference on Multimedia*, in MM' '14. New York, NY: Association for Computing Machinery, pp. 675–678, 2014. doi: 10.1145/2647868.2654889.

[86] The Theano Development Team *et al.*, "Theano: A Python framework for fast computation of mathematical expressions," 2016, [Online]. Available: https://arxiv.org/abs/1605.02688.

[87] F. Bastien *et al.*, "Theano: new features and speed improvements SpeechBrain view project parsing view project theano: New features and speed improvements," 2012. [Online]. Available: https://www.researchgate.net/publication/233753224.

[88] M. M. Yapıcı and N. Topaloğlu, "Performance comparison of deep learning frameworks," 2021. [Online]. Available: https://dergipark.org.tr/tr/pub/ci.

[89] C. Boufenar and M. Batouche, "Investigation on deep learning for off-line handwritten arabic character recognition using theano research platform," in *2017 Intelligent Systems and Computer Vision (ISCV)*, pp. 1–6, 2017. doi: 10.1109/ISACV.2017.8054902.

[90] Q. Wang, M. Ihme, Y.-F. Chen, and J. Anderson, "A TensorFlow simulation framework for scientific computing of fluid flows on tensor processing units," *Computer Physics Communications*, vol. 274, p. 108292, 2022, doi: 10.1016/j.cpc.2022.108292.

[91] J. Rokai, I. Ulbert, and G. Márton, "Edge computing on TPU for brain implant signal analysis," *Neural Networks*, vol. 162, pp. 212–224, 2023, doi: https://doi.org/10.1016/j.neunet.2023.02.036.

[92] A. Carreon, S. Barwey, and V. Raman, "A generative adversarial network (GAN) approach to creating synthetic flame images from experimental data," *Energy and AI*, vol. 13, p. 100238, 2023, doi: https://doi.org/10.1016/j.egyai.2023.100238.

[93] I. Ullah, U. U. Hassan, and M. I. Ali, "Multi-level federated learning for industry 4.0- A crowdsourcing approach," *Procedia Computer Science*, vol. 217, pp. 423–435, 2023, doi: https://doi.org/10.1016/j.procs.2022.12.238.

[94] E. J. Y. Koh *et al.*, "An automated machine learning (AutoML) approach to regression models in minerals processing with case studies of developing industrial comminution and flotation models," *Minerals Engineering*, vol. 189, p. 107886, 2022, doi: https://doi.org/10.1016/j.mineng.2022.107886.

[95] M. Schmitt, "Automated machine learning: AI-driven decision making in business analytics," *Intelligent Systems with Applications*, vol. 18, p. 200188, 2023, doi: https://doi.org/10.1016/j.iswa.2023.200188.

[96] M. Francia, J. Giovanelli, and G. Pisano, "HAMLET: A framework for Human-centered AutoML via Structured Argumentation," *Future Generation Computer Systems*, vol. 142, pp. 182–194, 2023, doi: https://doi.org/10.1016/j.future.2022.12.035.

[97] D. Chawla and P. S. Mehra, "A survey on quantum computing for internet of things security," *Procedia Computer Science*, vol. 218, pp. 2191–2200, 2023, doi: https://doi.org/10.1016/j.procs.2023.01.195.

[98] W. Zhao, T. Guo, X. Yu, and C. Han, "A learnable sampling method for scalable graph neural networks," *Neural Networks*, 2023, doi: https://doi.org/10.1016/j.neunet.2023.03.015.

[99] X. Fan, M. Gong, and Y. Wu, "Markov clustering regularized multi-hop graph neural network," *Pattern Recognition*, p. 109518, 2023, doi: https://doi.org/10.1016/j.patcog.2023.109518.

[100] F. Behrad and M. Saniee Abadeh, "An overview of deep learning methods for multimodal medical data mining," *Expert Systems with Applications*, vol. 200, p. 117006, 2022, doi: https://doi.org/10.1016/j.eswa.2022.117006.

9 Hard and Soft Computing

9.1 INTRODUCTION

Advancement in computing technology has opened the space for more ideas and understanding of computational complexities. Hence, the emergence of two complementary techniques vis-à-vis hard computing and soft computing. While the former works based on certainty, accuracy, and rigidity, the latter relies on approximation, uncertainty, and flexibility. The upsurge in the adoption of machine learning has witnessed significant advancements while creating a novel approach called soft computing for addressing several complex problems including thin film and atomic layer deposition (ALD). Intelligent decision-making has necessitated a shift towards the soft computing approach owing to the growth in data volume in recent years, particularly in thin film technology [1]. In this chapter, we examine the key distinguishing features of these complementary approaches while drawing insights from their strength and limitations. More attention is drawn to the soft computing approach as it influences the trajectories of computing in thin film technology.

9.2 WHAT ARE HARD COMPUTING AND SOFT COMPUTING?

Using exact mathematical models and methods to solve problems is the core of the hard computing paradigm. It is frequently applied in situations where the problem can be precisely modeled using mathematical models and the data are well-defined. Logic-based methods, traditional optimization, and numerical analysis are a few examples of hard computing techniques. Hard computing has a solid theoretical foundation and seeks to offer precise solutions to problems. These methods can be unnecessarily complex and time-consuming to compute since they demand clean, noise-free data [2]. Real-life problems have become more complex recently, and data collected from manufacturing and production processes tend to be noisy because of fluctuating process conditions and subjective evaluations. Moreover, because hard computing techniques lack intelligence, it may be difficult to reveal hidden patterns in real-time data handling for decision-making. Hence, researching imprecision and uncertainty was the driving force behind soft computing systems, which trade off certainty and precision for tractability and durability [1].

Since its inception, computing has advanced significantly. Soft computing is one of the computer concepts that has been essential to this change. A paradigm known as "soft computing" is oriented on problem-solving strategies that imitate human-like reasoning [3]. It is frequently employed in situations when the data are ambiguous or inaccurate and where it is challenging to formulate the issue using conventional mathematical models. Fuzzy logic (FL), artificial neural networks, and evolutionary algorithms are a few examples of soft computing techniques [3–6]. These methods are based on social and biological behavioral trends that form the basis of the

DOI: 10.1201/9781003346234-11

emergence of the prominent soft computing techniques mentioned earlier. With the use of soft computing, problems can be solved that are reliable and tolerant of data noise and ambiguity. Since its start, the discipline of soft computing has seen a great amount of application and model refinement. Its use has been noted in both the manufacturing and service sectors to increase product and process efficiency and effectiveness at a lower cost by making a small sacrifice in precision and certainty [1]. Soft computing techniques have found applications in predictive modeling and optimization problems with laudable outcomes. The fact that soft computing only offers approximations of solutions rather than exact ones is one of its critical drawbacks. In applications where accurate solutions are required, this might be a drawback. Understanding how the system came up with its solution might also be tricky because soft computing techniques can be challenging to interpret. Nonetheless, it is noteworthy that hard computing and soft computing techniques have distinct characteristics that make them stand out on their own. It is therefore significant to have a good grasp of the problem for a proper selection of the suitable technique to deploy.

9.3   CHARACTERISTICS OF HARD COMPUTING AND SOFT COMPUTING

9.3.1   HARD COMPUTING

9.3.1.1   Precision

Hard computing depends heavily on precision since it defines how accurate and trustworthy a system's output will be [7]. In order to create results that are valuable and reliable, hard computing frequently calls for a great deal of precision, especially in scenarios where minor errors might have massive implications [2]. It frequently employs specialized numerical techniques to reduce computational error and uncertainty.

High-precision arithmetic is often employed in hard computing toward high degree of precision. Higher digit counts than those used in ordinary arithmetic are used to represent numbers in high-precision arithmetic, which can increase calculation accuracy. Unlike floating-point arithmetic, high-precision may express integers with any number of digits. However, high-precision calculation requires expensive software and equipment [8]. Hard computing involves careful consideration of numerical methodologies, processing resources, and error-correcting algorithms in order to achieve high levels of precision [9].

9.3.1.2   Determinism

Deterministic features in hard computing relate to a system's capacity to provide reliable results based on a given input or group of inputs. This implies that the system will always generate the same result, devoid of unpredictability or uncertainty, given the same input or set of inputs [7]. In order to produce reliable outcomes, hard computing systems rely on exact algorithms and logical principles [9,10].. Every time these systems are deployed, their outputs are intended to be reliable and predictable because these systems are created to carry out specified duties using explicit information and regulations. Hard computing systems' deterministic nature provides various benefits. Due to the fact that they are created to function in a specified and predictable manner, deterministic systems are very dependable. Because of this, customers can count on the system to deliver consistent results each and every instance they are deployed.

Results can be replicated because hard computing systems are deterministic. This means that tests and simulations may be repeated, which is crucial for confirming scientific findings and progressing research in a variety of domains. Deterministic systems could not work effectively, for instance, in simulation or financial modeling applications where unpredictability or uncertainty is essential. Soft computing techniques like neural networks or fuzzy logic may be more suitable in certain circumstances. The nondeterministic characteristics of soft computing systems, such as artificial neural networks and fuzzy logic systems, allow them to learn from data and modify their inputs over time [1]. The trade-off is that, in comparison to hard computing systems, their results may prove less predictable and harder to interpret.

9.3.1.3   Symbolic Computation

A key element of hard computing that employs exact mathematical techniques for addressing issues is symbolic computation involving equations, functions, and algebraic expressions [11]. Further to this, because algebraic expressions can be altered in a variety of ways, symbolic computation is more adaptable than numerical approaches [8]. This is especially helpful in applications like optimization where working with algebraic formulas to find a solution may be necessary. In addition, analytical formulations for issue solutions can be derived via symbolic computation, which can reveal details about the underlying mathematical relationships and structures [2,7]. Engineering, physics, and computer science use symbolic computation extensively. For performance analysis and optimization of complex systems like control systems, analytical solutions are often applied. Computer vision and machine learning techniques use symbolic computation to create closed-form optimization solutions.

9.3.2   Soft Computing

9.3.2.1   Flexibility

A crucial characteristic that sets soft computing methods apart from conventional computing methods is their flexibility. Soft computing techniques can deal with data that are imprecise, ambiguous, or uncertain, which enable them to solve issues that could be challenging to identify or address with conventional computer techniques [3]. Soft computing systems can analyze erroneous and uncertain input and generate trustworthy conclusions by utilizing fuzzy logic, neural networks, and genetic algorithms (GA) [12].

The flexibility of these techniques comes from their capacity to analyze and make sense of incomplete and ambiguous data. Because traditional computing techniques require exact and correct data, they are inappropriate for issues with ambiguity and uncertainty [1,4].

Another soft computing technique that can handle ambiguous or imprecise input is neural networks [1]. Because they mimic the brain, neural networks can learn and adapt. Neural networks can process large volumes of data and identify patterns or

generate predictions from defective or unclear data. This is accountable for its effectiveness and robustness in applications like image and speech recognition, where
the input data may be flawed or missing [6]. A soft computing technique called GAs
mimics the process of natural selection. Genetic algorithms are highly suited for
applications like scheduling and resource allocation because they can tackle optimization problems with many restrictions or objectives [13]. By employing stochastic
search techniques that explore the search space in a probabilistic manner, GAs may
also handle noisy or incomplete data.

9.3.2.2 Bio-Inspiration

Soft computing techniques use methods that imitate the behavior of natural systems
to tackle complicated problems [3]. They are inspired by biological systems. Bio-
inspiration is the process of developing new algorithms, models, and approaches by
taking inspiration from biological systems [14]. In order to address complex issues,
soft computing techniques employ a variety of bio-inspired strategies, including neural networks, GAs, and swarm intelligence. For instance, neural networks, which
use artificial neurons to learn from instances and draw generalizations from them,
are modeled after the structure and operation of the human brain. Natural selection
served as the inspiration for GAs, which use a set of potential solutions to address
optimization issues [6,15].

There are many benefits of using bio-inspired methods in soft computing. Firstly,
these methods are quite flexible and may be used to address a wide range of issues
in numerous fields. Furthermore, they frequently produce excellent results and can
resolve issues that conventional approaches find challenging or impossible. Also,
solutions derived from bio-inspired methodologies are frequently quite strong and
adaptable to changes in the environment or input data. Soft computing techniques
can handle complicated issues that are challenging or impossible to answer using
conventional methods by utilizing the advantages of bio-inspired techniques and
addressing their shortcomings [3].

The immune system of the human body serves as a model for artificial immune
systems [16]. They involve a collection of synthetic immune cells that can identify
and react to antigens like germs and viruses [17]. The foraging habits of ants serve
as a model for ant colony optimization (ACO) [18,19]. A group of synthetic ants are
used, and they leave pheromones on a graph that represents an issue. The difficulty
is solved when a trail forms as a result of the pheromones' ability to draw in other
ants. Routing, planning, and allocation of resources are examples of optimization
problems that are solved using ant colonies..

9.3.2.3 Approximation

Soft computing, which focuses on the use of adaptable, heuristic methods for tackling complicated problems, is fundamentally characterized by approximation [3]. Its
approximation-based character results from the reality that many real-world issues
are either too complex or insufficiently understood to be resolved by accurate mathematical techniques. The interactions between many variables are frequently complicated and nonlinear, and the data that are provided are frequently inaccurate or
inadequate [1].

A prominent soft computing method is fuzzy logic, which is founded on the idea of using fuzzy sets and fuzzy rules to describe uncertain or inaccurate information [20]. In instances when precise numerical data are unavailable or challenging to get, fuzzy logic enables the use of continuous truth values that vary from fully false to completely true [21]. Neural networks, which are computer models created to imitate the structure and operation of the human brain, are a key component of soft computing. Neural networks are particularly effective at solving nonlinear issues that are challenging to model using conventional mathematical techniques [22]. In order to make predictions based on the data at hand, neural networks employ approximate reasoning. To increase their accuracy over time, they are trained using big datasets.

Another key tactic in soft computing is probabilistic reasoning, which uses statistical models and methods to calculate the likelihood of various outcomes or events [23]. When there is insufficient or uncertain data, probabilistic reasoning is very helpful since it enables the use of probabilities to express the system's level of uncertainty.

9.4   HARD AND SOFT COMPUTING TECHNIQUES

A more adaptable, intuitive strategy, based on experience and instincts could be a viable alternative to the conventional and logical strategy when attempting to address a challenge. Hard computing and soft computing have proffered remarkable solutions in the world of computers. While soft computing uses more flexible, variable ways to deal with ambiguity and incomplete data, hard computing focuses on mathematical models and algorithms to produce exact calculations and judgments. Both strategies have pros and cons, so your choice depends on the issue. In this section, we explore the fascinating world of computational problem-solving using hard and soft computing methods, from cutting-edge machine learning algorithms to traditional rule-based systems, and discuss their benefits.

9.4.1   HARD COMPUTING TECHNIQUES

9.4.1.1   Linear Programming

A linear programming is a mathematical method used in hard computing to optimize a linear objective function under a set of linear constraints. It is commonly utilized in disciplines including resource allocation, logistics, and operations research. Finding the optimum answer that maximizes or minimizes a linear objective function while taking into account a number of linear constraints is the core tenet of linear programming. The goal or objective of the problem is represented by the objective function, a mathematical expression, and the constraints are the requirements that must be met. In order to use linear programming, variables must be defined, constraints must be identified, and the objective function, or the desired outcome, must be identified [24]. Inequalities must be created and then graphed in order to solve problems using linear programming. Linear programming can be done manually in some cases, but complex variables and computations require computer software. A and B are sold to increase profits. They aim for $5,000 and $6,500 earnings but can only create a specific amount of product A and B due to resource limits. The corporation can only make 600 A and 800 B with a $120,000 expenditure. To optimize earnings, the corporation is determining A and B unit production. This issue can be expressed using a linear programming function with objective function as in Equation 9.1 by giving the objective function and limits:

$$\text{Maximize } C = 5,000\ A + 6,500\ B \qquad (9.1)$$

If C denotes the overall profit, A is the quantity of Product A while B is the quantity of B produced. The constraint would be:

$$A \leq 600 \qquad (9.2)$$

$$B \leq 800 \qquad (9.3)$$

$$600A + 800B \leq 120,000\ 9.4 \qquad (9.4)$$

Equations 9.2 and 9.3 represent the constraint on production while Equation 9.4 represents the constraints on budget.

Real-life application of linear programming is practicable. The methods of linear programming are used by farmers. By selecting what crops they should cultivate, the volume, and as well as their optimal utilization, farmers may enhance their earnings. In nutrition, linear programming offers a potent technique to help with nutritional planning. Nutritionists can deploy it to give impoverished households food baskets that are affordable and nutritious. Dietary recommendations, nutritional advice, cultural acceptance, or some mix of these may all be the constraints. In an assembly line, raw materials must pass through various machinery for predetermined lengths of time. A business can use a linear formula to determine how much raw material to employ in order to optimize profit. The amount of time spent on each computer is one restriction. Any equipment causing bottlenecks must be eliminated [24].

The literature is replete with sufficient amount of application of linear programming techniques in solving real-world problems. By taking into consideration the reliability of a probabilistic variable, it has been used in the field of hydrology to optimize firm output at the design stage of a hydro plant toward a protracted hydropower reserve management [25]. The research by Hyun [26] generated a tracking profile for an antenna subjected to system constraints toward maximizing transmission duration. The waste management field has greatly benefited from the use of linear programming to address a variety of issues, including paint waste management optimization [27] and waste management system planning [28–30], health monitoring [31], energy saving optimization [32], and maintenance planning and schedules [33], amongst others.

As a hard computing method, linear programming offers many benefits. It can handle a wide range of problems, from basic to complex, and can locate the optimum solution in a workable area. It is also particularly effective because there are numerous algorithms available to swiftly and precisely handle problems involving linear programming. However, linear programming can have some drawbacks. Only linear relationships between variables can be handled, which may not be suitable for all issues. Also, it makes the unquestionable assumption that the problem parameters are known, which may not always be the case in real-world scenarios.

9.4.1.2 Expert-System

Expert systems are computer programs that simulate human decision-making in a given domain by using AI and machine learning techniques [34]. To reach a conclusion or make a suggestion, they frequently use a set of rules, information, and inference engines [35]. Expert systems use a deterministic method of problem-solving, making them a subset of hard computing [1,36]. They don't use the statistical or probabilistic techniques used in soft computing; instead, they rely on rules and logical reasoning to reach a conclusion. They can collect and preserve specialist knowledge, increase the accuracy of decisions, and lower expenses, among other benefits. They do, however, have drawbacks, such as the requirement for in-depth topic expertise and the incapacity to deal with ambiguity and uncertainty. A rule-based or knowledge-based system is a kind of expert system used in hard computing that makes decisions or solves problems using a set of rules and logical reasoning. Another name for it is a knowledge-based system.

A collection of rules is formulated by a single expert or by a group of experts in a rule-based system [37]. Their field expertise informs these rules. Each rule has an antecedent (conditions) and a consequence. Based on the input data, the conditions are assessed, and if they are true, the consequent is carried out. The transparency and simplicity of rule-based systems is one of their benefits. Each guideline is clearly described, which makes it simple to follow the thought and selection processes. Human specialists will find it simpler to verify and improve the system's performance as a result [38]. Rules-based systems have major limitations. Since human professionals must program the system, they are difficult to design and maintain. They may also struggle with unexpected situations not covered by the current rules. Because they offer a strong and adaptable tool for encoding knowledge in a formal and structured way, logic-based approaches are frequently utilized in the creation of expert systems [39,40]. Formal logic is a mathematical framework for inferring conclusions from a set of premises and is the foundation of logic-based approaches [41]. Expert systems can utilize formal reasoning techniques to reach conclusions and make judgments based on the knowledge at their disposal by encoding knowledge in a logical manner.

9.4.1.3 Numerical Methods

Mathematical and computational problems that cannot be addressed analytically are solved using numerical methods. These techniques rely on mathematical algorithms that approximate answers to challenging issues using numerical analysis. In disciplines like engineering, physics, and computer science where exact mathematical solutions are necessary, numerical approaches are frequently used. The fact that numerical approaches rely on precise and deterministic algorithms is one of its key characteristics. This implies that the numerical methods' solutions are exact and precise, offering a degree of confidence in the outcomes. By offering precise solutions that may be utilized to make predictions and wise decisions, numerical approaches can aid researchers and engineers in understanding complicated systems and processes. These numerical approaches are frequently computationally demanding in that in order to do calculations, they need substantial computational resources, such as high-performance computer systems. However, improvements in processing capability and algorithms have made it possible to execute numerical calculations on desktop and laptop computers, opening up these techniques to a wider variety of academics and professionals.

Engineering uses numerical methods to simulate physical systems, develop and analyze structures, and optimize designs [42–45]. Engineering uses numerical methods for computational fluid dynamics [46,47], finite element analysis [48], and optimization [45]. Numerical approaches model electromagnetism, fluid dynamics, and quantum mechanics [49,50]. Astrophysics models stars and galaxies numerically [51]. Numerical methods in computer science model complex systems, develop algorithms, and solve machine learning and AI concerns [52]. Numerical methods are used in Monte Carlo simulations, numerical linear algebra, and optimization [9,53–55].

9.4.2 Soft Computing Techniques

9.4.2.1 Artificial Neural Network

Artificial neural networks (ANN) are machine learning models that mimic the human brain. They are multilayer neural networks with connected nodes. Each neuron receives inputs, applies a weighted sum, and produces an output after passing the output via an activation function. The ability to learn from ambiguous or imprecise data is one way that ANNs are employed in soft computing [56]. Each example in the set of examples used to train ANNs contains a set of input values and an associated output value. In order to reduce the disparity between the expected output and the actual output, the network modifies the weights between neurons during the training phase. The ability to represent intricate nonlinear interactions between input and output variables is another way ANNs are employed in soft computing [57]. ANNs are ideal for simulating complicated systems where the underlying linkages are unclear since they can approximate any continuous function.

ANN is built on the principles of the structure and operation of the human brain. Neurons, the interconnected nodes that make up ANNs, are arranged into layers. The input layer with inputs $x_j : (j = 1,2...n)$, the hidden layer(s) with neurons $n_j : (j = 1,2...n)$, and the output layer with output $o_j : (j = 1,2...n)$ are the three types of layers that are commonly present in an ANN as represented in Figure 9.1. The neurons in the hidden layers and output layer are in charge of processing and turning the input data into useful output, whereas each neuron in the input layer corresponds to a feature or input variable of the data.

Weights are used to depict the connections between neurons in various layers and to quantify how strong they are [58]. The weights between neurons are changed during training in order to reduce the discrepancy between expected and actual output. Initial weight for ANN models is updated during training until a stopping condition is satisfied; if otherwise, the new weight is ignored, and the learning rate parameters are reduced by lowering the scalar parameters [59–61]. For each training set, the biases and weights are updated using Equation 9.5:

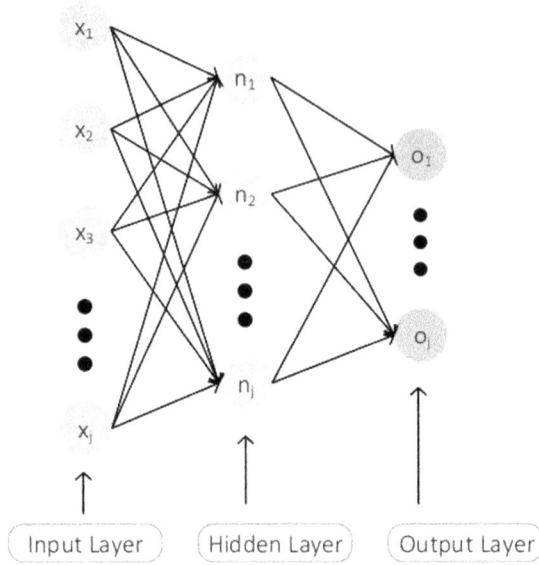

FIGURE 9.1   Artificial neural network (ANN) architecture [1].

$$\theta_{t+1} = \theta_t - \eta . \nabla_{\theta t} E\left(\theta_t; x^{(i)}; y^{(i)}\right) \tag{9.5}$$

where θ represents the weights and biases, $x^{(i)}$ is the input of the training sample, $y^{(i)}$ is the target label, η is the learning rate, and E is the loss function.

The multilayer perceptron (MLP) is a common ANN architecture comprising several hidden layers. This causes an increase in the complexity of the system. The output of the first hidden layer neurons in a network with two hidden layers, where the first hidden layer has m_1 neurons and the second hidden layer has m_2 neurons, first and second hidden layer weights are w_{il}^1 and w_{ij}^2, respectively, and first and second hidden layer activation functions are ϕ_i and ψ_j, respectively, can be expressed in equation. According to Equation 9.7, the second hidden layer's input is provided by the output of the neurons in the first hidden layer as in Equation 9.6.

$$\xi_i = \psi_j\left(\sum_{i=0}^{p} w_{il}^1 U_l\right), \; u_0 \text{ and } \psi_j\left(\cdot\right) = 1 \tag{9.6}$$

$$y = \sum_{i=0}^{m_2} w_1 \phi_i\left(\sum_{j=0}^{m_1} w_{ij}^2 \xi_j\right), \; \phi_0\left(\cdot\right) = 1 \tag{9.7}$$

9.4.2.2 Fuzzy Logic

FL is one of the core aspects of soft computing which is essential to systems with high MIQs (machine intelligence quotients) [12]. The FL approach is a variant of multivalued logic, often recognized as the logic of approximation reasoning. The idea of the fuzzy logic is comparable to the fuzzy set theory, in which nondistinct borders within classes persist. The approach establishes the membership function that emerges from human logic that all events' validity is a function of degree [1]. When there is insufficient or unclear information, fuzzy logic can be used to reason in cases where classic Boolean logic cannot. The fuzzy logic is a mathematical approach which addresses ambiguity and imprecision. There is no third option in conventional Boolean logic; variables can only be true or false. Contrarily, fuzzy logic permits variables to have degrees of truth ranging from 0 to 1, which permits more subtle reasoning [62].

In fuzzy logic, two ideas are fundamental to its uses. The first is a linguistic variable, whose values are phrases or clauses in a language, natural or artificial. The other is called fuzzy-if-then-rules, and it has premises with linguistic terms as its antecedent and consequents. The granularity of elements and their interactions is the sole goal of linguistic terms. By granulating, the employment of linguistic variables and fuzzy if-then rules produces lossy data compression. In this manner, fuzzy logic replicates the amazing potential of the human mind to focus on information that is crucial to decisions and synthesize information.

A "fuzzy set" is the main idea behind fuzzy logic. When reasoning about uncertain or imperfect information, fuzzy logic uses fuzzy sets to symbolize the information as well as fuzzy logic operations [63]. A value set with degrees of membership ranging from 0 to 1 is referred to as a fuzzy set. For example, the fuzzy set "tall people" can include individuals who are very tall (membership degree: 0.9), moderately tall (membership degree: 0.5), and individuals who are only slightly taller than normal (membership degree: 0.1) [64,65]. In the fuzzy set theory, x belongs to a set A, such that $\mu_A(x)$ comparable to the crisp $\mu_A(x) = 1$ could be partial provided $x \in A$, and $\mu_A(x) = 0$ if $x \notin A$. Sequel to this, we can thus define and mathematically express different fuzzy logic operators as follows:

i. **Union ("OR"):** By obtaining the maximum membership degree at each point, two or more fuzzy sets are merged into one fuzzy set using union fuzzy operator:

$$\mu_{A \cup B}(x) = \max\left[\mu_A(x), \mu_B(x)\right] \tag{9.8}$$

ii. **Intersection ("AND"):** This merges two or more fuzzy sets into a single fuzzy set by calculating the minimum membership degree at each location:

$$\mu_{A \cap B}(x) = \min\left[\mu_A(x), \mu_B(x)\right] \tag{9.9}$$

iii. **Complement ("NOT"):** Transforms a fuzzy set's membership degrees so that a membership degree of 0 is now 1 while a membership degree of 1 becomes 0:

$$\mu_{\bar{A}}(x) = 1 - \mu_A(x) \tag{9.10}$$

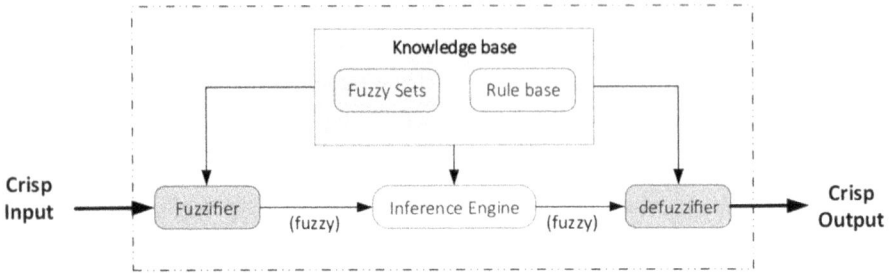

FIGURE 9.2   Fuzzy inference system [1].

A decision-making (inference) system known as a fuzzy inference system (FIS) employs fuzzy logic with a sequential sequence of commands whose implementation produces fuzzy (approximate) responses to take actions based on uncertain or incomplete information [66]. When utilizing the FIS, the system reads in the input variables and assigns them to the relevant fuzzy sets. The degree of membership of each rule is then calculated by applying the fuzzy rules. Lastly, it combines the fuzzy rule outputs to determine the output variable. A typical FIS structure is represented in Figure 9.2.

9.4.2.3   Swarm Intelligence

The flocking of birds and bee swarms are two examples of how complex systems may coexist harmoniously in the natural world. These social animals have developed cooperative and problem-solving abilities over several years, resulting in emergent behaviors that are more than the sum of their parts [67]. This idea serves as the foundation for swarm intelligence. Swarm intelligence, which uses the power of collective intelligence, is setting the bar for creating fresh, ground-breaking answers to some of the most difficult problems the world has ever faced [68]. Swarm intelligence is a soft computing approach which is a collection of computational methods that let computers gain knowledge from past data and experience. Problems requiring optimization, search, and decision-making are particularly well-suited for swarm intelligence [69]. Swarm intelligence presents a promising technique to problem-solving in this age of big data and complicated systems. We can design algorithms that can solve issues that are outside the purview of conventional computing techniques by taking inspiration from the group behavior of social animals. Few of the prominent swarm intelligence algorithms are discussed as follows:

9.4.2.3.1   Genetic Algorithm

Genetic Algorithm (GA) is a ubiquitous optimization method which draws its inspiration from genetics and the concept of natural selection [70]. GA belongs to the broader category of Evolutionary Algorithms (EA), a group of computational optimization methods that mimic the process of natural selection in order to address challenging issues [1]. Due to its ease of use and effectiveness in locating the best solutions, GA is one of the EA approaches that is most frequently utilized [71]. With the use of a set of parameters or variables, GA creates a population of probable answers to a problem. The effectiveness of each solution's fitness is then assessed in relation to how effectively it resolves the issue at hand. An objective function or fitness function that measures the quality of the solution is often used in the fitness evaluation [72]. The subset of the population chosen for the breeding process is then chosen using the selection procedure. In order to create a new population of prospective solutions, the genetic components of the chosen solutions are combined in this procedure. After evaluating this new population, the procedure is repeated until a workable solution is identified. GA uses three operators: crossover, mutation, and reproduction. The GA starts a certain population of the chromosomes at random. Figure 9.3 presents the process flowchart for the GA process. The sequential steps carried out for each iteration in GA based on the three operators are as follows:

    I. *Selection*: The best chromosomes in the population that represent the best solutions are identified through an evaluation of the fitness function for each chromosome. These chromosomes serve as parents for the new generation of offspring.
   II. *Crossover:* By using this method, offspring arising from the hybridization of two parents are created. As a result, the offspring are more physically fit than their parents. Many methods, including diagonal crossover, cycle order, partially mapped, uniform, tournament, ranking selection, amongst others, are used to carry out the crossover technique, which establishes the structure and child-to-parent chromosome [73].
  III. *Mutation*: This strategy avoids taking the local optimal solutions as the global optimum solutions by looking for fresh solutions inside the search area that is accessible. To accomplish this, chromosomes' genes are modified in a randomized manner [1].

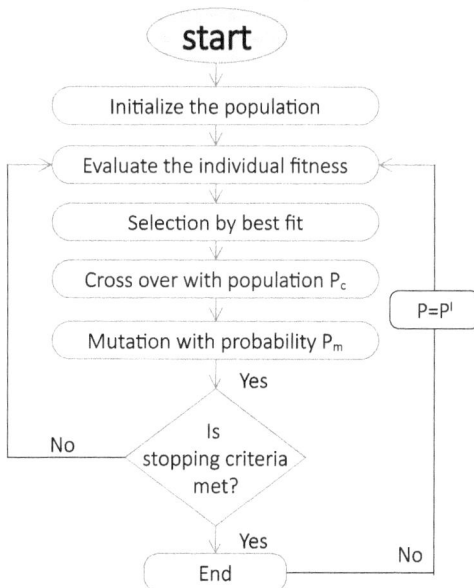

FIGURE 9.3 Process flowchart of genetic algorithm.

9.4.2.3.2   Particle Swarm Optimization

The prominent swarm intelligence technique known as particle swarm optimization (PSO) has been successfully applied to a variety of challenging optimization problems. PSO is a soft computing approach that draws inspiration from animal social behavior, particularly the insects, bird, and fishes. To find the best answer to a problem, PSO imitates these tendencies [74]. Kennedy and Eberhart first proposed PSO in 1995 [75], and since then it has developed into a well-known and incredibly effective optimization method. PSO's fundamental goal is to imitate a group of particles moving through a search space in quest of the best answer [76]. Each particle in the search space represents a potential solution, and it moves according to its own experience as well as the experiences of its neighbors [13]. The PSO method begins by initializing a collection of particles in the search space at random. Every particle has a position and velocity vector. A particle's position represents a potential solution, and its velocity vector shows the direction and speed with which it is moving across the search space. Based on the objective function of the problem being solved, each particle's fitness value is assessed. Based on its own experience and the experience of its neighbors, each particle's position and velocity are updated during each iteration of the algorithm. The following equations establish how the algorithm updates:

$$v_i(t+1) = wv_i(t) + c_1 r_1 \left(x_{Pbest} - x_i(t) \right) + r_2 c_2 \left(x_{Gbest} - x_i(t) \right) \tag{9.11}$$

$$x_i(t+1) = x_i(t) + v_i(t+1) \tag{9.12}$$

where $v_i(t)$ is the particle's velocity at time t, w denotes the inertia weight, *Pbest* denotes the particle's best solution so far, *Gbest* denotes the best solution for all particles, parameters c_1 and c_2 denote acceleration coefficients, while r_1 and r_2 denote random values between 0 and 1. Exploration and exploitation are balanced according to the inertia weight w. These control parameters are significant to the performance of the algorithm. The effects of the particle's own experience and the experiences of its neighbors on its movement are controlled by the acceleration coefficients c_1 and c_2. A stopping criterion, such as a maximum number of iterations or a desirable level of fitness, must be satisfied before the algorithm can stop. In addition, the PSO technique's effectiveness is improved by the five swarm intelligence principles of proximity, quality, diversified response, stability, and adaptability [77]. The process flowchart for the PSO is presented in Figure 9.4.

9.4.2.3.3   Ant Colony Optimization

Picture an active ant colony where many tiny insects collaborate to discover the quickest route from their nest to a food supply. They accomplish this by leaving pheromone trails that direct other ants to the source of food. The Ant Colony Optimization (ACO) algorithm was developed as a result of this extraordinary behavior. ACO makes a group of artificial ants that travel the problem space while leaving pheromones on the solutions they stop at by modeling their behavior after that of ants [78]. The pheromones serve as a route of communication that enables the ants to inadvertently exchange details about the caliber of the solutions they have discovered [19]. The working principle of the ACO is based on the discovery that ants may leave

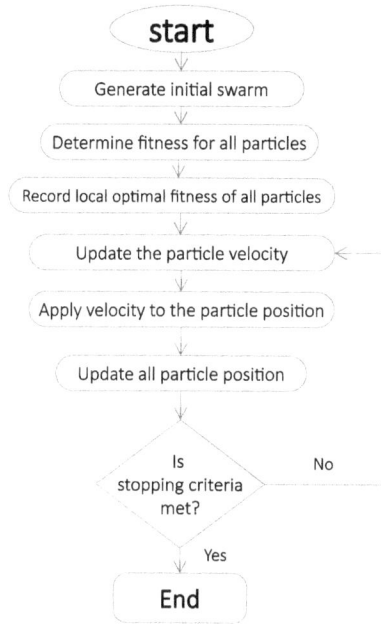

FIGURE 9.4   Process flowchart for the particle swarm optimization (PSO).

and follow pheromone trails to determine the shortest route between their nest and a food source [79]. As they migrate toward the food source, ants leave pheromone trails, which get stronger while more ants utilize them. Ants can indirectly communicate with one another by pheromone trails in this manner, and the entire colony can work together to determine the quickest route from the nest to the food source.

A set of artificial ants that move through the problem space and leave pheromones on the solutions they stop at are created by the ACO algorithm to simulate this behavior. Depending on how well the solutions the pheromone trails lead to turn out, they are then either strengthened or eliminated. By utilizing the data present in the pheromone trails, the ACO algorithm is able to use the information to converge toward the best solution to the issue [80]. The performance of ACO has been further improved as a result of the creation of numerous upgrades such as local search and multiple colony techniques, which were inspired by the success of ACO. Being one of the most potent optimization methods now in use, ACO continues to excite researchers and scientists across the globe with its ability to resolve challenging issues. The conventional ACO's state transition rule is a pseudorandom rule, which means that the roulette method is used to calculate the likelihood of a transition from the current node to the next viable node [81]. Both local and global search are the foundations of the continuing ACO. Local ants have the ability to go toward latent area with best solution in terms of transition probability of location k, as Equation 9.13 depicts [80]; Figure 9.5 represents the process flowchart for the ACO:

$$P_k(t) = \frac{t_k(t)}{\sum_{j=1}^{n} t_j(t)} \tag{9.13}$$

```
                    ( start )
                         |
           ( Assign the values of ACO parameters )
                         |
      ( Initialize concentration of pheromone at each region )
                         |
         ( Generate region to explore memory )
                         |
           ( Determine objective function )
                         |
                    / Is \
                   / stopping criteria \ ──────── No
                    \ met? /
                         |
                       Yes
              ( Repeat for all ants )
                         |
             ( Pheromone evaporation )
                         |
             ( Local optimum obtained )
                         |
                    ( End )
```

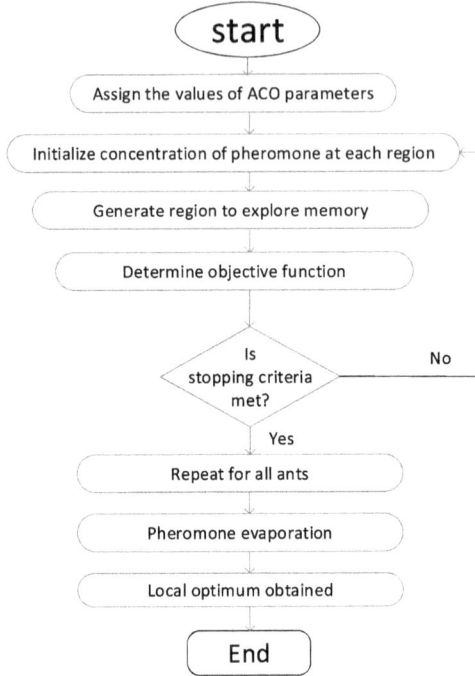

FIGURE 9.5 Process flowchart for ant colony optimization (ACO).

If n is the total number of global, $t_k(t)$ is the total pheromone in region k. The following equation updates the pheromone:

$$t_i(t+1) = (1-r)t_i(t) \tag{9.14}$$

9.4.2.3.4 Firefly Algorithm

The firefly algorithms (FAs) imitate the fascinating attributes of the biolumines-cent fireflies to address optimization problems. A group of fireflies try to attract a mate by flickering and flashing their lights in the darkness of the night. The FA leverages the flashing behavior of fireflies to solve a variety of optimization issues, and this is exactly what it accomplishes [82]. FA is a population-based algorithm, and each firefly symbolizes a potential answer to the optimization issue while the quality of the answer is shown by the firefly's flashing activity. The program employs firefly attraction to direct the search in the direction of the optimal solution [83]. An initial population of fireflies is randomly distributed around the problem space at the beginning of the algorithm's procedures. It then updates each firefly's position and brightness iteratively in accordance with their attraction to one another [84].

Three elements—distance, brightness, and a random parameter—are the foundations of firefly attraction to one another. The algorithm's random parameter, which adds stochasticity, has an impact on the attraction, which increases with brightness and diminishes with distance. Based on the attraction between fireflies and each firefly's prior position, the algorithm changes each firefly's position [83]. The FA is a potent optimization method thanks to a number of its advantages. The capability of the method to quickly converge on the optimal solution is one of its main advantages. It is an appealing choice for researchers and practitioners because it is also reasonably easy to use. Also, in a number of optimization problems, firefly algorithm has been found to perform better than other metaheuristic algorithms like GAs and Particle Swarm Optimization [83].

9.5 HYBRID SOFT COMPUTING TECHNIQUES

Hybrid model produces a more powerful and reliable system that can overcome the shortcomings of the standalone approaches. An upsurge in the applications of hybridized soft computing technique has been experienced in recent years owing to the substantial advantages it provides over single-method techniques. Due to common inadequacies in the critical aspect of soft computing such as parameter estimation or data fitting, single-method techniques have been proven to be less effective [1]. Compared to single-method approaches, hybrid systems are more easily able to adapt to varying conditions and settings. This is due to the system's ability to transition between various strategies in response to the scenario, enabling it to manage a wider range of circumstances. Many different methods, such as boosting, voting, bagging, or stacking, are used to create hybrid models [85]. Few hybrid models in soft computing are discussed briefly in this section.

9.5.1 ADAPTIVE NEURO-FUZZY INFERENCE SYSTEM

Adaptive neuro-fuzzy inference system (ANFIS) is a hybrid machine learning algorithm and a soft computing model which incorporates fuzzy logic system and ANN to learn from experience and make decisions in a manner that mimics that of humans [86]. The antecedent and consequence components of the Takagi-Sugeno FIS are coupled by fuzzy rules in the ANFIS modeling approach [87]. ANFIS is a suitable hybridized soft computing approach as it integrates two important soft computing approach via-a-vis ANN and fuzzy logic in a complimentary manner. The least-square approach and the backpropagation gradient-descent of the hybrid learning algorithm are combined to create the ANFIS, which optimizes the output's linear consequent parameters and the nonlinear premise parameter through fuzzy membership [88,89]. ANFIS comprises five layers vis-à-vis input, fuzzification, normalization, defuzzification, and output layer. The architectural framework of the ANFIS model is presented in Figure 9.6. These layers are described as follows.

Input layer: The system gets input variables or features from the outside environment at the input layer. To enable uniform processing across various inputs, the variables are typically standardized to a range of 0 to 1. This layer contains fuzzy

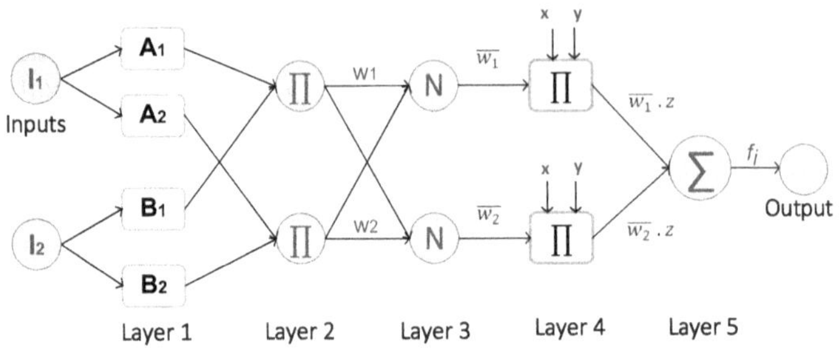

FIGURE 9.6 Architectural framework of adaptive neuro-fuzzy inference system (ANFIS) model.

membership functions $\tau A_i(x)$ and $\tau B_i(y)$, while its output function defined by O_{1j} at each node is presented in Equation 9.15:

$$O_{1j} = \tau A_i(x) \text{ and } F_{1j} = \tau B_i(y) \tag{9.15}$$

Fuzzification layer: This maps the crisp input variables to fuzzy sets, where each fuzzy set is represented by the degree of membership of the input variables. Using membership functions, this layer determines the extent to which each input variable belongs to the fuzzy sets. Equation 9.16 presents the output of the fuzzy layer which are the products of the input signals:

$$O_{2j} = \tau A_i(x) \times \tau B_i(y) \tag{9.16}$$

Normalization layer: The fuzzy logic is applied at the layer. Based on the degree of membership of the input parameters to the antecedents of the rules, it employs the rules provided by the system designer to calculate the degree of activation of each rule. As seen in Equation 9.17, the output function is a fraction of the node's firing strength to the sum of firing strength of the other nodes:

$$O_{3j} = \bar{w}_i = \frac{w_1}{w_1 + w_2 + w_3} \tag{9.17}$$

Defuzzification layer: The defuzzification layer changes the inference layer's fuzzy output value into a crisp output value. To do this, the outputs ε of each rule are averaged using a weighted formula, where the weights represent the rules' firing strength as shown in Equation 9.18.

$$O_{4j} = \bar{w}_i f_i = w_1(p_1x + q_1y + r_1) \tag{9.18}$$

Output layer: The ANFIS model's output layer produces the precise output value determined by the defuzzification layer. At this layer, a single node adds all the signals from all the layers using the summing function shown in Equation 9.19 to determine the overall output:

$$O_{5j} = \sum_{i=1}^{n} \overline{w}_i f_i = \frac{\sum_{i=1}^{n} w_i f_i}{\sum_{i=1}^{n} w_i} \qquad (9.19)$$

9.5.2 Hybrid Neuro-Fuzzy Models

In earlier applications, ANFIS has remarkably solved a number of nonlinear issues. It has the advantage of integrating the linguistic and numerical benefits of fuzzy logic and ANN. The exciting attributes of ANFIS such as less memory error, rapid learning capability, flexible computational framework, and adaptive capacity than the ANN has established the significance of its hybridization [90]. However, there are certain restrictions with the classical ANFIS model in its standalone form. It has some flaws like overfitting, artificialities, irregularities, amongst others [91,92]. It could be quite sophisticated, as it requires a lot of hyperparameter settings and tuning, consequently making them computationally intensive particularly when handling complicated systems or large amounts of data. Due to the growth in the research space in neuro-fuzzy model application, hybridization of neuro-fuzzy models with optimization algorithms such as evolutionary algorithms as an approach for overcoming some of its limitations in standalone form has gained traction [13]. Evolutionary algorithms are stochastic multi-objective optimization techniques which are contingent on selection and variation principle to replicate the process of natural evolution [93]. The antecedent and consequent parameters of the conventional ANFIS and the ideal training loss function are tuned by these evolutionary algorithms to facilitate a better convergence speed. The literature is replete with several evolutionary algorithms for tuning the parameters of neuro-fuzzy models for a better performance in several applications. Examples of these algorithms are PSO, GAs, and differential evolution (DE). Some new-generation meta-heuristic algorithms such as FFA, gray wolf optimization (GWO), teaching learning-based optimization (TLBO), and ACO, amongst others, have also been developed to achieve the same benefit [91,94].

9.6 MERITS AND DEMERITS OF SOFT COMPUTING METHODS

The real-life scenarios have benefitted from the applications of diverse soft computing approaches. Instead of competing with one another, these tools work in tandem [2,4]. Although they have shown considerable success, they have certain innate benefits as well as drawbacks that should guide their selection. Some of the generic and model-specific merits and demerits of the soft computing techniques studied in this chapter are highlighted as follows:

9.6.1 MERITS

   i. They are particularly tolerant of noise, uncertainty, and inadequate data. Even when the data are noisy, imperfect, or inconsistent, they can nevertheless produce reasonable results.
   ii. They can learn from experience and develop over time since they are adaptive. Their suitability for solving wide range of problems is attributed to their incredibly adaptable computing ability.
   iii. When compared to other machine learning techniques, many soft computing techniques are more transparent and explanatory since they can reveal the reasoning behind their decisions.
   iv. They are suited for applications that require real-time processing since they are typically less computationally intensive than classic machine learning approaches like deep learning.
   v. Soft computing techniques allow for hybridization with other systems such that they leverage the strength of each method.

9.6.2 DEMERITS

Soft computing methods have a number of benefits over conventional methods, but they also have certain drawbacks including but not limited to the following:

   i. This technique lacks a rigorous rational or mathematical foundation because they are dependent on heuristic or empirical methods. Thus, they face challenges to assess or comprehend how they operate.
   ii. The models that soft computing approaches develop can occasionally be challenging to interpret, despite the fact that they might offer insights into their rationale. This may be a challenge when the model's conclusions must be justified to others.
   iii. They might not always yield the greatest outcomes, especially when compared to other methods like deep learning. For instance, deep learning methods can outperform soft computing methods in picture identification tasks.
   iv. Overfitting, which happens when a model is too complicated and matches the training data too closely, can be problematic for soft computing techniques. This leads to poor generalization to new data.
   v. It can take a lot of effort and experience to fine-tune the many factors that soft computing approaches need for, such as the quantity and kind of fuzzy sets.

They lack a defined framework for model creation and assessment, which can make it challenging to compare outcomes across research or applications.

Although we have highlighted the holistic and general demerits and limitations of soft computing, the drawbacks are more unique to the method than general because of their diverse frameworks. Presented in Table 9.1 are some of the disadvantages which are peculiar to the soft computing approaches considered in this study.

TABLE 9.1
Specific Demerits of Soft Computing Techniques [1]

S/N	Soft Computing Method	Specific Demerits
1	Artificial neural network	1. It is not deterministic to arrive at the ideal number of neurons, hidden layers, or topology. 2. The quantity of the data used for training the ANN model affects how well the resultant model performs on novel datasets. 3. ANN is vulnerable to both over- and underfitting. 4. Training with complicated models and huge amounts of data takes more time. 5. Its design as a black box results in intriguing network activity.
2	Fuzzy logic	1. To create a rule base, human expertise is mandatory. 2. Its capacity for generalization is limited. It is frequently peculiar to the problem. 3. It is limited in application.
3	Swarm intelligence	1. They many not scale well owing to intricate challenges and the number of agents may cause swarm intelligence algorithms to not scale adequately. 2. They are also problem-specific and may not be suitable for general problems.
4	Adaptive neuro-fuzzy inference systems	1. It could be fairly complicated, requiring numerous settings and parameters to be tuned. 2. It can be computationally expensive, especially when working with complex systems or significant volumes of data. This may restrict their usefulness, especially in real-time or online settings where responsiveness is essential.

ANN, artificial neural networks.

9.7 CONCLUSION

This chapter investigated two complementary computing techniques exhibiting great strength and weaknesses. While hard computing relies on exactness and accuracy as in mathematical equations, soft computing techniques are more adaptive and flexible and thus suited for uncertainty and ambiguity. Moreover, this chapter demonstrated the deficiency of hard computing techniques in handling some complex real-life problems delineated by noise, and uncertainties thus failing in mimicking intelligent human-like decision makings. However, the integration of these computing approaches could transform the thin film deposition space, causing a paradigm shift in the development of novel materials and fostering sustainable innovations in the near future.

REFERENCES

[1] P. A. Adedeji, O. O. Olatunji, N. Madushele, and A. O. Ajayeoba, "Soft computing in renewable energy system modeling," *Design, Analysis and Applications of Renewable Energy Systems*, pp. 79–102, 2021. doi: 10.1016/B978-0-12-824555-2.00026-5.

[2] A. Kamiya, S. J. Ovaska, R. Roy, and S. Kobayashi, "Fusion of soft computing and hard computing for large-scale plants: a general model," *Applied Soft Computing*, vol. 5, no. 3, pp. 265–279, 2005, doi: 10.1016/j.asoc.2004.08.005.

[3] N. Bharat, A. Kumar, and P. S. C. Bose, "A study on soft computing optimizing techniques," In *Materials Today: Proceedings*, Elsevier Ltd, pp. 1193–1198, 2021. doi: 10.1016/j.matpr.2021.08.068.

[4] P. Chen and J. Zhang, "Research on applications of soft computing agents," In *2008 International Symposium on Intelligent Information Technology Application Workshops*, pp. 259–262, 2008. doi: 10.1109/IITA.Workshops.2008.132.

[5] J. S. M. Hanwen Huang, Yufeng Liu, Ming Yuan, "Statistical significance of clustering using soft thresholding," *Journal of Computational and Graphical statistics*, vol. 24, no. 4, pp. 975–993, 2015.

[6] M. Guermoui and A. Rabehi, "Soft computing for solar radiation potential assessment in Algeria," *International Journal of Ambient Energy*, vol. 41, no. 13, pp. 1524–1533, 2020, doi: 10.1080/01430750.2018.1517686.

[7] M. J. Schaefer, "Precise optimization using range arithmetic," *Journal of Computational and Applied Mathematics*, vol. 53, no. 3, pp. 341–351, 1994, doi: 10.1016/0377-0427(94)90062-0.

[8] H. Doraisamy, T. Ertekin, and A. S. Grader, "Field development studies by neuro-simulation: an effective coupling of soft and hard computing protocols," *Computers & Geosciences*, vol. 26, no. 8, pp. 963–973, 2000, doi: 10.1016/S0098-3004(00)00032-7.

[9] E. Lorin and S. Tian, "A numerical study of fractional linear algebraic systems," *Mathematics and Computers in Simulation*, vol. 182, pp. 495–513, 2021, doi: 10.1016/j.matcom.2020.11.010.

[10] D. W. Braithwaite, "Relations between geometric proof justification and probabilistic reasoning," *Learning and Individual Differences*, vol. 98, p. 102201, 2022, doi: 10.1016/j.lindif.2022.102201.

[11] Manuel Bronstein, "Symbolic integration I. Transcendental functions," In *Algorithms and Computation in Mathematics,* E. Becker, M. Bronstein, H. Cohen, D. Eisenbud, and R. Gilman, Eds., Springer-Verlag Berlin Heidelberg GmbH, 1997.

[12] L. A. Zadeh, "Soft computing and fuzzy logic," *IEEE Software*, vol. 11, no. 6, pp. 48–56, 1994, doi: 10.1109/52.329401.

[13] O. Adeleke, S. Akinlabi, T. C. Jen, P. A. Adedeji, and I. Dunmade, "Evolutionary-based neuro-fuzzy modelling of combustion enthalpy of municipal solid waste," *Neural Computing and Applications*, vol. 2, 2022, doi: 10.1007/s00521-021-06870-2.

[14] A. Darwish, "Bio-inspired computing: Algorithms review, deep analysis, and the scope of applications," *Future Computing and Informatics Journal*, vol. 3, no. 2, pp. 231–246, 2018, doi: 10.1016/j.fcij.2018.06.001.

[15] K. Jithesh, P. B. Anto, P. K. Reshma, and M. Aravindhan, *Proceedings of Fourth International Conference on Soft Computing for Problem Solving*, vol. 336, 2015, doi: 10.1007/978-81-322-2220-0.

[16] B. J. Bejoy, G. Raju, D. Swain, B. Acharya, and Y.-C. Hu, "A generic cyber immune framework for anomaly detection using artificial immune systems," *Applied Soft Computing*, vol. 130, p. 109680, 2022, doi: 10.1016/j.asoc.2022.109680.

[17] N. Arab, H. Nemmour, and Y. Chibani, "A new synthetic feature generation scheme based on artificial immune systems for robust offline signature verification," *Expert Systems with Applications*, vol. 213, p. 119306, 2023, doi: 10.1016/j.eswa.2022.119306.

[18] Q. Duan and T. W. Liao, "Improved ant colony optimization algorithms for determining project critical paths," *Automation in Construction*, vol. 19, no. 6, pp. 676–693, 2010, doi: 10.1016/j.autcon.2010.02.012.

[19] M. Dorigo and L. M. Gambardella, "Ant colony system: A cooperative learning approach to the traveling salesman problem," *IEEE Transactions on Evolutionary Computation*, vol. 1, no. 1, pp. 53–66, 1997.

[20] H. Kuşan, O. Aytekin, and I. Özdemir, "The use of fuzzy logic in predicting house selling price," *Expert Systems with Applications*, vol. 37, no. 3, pp. 1808–1813, 2010, doi: 10.1016/j.eswa.2009.07.031.

[21] A. Karami, G. H. Roshani, A. Salehizadeh, and E. Nazemi, "The fuzzy logic application in volume fractions prediction of the annular three-phase flows," *Journal of Nondestructive Evaluation*, vol. 36, no. 2, pp. 1–9, 2017, doi: 10.1007/s10921-017-0415-7.

[22] E. Grossi and M. Buscema, "Introduction to artificial neural networks," *European Journal of Gastroenterology & Hepatology*, vol. 19, no. 12, pp. 1046–1054, 2007, doi: 10.1097/MEG.0b013e3282f198a0.

[23] A. Zakharova and A. Podvesovskii, "Application of probabilistic reasoning for risk assessment and mitigation in hydrocarbon field development," *IFAC-PapersOnLine*, vol. 55, no. 9, pp. 210–215, 2022, doi: 10.1016/j.ifacol.2022.07.037.

[24] J. Dianne Dotson, "Five areas of application for linear programming techniques," 2018, https://sciencing.com/careers-use-linear-equations-6060294.html.

[25] C. Chen, S. Feng, S. Liu, H. Zheng, H. Zhang, and J. Wang, "A stochastic linear programming model for maximizing generation and firm output at a reliability in long-term hydropower reservoir operation," *Journal of Hydrology*, vol. 618, p. 129185, 2023, doi: 10.1016/j.jhydrol.2023.129185.

[26] J. H. Hyun, "Antenna tracking profile generation using mixed-integer linear programming," *Advances in Space Research*, 2023, doi: 10.1016/j.asr.2023.02.040.

[27] J. Wang, M. Cevik, S. H. Amin, and A. A. Parsaee, "Mixed-integer linear programming models for the paint waste management problem," *Transportation Research Part E: Logistics and Transportation Review*, vol. 151, p. 102343, 2021, doi: 10.1016/j.tre.2021.102343.

[28] S. Cheng, C. W. Chan, and G. H. Huang, "An integrated multi-criteria decision analysis and inexact mixed integer linear programming approach for solid waste management," *Engineering Applications of Artificial Intelligence*, vol. 16, no. 5, pp. 543–554, 2003, doi: 10.1016/S0952-1976(03)00069-1.

[29] H. Zhu and G. H. Huang, "SLFP: A stochastic linear fractional programming approach for sustainable waste management," *Waste Management*, vol. 31, no. 12, pp. 2612–2619, 2011, doi: 10.1016/j.wasman.2011.08.009.

[30] M. E. Batur, A. Cihan, M. K. Korucu, N. Bektaş, and B. Keskinler, "A mixed integer linear programming model for long-term planning of municipal solid waste management systems: Against restricted mass balances," *Waste Management*, vol. 105, pp. 211–222, 2020, doi: 10.1016/j.wasman.2020.02.003.

[31] T. Masino, P. E. Colombo, K. Reis, I. Tetens, and A. Parlesak, "Climate-friendly, health-promoting, and culturally acceptable diets for German adult omnivores, pescatarians, vegetarians, and vegans - a linear programming approach," *Nutrition*, vol. 109, p. 111977, 2023, doi: 10.1016/j.nut.2023.111977.

[32] Z. Xiong, M. Zhao, Z. Yuan, J. Xu, and L. Cai, "Energy-saving optimization of application server clusters based on mixed integer linear programming," *Journal of Parallel and Distributed Computing*, vol. 171, pp. 111–129, 2023, doi: 10.1016/j.jpdc.2022.09.009.

[33] J. Hu, Y. Wang, Y. Pang, and Y. Liu, "Optimal maintenance scheduling under uncertainties using linear programming-enhanced reinforcement learning," *Engineering Applications of Artificial Intelligence*, vol. 109, p. 104655, 2022, doi: 10.1016/j.engappai.2021.104655.

[34] C. B. R. Ng, C. Bil, S. Sardina, and T. O'bree, "Designing an expert system to support aviation occurrence investigations," *Expert Systems with Applications*, vol. 207, p. 117994, 2022, doi: 10.1016/j.eswa.2022.117994.

[35] R. H. Faisal *et al.*, "A modular fuzzy expert system for chemotherapy drug dose scheduling," *Healthcare Analytics*, vol. 3, p. 100139, 2023, doi: 10.1016/j.health.2023.100139.

[36] M. R. Nayeri, B. N. Araabi, M. Yazdanpanah, and B. Moshiri, "Design, implementation and evaluation of an expert system for operating regime detection in industrial gas turbine," *Expert Systems with Applications*, vol. 203, p. 117332, 2022, doi: 10.1016/j.eswa.2022.117332.

[37] L.-H. Yang, F.-F. Ye, J. Liu, and Y.-M. Wang, "Belief rule-base expert system with multilayer tree structure for complex problems modeling," *Expert Systems with Applications*, vol. 217, p. 119567, 2023, doi: 10.1016/j.eswa.2023.119567.

[38] J. Kołodziejczyk, N. Grzegorczyk-Dłuciak, and E. Kuliga, "Rule-based expert system supporting individual education-and-therapeutic program composition in SYSABA," *Procedia Computer Science*, vol. 207, pp. 4535–4544, 2022, doi: 10.1016/j.procs.2022.09.517.

[39] A. Trivedi and P. Kumar Gurrala, "Fuzzy logic based expert system for prediction of tensile strength in fused filament fabrication (FFF) process," *Materials Today: Proceedings*, vol. 44, pp. 1344–1349, 2021, doi: 10.1016/j.matpr.2020.11.391.

[40] D. Rajamani, B. Esakki, P. Arunkumar, and R. Velu, "Fuzzy logic-based expert system for prediction of wear rate in selective inhibition sintered HDPE parts," *Materials Today: Proceedings*, vol. 5, no. 2, Part 1, pp. 6072–6081, 2018, doi: 10.1016/j.matpr.2017.12.212.

[41] S. A. Abdul-Wahab, A. Elkamel, M. A. Al-Weshahi, and A. S. Al Yahmadi, "Troubleshooting the brine heater of the MSF plant fuzzy logic-based expert system," *Desalination*, vol. 217, no. 1, pp. 100–117, 2007, doi: 10.1016/j.desal.2007.01.014.

[42] N. Atabaki, N. Jesuthasan, and B. R. (Rabi) Baliga, "Chapter Five - Hybrid numerical models for predictions of thermofluid phenomena in engineering systems, with application to a loop heat pipe in steady operation," In *Advances in Heat Transfer*, J. P. Abraham, J. M. Gorman, and W. J. Minkowycz, Eds., Elsevier, pp. 179–239, 2022. doi: 10.1016/bs.aiht.2022.07.005.

[43] J. Náprstek, "Combined analytical and numerical approaches in Dynamic Stability analyses of engineering systems," *Journal of Sound and Vibration*, vol. 338, pp. 2–41, 2015, doi: 10.1016/j.jsv.2014.06.029.

[44] A. H. B. Duffy, A. Persidis, and K. J. MacCallum, "NODES: a numerical and object based modelling system for conceptual engineering design," *Knowledge-Based Systems*, vol. 9, no. 3, pp. 183–206, 1996, doi: 10.1016/0950-7051(95)01027-0.

[45] W. Li, X. Li, Y. Mei, G. Wang, W. Yang, and H. Wang, "A numerical simulation approach of energy-absorbing anchor bolts for rock engineering," *International Journal of Rock Mechanics and Mining Sciences*, vol. 158, p. 105188, 2022, doi: 10.1016/j.ijrmms.2022.105188.

[46] K. C. Lin and C.-C. Liao, "Simulation of slurry residence time during chemical-mechanical polishing using 3-D computational fluid dynamics," *Chemical Engineering Research and Design*, vol. 191, pp. 375–386, 2023, doi: 10.1016/j.cherd.2023.01.025.

[47] B. Wang, W. Wang, Y. Li, and F. Lan, "Aerodynamic characteristics study of vehicle-bridge system based on computational fluid dynamics," *Journal of Wind Engineering and Industrial Aerodynamics*, vol. 234, p. 105351, 2023, doi: 10.1016/j.jweia.2023.105351.

[48] D.-S. Wang *et al.*, "Finite element analysis of the main reactor vessel in the China initiative accelerator driven system," *Engineering Failure Analysis*, vol. 146, p. 107121, 2023, doi: 10.1016/j.engfailanal.2023.107121.

[49] A. M. Hissanaga, J. R. Barbosa Jr, and A. K. da Silva, "Numerical analysis of inorganic fouling with multi-physics turbulent models," *Applied Thermal Engineering*, vol. 220, p. 119624, 2023, doi: 10.1016/j.applthermaleng.2022.119624.

[50] A. Tabandeh, N. Sharma, L. Iannacone, and P. Gardoni, "Numerical solution of the Fokker-Planck equation using physics-based mixture models," *Computer Methods in Applied Mechanics and Engineering*, vol. 399, p. 115424, 2022, doi: 10.1016/j.cma.2022.115424.

[51] C. Klingenberg, "Chapter 17- Numerical methods for astrophysics," In *Handbook of Numerical Analysis*, R. Abgrall and C.- W. Shu, Eds., Elsevier, 2017, pp. 465–477. doi: 10.1016/bs.hna.2016.11.001.

[52] W. Zhen, H. Li, and Q. Wang, "Simulation of residual stress in aluminum alloy welding seam based on computer numerical simulation," *Optik*, vol. 258, p. 168785, 2022, doi: 10.1016/j.ijleo.2022.168785.

[53] K. Aledealat, B. Aladerah, and A. Obeidat, "Study of structural, electronic, and magnetic properties of L10-ordered CoPt and NiPt: An ab initio calculations and Monte Carlo simulation," *Solid State Communications*, vol. 363, p. 115112, 2023, doi: 10.1016/j.ssc.2023.115112.

[54] N. Saber, Z. Fadil, A. Mhirech, B. Kabouchi, and L. Bahmad, "Magnetic properties and thermal behavior of the monolayer Rubrene-like nano-island: Monte Carlo simulations," *Solid State Communications*, vol. 362, p. 115084, 2023, doi: 10.1016/j.ssc.2023.115084.

[55] L. Tan, S. Kothapalli, L. Chen, O. Hussaini, R. Bissiri, and Z. Chen, "A survey of power and energy efficient techniques for high performance numerical linear algebra operations," *Parallel Computing*, vol. 40, no. 10, pp. 559–573, 2014, doi: 10.1016/j.parco.2014.09.001.

[56] M. Hemmat, D. Toghraie, and F. Amoozad, "Prediction of viscosity of MWCNT-Al2O3 (20:80)/SAE40 nano-lubricant using multi-layer artificial neural network (MLP-ANN) modeling," *Engineering Applications of Artificial Intelligence*, vol. 121, p. 105948, 2023, doi: 10.1016/j.engappai.2023.105948.

[57] R. Langbauer, G. Nunner, T. Zmek, J. Klarner, R. Prieler, and C. Hochenauer, "Modelling of thermal shrinkage of seamless steel pipes using artificial neural networks (ANN) focussing on the influence of the ANN architecture," *Results in Engineering*, p. 100999, 2023, doi: 10.1016/j.rineng.2023.100999.

[58] F. Ahmed and W. Chen, "Investigation of steam ejector parameters under three optimization algorithm using ANN," *Applied Thermal Engineering*, vol. 225, p. 120205, 2023, doi: 10.1016/j.applthermaleng.2023.120205.

[59] A. J. Kilani, O. Adeleke, and C. A. Fapohunda, "Application of machine learning models to investigate the performance of concrete reinforced with oil palm empty fruit brunch (OPEFB) fibers," *Asian Journal of Civil Engineering*, vol. 23, no. 2, pp. 299–320, 2022, doi: 10.1007/s42107-022-00424-0.

[60] O. Adeleke, S. Akinlabi, T.-C. Jen, and I. Dunmade, "Prediction of the heating value of municipal solid waste: a case study of the city of Johannesburg," *International Journal of Ambient Energy*, pp. 1–12, 2020, doi: 10.1080/01430750.2020.1861088.

[61] O. Adeleke, S. A. Akinlabi, T. Jen, and I. Dunmade, "Application of artificial neural networks for predicting the physical composition of municipal solid waste : An assessment of the impact of seasonal variation," *Waste Management & Research*, vol. 39, no. 8, pp. 1058–1068, 2021, doi: 10.1177/0734242X21991642.

[62] S. F. Peixoto, A. M. Coimbra Horbe, T. M. Soares, C. A. Freitas, E. M. Dalat de Sousa, and E. R. Herrera de Figueiredo Iza, "Boolean and fuzzy logic operators and multivariate linear regression applied to airborne gamma-ray spectrometry data for regolith mapping in granite-greenstone terrain in Midwest Brazil," *Journal of South American Earth Sciences*, vol. 112, p. 103562, 2021, doi: 10.1016/j.jsames.2021.103562.

[63] G. Senthilkumar, R. Murugan, G. Gnanakumar, and N. Nithyanandan, "Fuzzy logic modelling of machining characteristics for CNC milling of EN24 using Ti-N coated tool," *Materials Today: Proceedings*, 2023, doi: 10.1016/j.matpr.2022.12.111.

[64] U. M. R. Paturi, D. G. Vanga, S. Cheruku, S. T. Palakurthy, and N. K. Jha, "Estimation of abrasive wear of nanostructured WC-10Co-4Cr TIG weld cladding using neural network and fuzzy logic approach," *Materials Today: Proceedings*, 2022, doi: 10.1016/j.matpr.2022.10.266.

[65] I. Tunc and M. T. Soylemez, "Fuzzy logic and deep Q learning based control for traffic lights," *Alexandria Engineering Journal*, vol. 67, pp. 343–359, 2023, doi: 10.1016/j.aej.2022.12.028.

[66] A. I. Provotar, A. V Lapko, and A. A. Provotar, "Fuzzy inference systems and their applications," *Cybernetics and Systems Analysis*, vol. 49, no. 4, pp. 517–525, 2013, doi: 10.1007/s10559-013-9537-9.

[67] P. Lu, F. Wen, Y. Li, and D. Chen, "Individual behaviors, social learning, and swarm intelligence: Real case and counterfactuals," *Expert Systems with Applications*, vol. 207, p. 117878, 2022, doi: 10.1016/j.eswa.2022.117878.

[68] I. Sudha *et al.*, "Pulse jamming attack detection using swarm intelligence in wireless sensor networks," *Optik*, vol. 272, p. 170251, 2023, doi: 10.1016/j.ijleo.2022.170251.

[69] S. Yelisetti, V. K. Saini, R. Kumar, R. Lamba, and A. Saxena, "Optimal energy management system for residential buildings considering the time of use price with swarm intelligence algorithms," *Journal of Building Engineering*, vol. 59, p. 105062, 2022, doi: 10.1016/j.jobe.2022.105062.

[70] A. A. Hassan, F. H. Fahmy, A. E.-S. A. Nafeh, and M. A. Abu-elmagd, "Genetic single objective optimisation for sizing and allocation of renewable DG systems," *International Journal of Sustainable Energy*, vol. 36, no. 6, pp. 545–562, Jul. 2017, doi: 10.1080/14786451.2015.1053393.

[71] O. Adeleke, S. Akinlabi, T. C. Jen, and I. Dunmade, "A machine learning approach for investigating the impact of seasonal variation on physical composition of municipal solid waste," *Journal of Reliable Intelligent Environments*, 2022, doi: 10.1007/s40860-021-00168-9.

[72] Y. Qin, Z. Li, J. Ding, F. Zhao, and M. Meng, "Automatic optimization model of transmission line based on GIS and genetic algorithm," *Array*, vol. 17, p. 100266, 2023, doi: 10.1016/j.array.2022.100266.

[73] H.-P. Schwefel, "Advantages (and disadvantages) of evolutionary computation over other approaches," *Evolutionary Computation*, vol. 1, pp. 20–22, 2000.

[74] A. Sajadi, A. Dashti, M. Raji, A. Zarei, and A. H. Mohammadi, "Estimation of cetane numbers of biodiesel and diesel oils using regression and PSO-ANFIS models," *Renewable Energy*, vol. 158, pp. 465–473, 2020.

[75] R. Eberhart and J. Kennedy, "A new optimizer using particle swarm theory," In *MHS'95. Proceedings of the Sixth International Symposium on Micro Machine and Human Science*, IEEE, pp. 39–43, 1995.

[76] M. Zanganeh, "Improvement of the ANFIS-based wave predictor models by the particle swarm optimization," *Journal of Ocean Engineering and Science*, vol. 5, pp. 84–99, 2020.

[77] P. A. Adedeji, S. Akinlabi, N. Madushele, and O. O. Olatunji, "Wind turbine power output very short-term forecast: A comparative study of data clustering techniques in a PSO-ANFIS model," *Journal of Cleaner Production*, vol. 254, 2020, doi: 10.1016/j.jclepro.2020.120135.

[78] M. Dorigo and G. Di Caro, "Ant colony optimization: a new meta-heuristic," In *Proceedings of the 1999 Congress on Evolutionary Computation-CEC99 (Cat. No. 99TH8406)*, vol. 2, pp. 1470–1477, 1999, doi: 10.1109/CEC.1999.782657.

[79] H. Zhao *et al.*, "VM performance-aware virtual machine migration method based on ant colony optimization in cloud environment," *Journal of Parallel and Distributed Computing*, vol. 176, pp. 17–27, 2023, doi: 10.1016/j.jpdc.2023.02.003.

[80] Z. E. Ahmed, R. A. Saeed, A. Mukherjee, and S. N. Ghorpade, "10- Energy optimization in low-power wide area networks by using heuristic techniques," In *LPWAN Technologies for IoT and M2M Applications*, B. S. Chaudhari and M. Zennaro, Eds., Academic Press, pp. 199–223, 2020. doi: 10.1016/B978-0-12-818880-4.00011-9.

[81] D. Zhao, A. Qi, F. Yu, A. A. Heidari, H. Chen, and Y. Li, "Multi-strategy ant colony optimization for multi-level image segmentation: Case study of melanoma," *Biomedical Signal Processing and Control*, vol. 83, p. 104647, 2023, doi: 10.1016/j.bspc.2023.104647.

[82] M. Ghasemi, S. kadkhoda Mohammadi, M. Zare, S. Mirjalili, M. Gil, and R. Hemmati, "A new firefly algorithm with improved global exploration and convergence with application to engineering optimization," *Decision Analytics Journal*, vol. 5, p. 100125, 2022, doi: 10.1016/j.dajour.2022.100125.

[83] J. Li, X. Wei, B. Li, and Z. Zeng, "A survey on firefly algorithms," *Neurocomputing*, vol. 500, pp. 662–678, 2022, doi: 10.1016/j.neucom.2022.05.100.

[84] X. Xue, "A compact firefly algorithm for matching biomedical ontologies," *Knowledge-Based Systems*, vol. 62, no. 7, pp. 2855–2871, 2020, doi: 10.1007/s10115-020-01443-6.

[85] J. Antonanzas, N. Osorio, R. Escobar, R. Urraca, F. J. Martinez-de-Pison, and F. Antonanzas-Torres, "Review of photovoltaic power forecasting," *Solar Energy*, vol. 136, pp. 78–111, 2016, doi: 10.1016/j.solener.2016.06.069.

[86] H. Fattahi, "Adaptive neuro fuzzy inference system based on fuzzy c - means clustering algorithm, a technique for estimation of TBM penetration rate," *Iran University of Science and Technology*, vol. 6, no. 2, pp. 159–171, 2016.

[87] D. Karaboga and E. Kaya, "Adaptive network based fuzzy inference system (ANFIS) training approaches: A comprehensive survey," *Artificial Intelligence Review*, vol. 52, no. 4, pp. 2263–2293, 2019, doi: 10.1007/s10462-017-9610-2.

[88] M. Mustapha, M. W. Mustafa, S. N. Khalid, I. Abubakar, and A. M. Abdilahi, "Correlation and wavelet-based short-term load forecasting using anfis," *Indian Journal of Science and Technology*, vol. 9, no. 46, 2016, doi: 10.17485/ijst/2016/v9i46/107141.

[89] V. Güldal and H. Tongal, "Comparison of recurrent neural network, adaptive neuro-fuzzy inference system and stochastic models in eğ irdir lake level forecasting," *Water Resources Management*, vol. 24, no. 1, pp. 105–128, 2010, doi: 10.1007/s11269-009-9439-9.

[90] M. Şahin and R. Erol, "A comparative study of neural networks and ANFIS for forecasting attendance rate of soccer games," *Mathematical and Computational Applications*, vol. 22, no. 4, p. 43, 2017, doi: 10.3390/mca22040043.

[91] A. Azad, M. Manoochehri, H. Kashi, S. Farzin, and H. Karami, "Comparative evaluation of intelligent algorithms to improve adaptive neuro- fuzzy inference system performance in precipitation modelling," *Journal of Hydrology*, vol. 571, no. February, pp. 214–224, 2019.

[92] A. Sarkheyli and A. Mohd, "Robust optimization of ANFIS based on a new modified GA," *Neurocomputing*, vol. 166, pp. 357–366, 2015, doi: 10.1016/j.neucom.2015.03.060.

[93] D. K. Roy, R. Barzegar, J. Quilty, and J. Adamowski, "Using ensembles of adaptive neuro-fuzzy inference system and optimization algorithms to predict reference evapotranspiration in subtropical climatic zones," *Journal of Hydrology*, vol. 591, p. 125509, 2020, doi: 10.1016/j.jhydrol.2020.125509.

[94] S. Ghordoyee Milan, A. Roozbahani, N. Arya Azar, and S. Javadi, "Development of adaptive neuro fuzzy inference system -Evolutionary algorithms hybrid models (ANFIS-EA) for prediction of optimal groundwater exploitation," *Journal of Hydrology*, vol. 598, no. March, p. 126258, 2021, doi: 10.1016/j.jhydrol.2021.126258.

Part III

---

*Machine Learning Applications
in Atomic Layer Deposition*

10 Why Machine Learning?

10.1 INTRODUCTION

Thin film technology has long led scientific progress as solar panels, computer screens, sensors, and implants use thin sheets [1,2]. The need for more efficient and effective manufacturing and characterization processes, however, is growing along with the demand for complex and sophisticated thin films [3]. Machine learning can optimize thin film characteristics and deposition procedures by utilizing intelligent algorithms, which can recognize patterns and relationships in large amounts of data [4,5]. Making sense of experimental data can be streamlined by machine learning, which enables scientists to make conclusions more quickly and accurately [6]. As thin film technology becomes increasingly sophisticated, machine learning offers a potent tool for modeling and simulating thin films at a degree of detail that was previously unachievable [7].

A robust method for depositing thin films with atomic accuracy is atomic layer deposition (ALD) [8–10]. ALD has turned into a crucial tool in a variety of industries, from microelectronics to energy storage, because of its capacity to create complex multilayered structures with a high degree of control over thickness, content, and morphology [11–13]. ALD is not without difficulties, despite its many benefits. The inherent complexity of the process is one of the main challenges owing to the many variables that can impact the quality and development of films during ALD [14]. Because of their frequent interdependence, it is challenging to optimize the procedure and obtain the desired deposited film qualities. Machine learning has the ability to revolutionize how we tackle ALD problems by utilizing the power of artificial intelligence, while assisting us in optimizing ALD procedures in ways that were previously not conceivable because of its capacity to evaluate enormous datasets and find intricate connections between process variables and film qualities [7].

Having examined the fundamental techniques of machine learning in previous chapters, we want to establish the significance and the need for machine learning–based modeling in thin film technology and ALD in this current chapter. While we are not undermining the uniqueness of other computational simulation techniques in thin film technology as examined in one of the preceding chapters, we shall examine the benefits of machine learning over these computational models in their standalone or hybridized form. The potentials of integrating machine learning algorithms with the computational simulation models will also be investigated.

DOI: 10.1201/9781003346234-13

10.2   LIMITATIONS OF NUMERICAL SIMULATIONS IN ATOMIC LAYER DEPOSITION

A critical challenge of the ALD process lies in comprehending the intricate physical and chemical processes involved in depositing atoms, one layer at a time. This has necessitated the need to optimize the deposition process for various materials and applications by simulating the behavior of atoms and molecules during the process using numerical computational simulations [3,14,15]. Several computational simulation approaches were studied in-depth in a previous chapter. More so, these numerical simulation approaches were described in a previous chapter as "hard computing techniques." If we wish to make the most of this cutting-edge technology, we must be aware of numerical simulations' shortcomings. Machine learning models can fill some of these gaps. We shall examine some of these limitations as follows.

10.2.1   Faulty Capture of the Physical Phenomena by the Model

The performance of the numerical simulations significantly depends on how accurately the physical models that explain the actions of thin film materials and the deposition process are established [16,17]. Despite tremendous advancements in our knowledge of the mechanisms underpinning the process of ALD, many fundamental physical and chemical processes remain mostly unexplained [14]. Creating precise ALD simulations may be limited due to these difficulties. The models are built around multiple assumptions, simplifications, and approximations that might not always actually apply in real-life situations. The quality and dependability of the simulation results can therefore be severely influenced by any mistakes or faults in these models. The use of empirical models, which are predicated on observed measurements and can sometimes not precisely capture the basic concepts of the physical processes, is a prominent reason for inadequacy and poor performance in modeling physical events [18,19]. For instance, empirical equations drawn from experimental data may be used to simulate the thin film deposition rate, but these equations might not take into account the intricate interactions that occur between precursor molecules, the substrate, and the reactant gases throughout the deposition process. By using empirical models, it's possible that simulations will be overly simplified or imprecise and won't accurately reflect how the deposition process actually works. A number of numerical simulations depend on abstracted representations of the process, which presuppose homogeneous deposition throughout the substrate layer and uniform distribution of the precursor molecule [3]. Several variables directly impact the deposition process ranging from the forms and position of the substrate surface, the existence of defects or contaminants, and the diffusion of precursor molecules across the surface [10,14]. Inaccurate simulations that default in accurately representing the ALD process can be attributed to ignoring these elements [20,21].

10.2.2   Handling the Atomic-Scale Processes of ALD is Challenging

The scale of the deposition process impacts the performance of the computational simulation process significantly [17]. The process involved in ALD is carried out at atomic-scale, thus necessitating a high degree of temporal and spatial accuracy. This operation scale poses a challenge to the simulation of the process. This challenge could be attributed to the complicated nature of the atoms and molecular relationships [16,17]. It is computationally costly, laborious, and not time-effective for the simulations to follow the motions of individual atoms and molecules. A further factor driving up the cost of computation is the magnitude of the deposition process, which necessitates broad simulation domains to accurately model substrate and environment impacts. Quantum mechanics, a sophisticated system of physical principles that describes how matter interacts at the atomic and subatomic scales, determines the interactions between atoms [22]. Modeling these interactions can be challenging since they depend on a number of variables, including the atoms' size, shape, and electrical configuration [23]. Inadequate experimental data may also be a contributing factor to the numerical computational simulation's shortcomings. Simulations can predict material behavior, but data quality affects accuracy. Without enough experimental data on a material's properties, simulations may not be accurate, limiting their application.

10.2.3 Limited Computational Resources

A high computational power and resources involving lots of memory and computing capacity are needed for accurate modeling of ALD processes. These resources might not be easily accessible, especially for complicated deposition procedures with numerous chemical and surface reactions. As a result, experts frequently adopt simplified models that might not fully reflect the process's genuine nature. Using coarse-grained models is one method for minimizing the amount of computing power needed for ALD simulations [24]. By lowering the quantity of particles that must be simulated, coarse-grained models make the system simpler while also using much less computer power. Yet, as they might not accurately represent the process's finer intricacies, coarse-grained models might sometimes result in inaccurate simulation results. To further overcome the challenge of limited resource in ALD simulation with numerical methods, machine learning–based modeling which would be extensively discussed in subsequent subsection and chapter would be viable. In order to create prediction models that can precisely forecast the behavior of the process, machine learning algorithms can be trained on massive datasets of ALD simulations [7]. By maintaining the ability to capture the crucial aspects of the process, these models can drastically cut the processing needs for ALD simulations.

10.2.4 Reliability of Experimental Data for Validation

The quantity and quality of experimental data for validation play a critical role in determining the effectiveness and reliability of numerical simulations in ALD procedures [14,23]. Experimental data provide crucial details on the operation and performance of ALD processes, which may be used to evaluate the simulation models and modify the input parameters to improve their performance [25]. In complex ALD processes such as those involving intricate chemical and surface reactions and interactions, getting experimental data for them might be a bit challenging [26]. Further to this, it might be difficult to collect precise data because the experiments may be constrained by the sensitivity and precision of the instruments utilized while the risky nature and complicated synthesis of certain materials might impede the practicality of an experimental research on them [25]. In cases where experimental data are absent, the simulation process is often predicated on assumptions which might not perfectly capture the real-life scenario of the deposition process, ultimately a flawed process design results. To abate this challenge, the simulation process can be validated using in situ characterization techniques like X-ray photoelectron spectroscopy and ellipsometry which can accurately present an insightful knowledge of the deposition process [27,28].

10.3   BENEFITS OF MACHINE LEARNING OVER COMPUTATIONAL SIMULATION APPROACH

The optimization of the ALD process and creation of novel materials has been built on the conventional computational simulations, which are predicated on mathematical models. These simulations can, however, be time-consuming, costly in terms of processing, and have accuracy limitations as noted earlier. Machine learning models have been a potentially viable strategy and a suitable alternative for examining and improving deposition processes in contemporary times. Although, the potential benefit of machine learning application in ALD has not been fully explored yet, but it is growing. A better comprehension of the behavior of ALD processes and improved prediction performance can be realized with less computational power and scalability by training machine learning models using experimental results. While both machine learning and numerical simulations have merits and demerits, machine learning models could be preferred over the computational approach due to the following reasons.

10.3.1   BETTER PERFORMANCE AND A MORE RELIABLE AND ACCURATE OUTCOME

The dependence of the classical numerical simulations on assumptions and approximations is a key factor limiting its performance in the ALD applications [3,29]. Another complementary approach that can enhance forecast accuracy and provide useful insights into the dynamics of the deposition processes is machine learning modeling [7]. In ALD, where the deposition process is contingent on multiple interconnected variables, machine learning models are built to capture complex interactions between input parameters and output variables. They can as well find patterns and hidden trends by learning from historical experimental data such that they are able to comprehend the subtleties and intricacies of ALD processes that the simulation models can't fathom [7]. Furthermore, the existence of complex nonlinear relationship between several variables of ALD process such as precursor chemistry, temperature, and pressure has made the classical numerical simulation difficult. The viability of the intelligent machine learning algorithms to capture these nonlinear relationships and their capability in detecting complex relationships within process parameters which might not be readily apparent makes them a suitable alternative to numerical simulation [30]. By identifying the crucial input variables that influence the process outputs, machine learning models may also be used to optimize deposition processes [7]. This assists in determining which processing parameters have the biggest effects on the deposition process by examining the learned weights of the model. With the help of this knowledge, ALD processes can be designed and optimized to produce higher-quality final products by altering deposition conditions or precursor flow rates.

10.3.2   COST-EFFECTIVE COMPUTING

The cost of computing has significantly increased with the upsurge in the demand for computational resources and power [31]. Consequently, there is a greater desire for more cost-effective computing [32]. The need for affordable computing solutions is greater than ever in this contemporary age of big data and increasing computational needs [31]. Numerical simulations might need a lot of computing power to operate, which would take both resources and time [18,19]. Machine learning can help in this situation. Machine learning can build reliable predictive models without using as much computational power as numerical models [33]. Machine learning can simply be trained on historical data to recognize patterns and trends in the dataset, unlike numerical computations which are based on complex systems, variables, and mathematical equations [34,35]. This accounts for the significant difference in their computational cost.

10.3.3   ADAPTATION AND FLEXIBILITY TO A NEW DATASET

It can be hard to forecast and control the features of the deposited material due to the high ambiguity and multidimensionality of variables involved in the processes [33]. The adaptability of machine learning algorithm models to an unseen and novel dataset is a crucial benefit it can offer in ALD process by learning from new data and modifying their predictions to better comprehend the process and its final outcomes [7,33]. They have the ability to modify their models real-time as new information emerges. This means that the algorithm can update its model as more data from the deposition process is gathered and make better predictions based on the updated data.

10.3.4   A DEEPER UNDERSTANDING AND ANALYSIS OF THE COMPLEX ALD DATA

The atomic layer deposition processes involve a complex inter-relationship between diverse complicated and high-dimensional variables which have a significant impact on the physical, chemical, and electrical properties of the deposited materials [14,25]. These massive and intricate datasets and parameters may be difficult for the conventional computational simulations approaches to handle, thus providing only limited insights into the process. This impedes an intelligent decision-making on the final deposited materials. However, machine learning algorithms are capable of swiftly analyzing huge, complex datasets and parameters while finding undiscovered patterns and connections between the independent parameters [36]. This may give a holistic comprehension of the deposition procedure, thus facilitating more correct estimates and more control over the properties of the deposited material. The robustness of machine learning algorithms in handling nonlinearity in the complex ALD variables fills up the gap opened by the limitation of the computational approaches to linearity. A useful technique in machine learning, called the feature selection, can be deployed in ALD process to choose the most crucial input factors that have an impact on the final properties of the deposited material. Owing to the influence of several input parameters on the behavior of the deposited material. Other machine learning algorithms such as classification and clustering can also be beneficial in providing a deeper insight into the complex deposition data by grouping and classifying materials toward finding new materials with attractive characteristics.

10.4   CHALLENGES AND LIMITATIONS OF MACHINE LEARNING APPLICATIONS IN ALD PROCESSES

Even though machine learning has demonstrated a great deal of potential in improving ALD procedures, its application is still at the infancy stage due to a number of challenges and shortfalls that must be overcome before it can be effectively used. The slow pace in the growth, applications, and full exploitation of the potential benefits of machine learning techniques in ALD is attributed to some limitations and challenges. The opportunities of machine learning applications in ALD are enormous, ranging from selecting the best precursors and fine-tuning process variables to forecasting deposition rates and film characteristics. We will delve deeper into these potential applications in subsequent chapters. However, in order to realize these potentials, it is critical to recognize its constraints and work to overcome them. We will examine some of the drawbacks to machine learning applications in ALD processes.

10.4.1   ACCESSIBILITY AND QUALITY OF ALD DATA

This is a prominent challenge for all machine learning applications. The success of machine learning models depends heavily on the quantity, integrity, and quality of available data [37]. For learning and making precise predictions, the algorithms primarily rely on reasonable amounts of high-quality data. Data integrity is crucial because the model's accuracy and efficiency depend on its training data. Noisy or inadequate data may lead to inaccurate predictions and poor model performance. The paucity of data in ALD process is due to the high cost of instrumentation and experimental procedures [7]. In addition, since there are fewer materials, precursors, and process conditions available due to the high cost of experimentation, the resulting data sets may be biased, resulting in models that do not accurately reflect the entire process space. This is also accountable for the lack of sufficient data in the literature. Despite the fact that there is a growing number of researches on ALD processes in the literature, a great deal of it is devoted to particular uses or types of materials, and the data collection and publishing are frequently incongruent. Due to this, combining datasets from various sources may be challenging, which would decrease the quantity of training data that machine learning algorithms could use.

10.4.2 Interpretability of Models' Outcome

It can be difficult to understand how a given conclusion was reached because many machine learning models are referred to as "black boxes" [31]. In the context of ALD, interpretability of the outcome of machine learning's outcome is critical in order to fully comprehend the behavior of the deposited materials resulting from complex physical and chemical reactions. The ALD process which produces the thin film comprises complex chemical processes, surface adsorption, and diffusion mechanism [17]. These inter-relationships are influenced by a number of variables, including but not limited to the concentration of the precursor, temperature, the characteristics of the substrate's wall, and pressure [25]. Machine learning may be unable to describe and interpret these underlying complex processes and causal relationship in its result. In order to improve the performance and generalization ability of the machine learning model in ALD, its interpretability and explainability must be established. One of such intervention is the use of explainable machine learning. Explainable machine learning research is emerging. It seeks to make machine learning models more transparent and interpretable to develop trust in them [38,39]. Decision trees and rule-based models may reveal deposition factors and film characteristics [40,41].

10.4.3 Minimal Knowledge of the Domain

The intrinsic physical and chemical reactions that control the deposition must be understood in order to develop machine learning for optimizing ALD processes [33]. To achieve accurate product design and optimal deposition process, the proper knowledge of the domain is required. However, machine learning models are stochastic and not deterministic; they have a limited knowledge of the physical domain of the process [42]. While this may not be a long-term limitation, machine learning can be incorporated with the physical and chemical processes that produce the thin film, otherwise an inaccurate model may result. It is crucial to know which physical and chemical variables significantly influence the deposition process and to decide on the best experimental setups to quantify those variables because finding the pertinent input features is one of the major hurdles in embedding domain knowledge into the ALD-based machine learning applications. Further to this, the proper choice of machine learning algorithm is contingent on the adequate knowledge of the domain and underlying processes. Sequel to this, neural networks may be better suited to forecast dynamic interplay between constituents and substrates than linear regression when it comes to predicting film thickness. Similarly, correct interpretation of the models' outcome relies largely on the succinct knowledge of the domain and underpinning processes. A sustainable approach to this challenge is the recommendation for team work and collective participation between machine learning programmer, ALD, and/or material science expert toward a common task while integrating the required statistical and computational simulation approaches.

10.4.4 Overfitting and Underfitting

When a machine learning model is overly complicated and learns the noise in the training data rather of the underlying patterns, this is known as overfitting while underfitting happens when a model is overly simplistic and unable to discern the fundamental trends in the data [43]. Because of this, the model does well with training data but poorly with novel, untried data. This is often experienced with most machine learning applications including ALD space. The use of a model that is too complex for the amount of the training dataset is the most frequent reason for overfitting. This may pose a significant setback to the performance and reliability of machine learning application in ALD owing to the complexity of the deposition process. The model must be complex enough to capture the inherent complex reaction in ALD to avoid underfittings, otherwise, there would be a need to increase the complexity of the model such as adding more features to the input data or adding more layers to the neural networks model. Optimal hyper-parameter tuning of models' parameters such as the learning rate, layer count, and regularization parameters is critical in this regard to enhance the model's performance on the test data [34,44]. Furthermore, it is crucial to have enough high-quality training data that reflect the fundamental behaviors of the deposition process in order to abate the drawbacks of overfitting and underfitting in ALD.

10.4.5 Flexibility with Changing Process Variables

Machine learning–based applications for analytical laboratory-based data such as ALD are created to predict outcomes based on a predetermined set of input and output parameters. But when it comes to adjusting to changes in input or output parameters, these models frequently lack flexibility. The inability of machine learning models to adjust to variations in input or output variables limits their ability to make accurate predictions from analytical laboratory data of ALD. New data, relabeling, and rigorous testing and validation make updating these models costly and time-consuming. Adding these changes to the model can require time and money. Retraining the model entails gathering fresh data reflecting adjustments to the input or output parameters, labeling the data, and then applying the labeled data to train a fresh model. The only way to confirm the accuracy of the updated model is through rigorous testing and validation. Noteworthily, if the changes are substantial, the updated model might not generalize well to other datasets. It might be necessary to create completely new models in some circumstances to account for adjustments to input or output parameters.

10.5   OPTIMAL SELECTION OF MACHINE LEARNING ALGORITHM FOR ALD MODELING

A proper understanding of machine learning algorithms and the deposition process is essential for choosing the best algorithm for a better model performance in ALD applications. It is crucial to take into account the type of data, the nature of the problem to be solved, the desired level of accuracy, and the algorithm's performance on the particular dataset being used when choosing the best algorithm for ALD modeling. It is possible to select an algorithm that can deliver precise predictions and ideal performance for the particular ALD problem at hand by taking these factors into account. Some of these important steps are explained as follows.

10.5.1 PROBLEM DEFINITION

Problem definition involves identifying and defining the precise task or problem that the machine learning model will attempt to solve. Creating a successful machine learning model depends on a definitive problem identification as the first step in optimal model selection for establishing the foundation for the subsequent steps in the development process [45]. Understanding the problem's specifics, the data at hand, and the ultimate goal are all part of the problem definition stage [46]. In the context of ALD modeling, predicting the growth rate, thickness, or composition of the deposited material as a function of different process parameters like precursor gas flow rate, temperature, pressure, and exposure time may be the goal. Figuring out the exact process parameters that will be applied to produce the data needed to train and validate the model's accuracy is a crucial step. The anticipated outcome of the model must then be specified. This could be a continuous or categorical output when a regression or classification model is used, respectively. In an ALD context, possible model outputs could be the rate of deposition or the kind of material deposited [7,33]. This type of output desired determines the type of algorithm selected. Another important consideration while defining the problem is the data availability and their quality. This could include data from experiments or data produced by simulations. The effectiveness of the predictive algorithm will be significantly influenced by the data's quality.

10.5.2 DATA COLLECTION AND PRE-PROCESSING

Machine learning techniques aim to identify and mathematically represent the patterns found in data. As a result, both the quantity and quality of the data used are essential for developing machine learning algorithms that are both effective and practical [47]. Data collection belongs to the difficult lifecycle of machine learning data processing phases. The robustness, accuracy, and performance of the model is contingent on the quality of the data. Another crucial and significant factor affecting the generalization capabilities of a supervised machine learning algorithm is data pre-processing. Most of the times, real-life data not excluding thin film deposition are unreliable and devoid of particular trends or patterns. They are also probably full of mistakes, such that they might require being pre-processed into an acceptable format for the algorithm [48,49]. Data pre-processing involves data cleaning, integration, and transformation. Data cleaning entails finding and fixing errors, discrepancies, and incomplete data which might entail applying methods like imputation, interpolation, or removing outliers. Data integration is the process of combining information from various sources to produce a single dataset that can be incorporated into the model while transforming the data into a format that the machine learning model can use is important.

10.5.3 Desired Level of Accuracy

The optimal machine learning approach to model ALD depends on precision. The model's forecast accuracy depends on the assignment. Accuracy can be achieved by considering the consequences of faulty forecasts. In ALD, inaccurate predictions can lead to film defects that affect device functioning. This circumstance requires great accuracy to reduce error risk. Data accessibility and computational power can affect accuracy. Sometimes data or computational resources limit accuracy. These conditions may require a less accurate, simpler model, more data, or more powerful computational resources.

10.5.4 Selection of Features

A crucial step in choosing the best machine learning model for ALD is feature selection. It is possible to increase the model's accuracy and performance while lowering its complexity and computational requirements by carefully choosing the most crucial features [50]. It aims to figure out which attributes or features are most significant and have the biggest effects on how the deposition process turns out [51]. The model becomes more precise, effective, and understandable by choosing only the most significant aspects. Feature selection techniques range from determining the characteristics that have the strongest correlation with the results of the deposition process to selecting the algorithm that repeatedly chooses subsets of features and assesses how well they affect the results of the deposition process before choosing the best subset. Feature selection becomes necessary as it purges redundant, irrelevant, and noisy features from the original feature space [51].

10.5.5 Data Scalability

Scalability is crucial when selecting a machine learning method for big datasets or real-time forecasting. "Scalability" is an algorithm's ability to handle more data or processing power without losing efficiency or performance [52]. The complexity of the machine learning algorithm increases with dataset size, which can result in longer training times and higher memory requirements [42]. By splitting up the processing workload among several nodes or employing algorithms made for distributed data, scalable algorithms are better able to handle large datasets. Scalability of an algorithm is a critical consideration especially when a real-time application is desired. In such applications, machine learning algorithm which can process data quickly and effectively is needed. Real-time predictions can be handled by scalable algorithms by utilizing speed-optimized algorithms like decision trees or linear regression.

10.6 INTEGRATION OF NUMERICAL SIMULATION AND MACHINE LEARNING MODELS IN ALD

The design and improvement of ALD processes must be guided by numerical simulations; however, machine learning integrations into these simulations is a promising strategy that might precipitate the discovery of novel materials and equipment.

In this section, we shall explore the contemporary developments in the integrations of machine learning and numerical simulations in ALD applications. The complementary approaches of the machine learning techniques and other numerical computational approaches in ALD can be utilized for exploring their combined strengths and benefits and overcoming their individual weaknesses in the modeling of significant properties of deposited film toward improved thin film deposition and emergence of novel materials. Machine learning can extract features, trends, patterns, and continuous data from the result of numerical simulation for developing an intelligence to predict models using several available algorithms to achieve beneficial outcomes for the deposited thin films.

10.6.1 Motivations for the Integration of Machine Learning and Numerical Simulations

Synthesis of thin films with specific properties for a range of applications in ALD process involves a complex inter-relationship between several nonlinear variables [18]. This complexity of the physical and chemical process involved in the deposition process has made the optimization of the process toward an improved performance of the deposited films a challenging task. As earlier examined, different simulations approaches are available for modeling the physical, electrical, morphological, and kinetic properties of the deposited films such as MD, density theory function, amongst others [19]. Although these simulations have offered insightful knowledge into the ALD process, they are frequently computationally intensive and demand a lot of computing power. Machine learning techniques, on the other hand, has offered remarkable solution to this challenge by improving the precision and effectiveness of numerical simulations in ALD without requiring the same level of computational resources and intensities. However, we have examined the limitations and challenges of both techniques in their standalone form. To maximize the potential of machine learning applications in ALD, we have identified some of their challenges such as interpretability of their outcome, minimal knowledge of the domain, overfitting and underfittings, and their non-flexibility with changing process variables, while some of the limitations with the numerical simulations are incorrect capture of the physical phenomena on which the model is built, restriction on simulation size, handling atomic-scale process, limited computing resources, amongst others. Despite the shortcomings of these approaches, they have been proven as viable and helpful modeling tools in the analysis and optimization of the ALD process.

To harness the benefits and strengths of these modeling and simulation techniques, and overcome their challenges, their integration would play a critical role. We can speed up the search for new materials, improve the deposition procedure, and lower the computational expense of simulations by combining machine learning with numerical simulations. We can extract useful attributes from the outcome of numerical models and accurately predict the attributes of thin films by using a machine-learning-enhanced computational simulation. Further to this, machine learning models can be used to identify the ideal deposition requirements and minimize the number of necessary experimental trials by training them on a sizable dataset of simulated or real-world data. This is achieved by optimizing the deposition process to identify the optimal set of input which produces the desired thin film properties. This concept of machine-learning-enhanced simulations has become a significant focus in ALD research with the potential to cause a paradigm shift in this space as motivated by the advances in computing technology.

10.6.2   A Quick Recap of Numerical Simulations in ALD

The cost of running experiments is high due to the increased cost of ALD instrumentations and equipment and other raw materials. They can also be dangerous because the gases used or released during chemical reactions can occasionally be highly flammable or harmful. Modeling and simulation of ALD processes with numerical simulation approaches such as density function theory (DFT), MD, Lattice Boltzmann simulations, Monte Carlo simulations, Knudsen number, and computational fluid dynamics (CFD), amongst others, are viable alternatives which are potent in reducing the cost of experimentation. These computational simulation techniques have been studied extensively in the previous chapters. We provide a quick recap here to enhance the understanding of its distinctions and integration with machine learning models in ALD applications. While the DFT simulates the interaction of molecules and atoms using the principles of quantum mechanics, MD uses Newton's laws of motion to model forces of attraction and the positions of molecules [3]. MD is also a computational method for calculating the transport characteristics and equilibrium of a classical many-body system [15,53]. In the condensed and liquid states, MD focuses on the dynamics of atoms, molecules, and clusters. ALD reactor gas flow and mixing are examined using CFD simulations. The concentration and distribution of precursor molecules inside the reactor can be predicted by the CFD simulations, and this can have an impact on the rate of deposition and the quality of the film [54]. Simulations using the finite element method (FEM) are used to examine the temperature distribution and heat transfer within the ALD reactor. These simulations can forecast the reactor's temperature profile and temperature gradients, which can have an impact on reaction rates and picture quality [55]. Table 10.1 summarizes some of these computational simulations with the film's properties which they attempt to model.

10.6.3   Potential Integration Techniques

The optimization of the ALD process can benefit significantly from the integration of machine learning and simulation data. It is possible to speed up the optimization process, lower the number of experimental trials, and lower the overall cost of the process by using machine learning models trained on simulation data. In addition, the machine learning models can offer perceptions into the physical and chemical mechanisms that control the ALD process, which can direct future study and development. The machine learning model can be utilized to draw knowledge from simulations and produce precise predictions of film characteristics. Figure 10.1 illustrates some of these integration approaches and the flow diagram for machine learning-based simulation. These approaches are discussed as follows.

TABLE 10.1

Summary of the Computation Simulation Methods in ALD

S/N	Simulation Technique	Summary	Target Film and/or Deposition Properties
1	Density function theory	A computational quantum mechanical modeling technique used in chemistry, physics, and materials science based on atomic interactions	Electronic structure and mechanical properties, energies and forces, reaction mechanism, deposition characteristics, overall film growth
2	Molecular dynamic simulation	Analyzing the physical movements of atoms and molecules based on particles' interactions	Thermodynamic, mechanical properties, animating process mechanism
3	Monte Carlo simulation	Stochastic computational method for estimating the probability of a process under different conditions and producing numerical results	Morphology, growth rate, kinetics of reactions, material processing
4	Lattice Boltzmann method	Mesoscopic modeling method used in Computational Fluid Dynamics (CFD) for complex fluid systems	Simulate flow of atomic layer deposition (ALD) gases, predicting optimal condition and ideal reactor design for ALD reactions
5	Computational fluid dynamics	CFD simulations are deployed in ALD processes to model and predict the reactant's behavior and the fluid flow within the ALD reactor	Pressure and temperature distribution, flow velocity, film thickness and uniformity

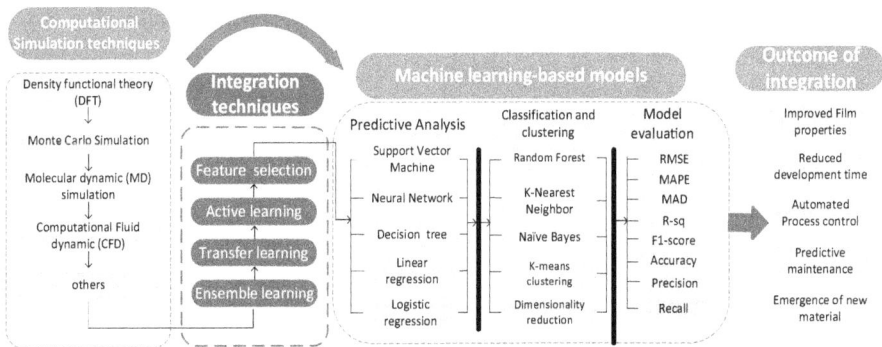

FIGURE 10.1 Integration of machine learning and computational simulation in atomic layer deposition (ALD) research.

10.6.3.1   Feature Extraction from Computational Simulations

A very important technique that can be deployed in machine-learning-enhanced computational simulation in ALD is the extraction of features from the outcome of the computational simulations such as DFT, MD, CFD, and Monte Carlo simulations. This method allows machine learning models to extract features which represent the physical and chemical scenario in the deposition process from simulation data in a technique called "feature engineering" for an accurate generalization of the film properties. Both supervised and unsupervised machine learning algorithms can be carefully selected for this task. This approach is based on the type of simulation from which features are being extracted from. In the context of dynamic simulations, attributes such as bond lengths, angles, and torsion angles and other mechanical properties which can reveal details about the process stages and the emergence of surface species can be drawn from the dynamics of the atoms and molecules engaged in the deposition process while features like electronic densities, energy, and forces, which speak about the reaction mechanism and the overall film growth can be extracted from DFT simulation and utilized by machine learning algorithms for optimal thin film deposition process. Details regarding the kinetics of the deposition process such as the rate of reactions and transition probability can be extracted from the Monte Carlo simulation process. Sequel to the feature extracted from the simulations, machine learning algorithms utilize these features as input and deploy a regression, classification, or clustering algorithm as the case may be to predict the film properties. To establish the performance and reliability of the machine learning model trained on simulation results, it must be validated against experimental data.

10.6.3.2   Feature Selection and Dimensionality Reduction from Simulated Data

The features and attributes of simulated datasets in ALD could be numerous, thus making it difficult to comprehend that such datasets' observations are located in high-dimensional space [56]. In order to select important features and minimize the computational complexity of the numerically simulated data in ALD, feature selection and dimensionality reduction are two machine learning techniques that can play a key role in this task. By doing so, we may be able to increase the precision of the simulation predictions, shorten the amount of time and resources needed to run the simulation analysis, and make it possible to explore the simulation parameter space more effectively. Dimensionality reduction is the technique of decreasing the number of features in data while retaining the same or more variance as is feasible in the original data [57]. These machine learning techniques could be particularly beneficial in computational simulations of ALD to locate the key attributes such as atomic species, surface locations, and other pertinent variables that influence the deposition process. As revealed in an earlier chapter, principal component analysis (PCA) is a prominent algorithm which can be used to achieve this task [58]. Dimensionality reduction helps to make the ALD simulated dataset less computationally complex and increases the effectiveness of the simulation analysis while preserving the relevant ALD details.

10.6.3.3 Active and Transfer Learning

While active learning is the process of actively prioritizing the high-impact data points that must be labeled in order to facilitate faster training and reduce computational time and intensity, transfer learning is aimed at applying the knowledge from an existing model that has already been trained to a new model [59,60]. This minimizes the quantity of data needed to train the new model and enhancing its accuracy and performance. Practical implication and benefits of active learning is that it reasonably minimizes the computational expenses associated with building machine learning models in the context of ALD. This is because fewer data points are needed to achieve a given level of accuracy because the most useful data points are chosen iteratively. This is especially helpful when working with simulation data, which can be expensive to generate by numerical computational simulations. Transfer learning can be applied to ALD to help machine learning models predict the characteristics of deposited films more accurately by transferring insights and information from MD, DFT, or other simulations. This strategy is especially helpful when obtaining training data for machine learning model is difficult or expensive.

10.6.3.4 Ensemble Learning

Ensemble learning is a machine learning approach which aims to improve predictive accuracy by integrating predictions from various models [61]. To build a more accurate prediction of the thin film characteristics in the deposition process, ensemble modeling is a technique which can integrate the outcomes of different machine learning algorithms which are separately trained on the results of various computational simulation methods or an experimental dataset. When the different models each do well on different aspects of the problem, or when the data used to train each model is different, this approach is especially helpful. The ability of ensemble learning to substantially enhance prediction performance over the use of a single model is one benefit of ensemble modeling. We can alleviate the impact of single model uncertainty and boost the prediction's overall robustness by incorporating the predictions of various models. The uncertainty associated with the prediction can also be estimated using ensemble modeling, which is crucial in ALD because the deposition conditions have a significant impact on the properties of deposited films. A weighted average or a more complex aggregation method can be used to combine the predictions of the individual models after they have been trained.

10.7 OUTCOME OF MACHINE LEARNING-ENHANCED COMPUTATIONAL SIMULATIONS IN ALD

We shall examine the fascinating results of combining computational simulations and machine learning in ALD and consider how this convergence might affect the field of thin film technology and materials science. These techniques have explored the benefits and strength of the standalone and individual approaches or a hybridized application to overcome their shortfalls and improve the properties of deposited thin film and the efficiency of the deposition process. While most of these techniques have potential benefits in ALD process, its application is still in the infancy stage.

However, some of the anticipated outcomes of the integration of machine learning and computational simulations approach are discussed as follows. Most of these results are intertwined.

10.7.1   IMPROVED DEPOSITED FILM PROPERTIES

The machine-learning-integrated computational simulation applications in the deposition process produces improved film properties compared to the single-method modeled deposition process. This feat is achieved by forecasting the ideal process conditions for depositing films with particular properties, using machine learning training fed with a large experimental and numerically simulated dataset. This enables the deposition processes to be optimized to produce expected film properties like thickness, uniformity, density, and crystallinity. Further to this, the incorporation of computational simulations and machine learning into ALD has made it possible to create new materials with enhanced properties. This is a paradigm in the research space as promising candidates for synthesis by using machine-learning-integrated computational simulations for the prediction of the properties of hypothetical material can be discovered toward optimizing the deposition of these materials. New materials with outstanding qualities like high thermal stability, high mechanical strength, and low electrical resistivity have resulted from this.

10.7.2   MINIMIZED DEVELOPMENT TIME

One of the important results of this integration has been a reduction in development time for ALD process which is known to be a complex interconnection of diverse variables. Extensive experimental investigation was previously required to optimize ALD processes and create new materials. However, this process has significantly improved in efficiency with the incorporation of machine learning and computational simulations. While computational simulations can be deployed to comprehend the mechanism of the growth of deposition film during the deposition process, machine learning algorithm determines the ideal process variables for depositing films with a particular set of properties. Further to this, the incorporation of machine learning and computational simulations in thin film deposition can also cut down on wasteful materials and experimentation expenses. We can minimize the number of requisite experimental procedures and optimize the process conditions by using simulations and predictive modeling, which leads to less waste production and a more economical process. This is also an extra benefit to the single method computational simulations

10.7.3   AUTOMATED AND IMPROVED DEPOSITION PROCESS CONTROL

The ability to optimize the deposition process by learning from sizable datasets produced by historical deposition experimental data is the key strength of utilizing machine learning algorithms in ALD. These datasets can then be used to train machine learning algorithms to find correlations and patterns between the process parameters and the final film properties. The evolution of a closed-loop control system that can modify the process parameters in real-time based on information from sensors measuring the film properties during deposition is another strategy that makes use of machine-learning-enhanced simulation approach. To achieve this, a control algorithm must be created that can modify the process variables in a way that reduces departures from the desired film properties. This closed-loop control system can lessen the need for manual adjustments by the operator while improving the uniformity and thickness control of deposited films. The application of machine-learning-enhanced simulation approach in ALD builds a closed-loop control system that can modify the process parameters in real-time using the input information of the film properties during deposition. This necessitates the creation of a control algorithm capable of modifying the process variables in a way that minimizes departures from the desired film properties. By lowering the need for manual tweaking by the operator, this closed-loop control system can increase the consistency and particle size control of deposited films. Machine learning can be used to anticipate when maintenance or calibration is required, minimizing risk, waste, and downtime, while boosting the deposition system's overall throughput. In addition, real-time process fault and deviation detection made possible by machine learning-enhanced simulations can eliminate the need for post-process evaluation and quality control assessments.

10.7.4 Predictive Maintenance

By utilizing the information gathered from deposition monitoring devices in thin film reactors, predictive maintenance is forecast when maintenance or service is required for a system. Machine learning-enhanced computational simulation in the context of ALD could create predictive maintenance strategies that can raise the efficiency and dependability of ALD systems. Monitoring device installed in the film deposition reactor can control and regulate various film parameters for implementing predictive maintenance in ALD. The devices collect data that can be used to train machine learning algorithms to find patterns and anomalies in the data that inform when upkeep or repair is required. Subsequently, the machine learning algorithm can be trained to predict the anticipated outcome of the real-time process for each condition by using computational simulation to model the behavior of the ALD system under various operating conditions. As a result, the algorithm is able to recognize unusual sensor readings that may be a sign of system or component failures. The simulation can also be used to pinpoint the fault's primary cause and offer suggestions for replacement or repair. System interruptions and failure can be reduced by foreseeing when maintenance or repairs are necessary, which boosts output and lowers costs.

10.8 CONCLUSION

This chapter has established the need and the motivations for the adoption of machine learning methods in ALD processes as a complementary technique to computational simulation approaches. Machine learning and computational simulations approaches have been explicitly applied in thin film deposition with viable outcomes. However, the combined benefits of the nexus of these two approaches have not been studied.

This chapter demonstrated exciting possibilities of these realms in thin films and ALD. Machine learning algorithms can discover intricate connections between deposition parameters and the properties of the deposited films by utilizing the massive volumes of data collected from ALD trials, leading to improved and more effective ALD processes. While machine learning models have drawbacks in ALD applications as examined in this chapter, its integration into the computational approaches can offer combined benefit for accurate and dependable forecasts. Moreover, application of machine learning in standalone form or integrated with other techniques in ALD research has the potential to revolutionize the entire spectrum of thin film materials towards a sustainable future with a higher level of desired success.

REFERENCES

[1] A. Maalouf, T. Okoroafor, S. Gahr, K. Ernits, D. Meissner, and S. Resalati, "Environmental performance of Kesterite monograin module production in comparison to thin-film technology," *Solar Energy Materials and Solar Cells*, vol. 251, p. 112161, 2023, doi: 10.1016/j.solmat.2022.112161.

[2] N. M. Badawy, H. S. El Samaty, and A. A. E. Waseef, "Relevance of monocrystalline and thin-film technologies in implementing efficient grid-connected photovoltaic systems in historic buildings in Port Fouad city, Egypt," *Alexandria Engineering Journal*, vol. 61, no. 12, pp. 12229–12246, 2022, doi: 10.1016/j.aej.2022.06.007.

[3] D. Sibanda, S. T. Oyinbo, and T. C. Jen, "A review of atomic layer deposition modelling and simulation methodologies: Density functional theory and molecular dynamics," *Nanotechnology Reviews*, vol. 11, no. 1, pp. 1332–1363, 2022, doi: 10.1515/ntrev-2022-0084.

[4] E. J. Y. Koh *et al.*, "An Automated Machine learning (AutoML) approach to regression models in minerals processing with case studies of developing industrial comminution and flotation models," *Minerals Engineering*, vol. 189, p. 107886, 2022, doi: 10.1016/j.mineng.2022.107886.

[5] M. Schmitt, "Automated machine learning: AI-driven decision making in business analytics," *Intelligent Systems with Applications*, vol. 18, p. 200188, 2023, doi: 10.1016/j.iswa.2023.200188.

[6] O. Adeleke, S. Akinlabi, T. C. Jen, P. A. Adedeji, and I. Dunmade, "Evolutionary-based neuro-fuzzy modelling of combustion enthalpy of municipal solid waste," *Neural Computing and Applications*, vol. 2, 2022, doi: 10.1007/s00521-021-06870-2.

[7] J. Jiang *et al.*, "The applications of Machine learning (ML) in designing dry powder for inhalation by using thin-film-freezing technology," *International Journal of Pharmaceutics*, vol. 626, p. 122179, 2022, doi: 10.1016/j.ijpharm.2022.122179.

[8] H.-Y. Choi, J. D. Jeon, S. E. Kim, S. Y. Jang, J. Y. Sung, and S. W. Lee, "Strained BaTiO3 thin films via in-situ crystallization using atomic layer deposition on SrTiO3 substrate," *Materials Science in Semiconductor Processing*, vol. 160, p. 107442, 2023, doi: 10.1016/j.mssp.2023.107442.

[9] L. F. Rasteiro, M. A. Motin, L. H. Vieira, E. M. Assaf, and F. Zaera, "Growth of ZrO2 films on mesoporous silica sieve via atomic layer deposition," *Thin Solid Films*, vol. 768, p. 139716, 2023, doi: 10.1016/j.tsf.2023.139716.

[10] L. Santinacci, "Atomic layer deposition: An efficient tool for corrosion protection," *Current Opinion in Colloid & Interface Science*, vol. 63, p. 101674, 2023, doi: 10.1016/j.cocis.2022.101674.

[11] L. Henderick *et al.*, "Plasma enhanced atomic layer deposition of a (nitrogen doped) Ti phosphate coating for improved energy storage in Li-ion batteries," *Journal of Power Sources*, vol. 497, p. 229866, 2021, doi: 10.1016/j.jpowsour.2021.229866.

[12] N. Navarrete *et al.*, "Improved thermal energy storage of nanoencapsulated phase change materials by atomic layer deposition," *Solar Energy Materials and Solar Cells*, vol. 206, p. 110322, 2020, doi: 10.1016/j.solmat.2019.110322.

[13] B. Ahmed, C. Xia, and H. N. Alshareef, "Electrode surface engineering by atomic layer deposition: A promising pathway toward better energy storage," *Nano Today*, vol. 11, no. 2, pp. 250–271, 2016, doi: 10.1016/j.nantod.2016.04.004.

[14] T. Muneshwar, M. Miao, E. R. Borujeny, and K. Cadien, "Chapter 11- Atomic layer deposition: Fundamentals, practice, and challenges," In *Handbook of Thin Film Deposition*, 4th edition, K. Seshan and D. Schepis, Eds. William Andrew Publishing, pp. 359–377, 2018. doi: 10.1016/B978-0-12-812311-9.00011-6.

[15] X. Hu, J. Schuster, S. E. Schulz, and T. Gessner, "Surface chemistry of copper metal and copper oxide atomic layer deposition from copper(ii) acetylacetonate: a combined first-principles and reactive molecular dynamics study," *Physical Chemistry Chemical Physics*, vol. 17, no. 40, pp. 26892–26902, 2015, doi: 10.1039/C5CP03707G.

[16] T. Justin Kunene, L. Kwanda Tartibu, K. Ukoba, and T.-C. Jen, "Review of atomic layer deposition process, application and modeling tools," *Materials Today: Proceedings*, vol. 62, pp. S95–S109, 2022, doi: 10.1016/j.matpr.2022.02.094.

[17] S. Yun, F. Ou, H. Wang, M. Tom, G. Orkoulas, and P. D. Christofides, "Atomistic-mesoscopic modeling of area-selective thermal atomic layer deposition," *Chemical Engineering Research and Design*, vol. 188, pp. 271–286, 2022, doi: 10.1016/j.cherd.2022.09.051.

[18] L. Angermann, *Numerical Simulations*, Rijeka: IntechOpen, 2010. doi: 10.5772/545.

[19] F. Bulnes and J. P. Hessling, *Recent Advances in Numerical Simulations*, Rijeka: IntechOpen, 2021. doi: 10.5772/intechopen.91589.

[20] Y. Abu-Zidan, P. Mendis, and T. Gunawardena, "Optimising the computational domain size in CFD simulations of tall buildings," *Heliyon*, vol. 7, no. 4, p. e06723, 2021, doi: 10.1016/j.heliyon.2021.e06723.

[21] M. Kisielewski, A. Maziewski, T. Polyakova, and V. Zablotskii, "Models and simulations for analysis of domain sizes in ultrathin magnetic films," *Journal of Magnetism and Magnetic Materials*, vol. 272–276, pp. E825–E826, 2004, doi: 10.1016/j.jmmm.2003.12.162.

[22] G. Kunstatter and J. Ziprick, "Quantum mechanics does not allow violation of the pigeon counting principle," *Physics Letters A*, vol. 415, p. 127642, 2021, doi: 10.1016/j.physleta.2021.127642.

[23] M. Becker and M. Sierka, "Atomistic simulations of plasma-enhanced atomic layer deposition," *Materials*, vol. 12, no. 16, 2019, doi: 10.3390/ma12162605.

[24] Y. Kosaku, Y. Tsunazawa, and C. Tokoro, "A coarse grain model with parameter scaling of adhesion forces from liquid bridge forces and JKR theory in the discrete element method," *Chemical Engineering Science*, vol. 268, p. 118428, 2023, doi: 10.1016/j.ces.2022.118428.

[25] P. O. Oviroh, R. Akbarzadeh, D. Pan, R. A. M. Coetzee, and T. C. Jen, "New development of atomic layer deposition: processes, methods and applications," *Science and Technology of Advanced Materials*, vol. 20, no. 1, pp. 465–496, 2019. doi: 10.1080/14686996.2019.1599694.

[26] R. W. Johnson, A. Hultqvist, and S. F. Bent, "A brief review of atomic layer deposition: from fundamentals to applications," *Materials Today*, vol. 17, no. 5, pp. 236–246, 2014, doi: 10.1016/j.mattod.2014.04.026.

[27] S. Chaoudhary *et al.*, "X-ray photoelectron spectroscopy and spectroscopic ellipsometry analysis of the p-NiO/n-Si heterostructure system grown by pulsed laser deposition," *Thin Solid Films*, vol. 743, p. 139077, 2022, doi: 10.1016/j.tsf.2021.139077.

[28] M. Barclay, S. B. Hill, and D. H. Fairbrother, "Use of X-ray photoelectron spectroscopy and spectroscopic ellipsometry to characterize carbonaceous films modified by electrons and hydrogen atoms," *Applied Surface Science*, vol. 479, pp. 557–568, 2019, doi: 10.1016/j.apsusc.2019.02.122.

[29] J. B. Cunningham, "Assumptions underlying the use of different types of simulations," *Simulation & Games*, vol. 15, no. 2, pp. 213–234, 1984, doi: 10.1177/0037550084152004.

[30] Jason Brownlee, "How machine learning algorithms work (they learn a mapping of input to output)," 2016, https://machinelearningmastery.com/how-machine-learning-algorithms-work/.

[31] D. Dandolo, C. Masiero, M. Carletti, D. Dalle Pezze, and G. A. Susto, "AcME-Accelerated model-agnostic explanations: Fast whitening of the machine-learning black box," *Expert Systems with Applications*, vol. 214, 2023, doi: 10.1016/j.eswa.2022.119115.

[32] C. Wu, Q. Peng, Y. Xia, Y. Jin, and Z. Hu, "Towards cost-effective and robust AI microservice deployment in edge computing environments," *Future Generation Computer Systems*, vol. 141, pp. 129–142, 2023, doi: 10.1016/j.future.2022.10.015.

[33] Y. Ding, Y. Zhang, Y. M. Ren, G. Orkoulas, and P. D. Christofides, "Machine learning-based modeling and operation for ALD of SiO2 thin-films using data from a multiscale CFD simulation," *Chemical Engineering Research and Design*, vol. 151, pp. 131–145, 2019, doi: 10.1016/j.cherd.2019.09.005.

[34] M. Mohammed, M. B. Khan, and E. B. M. Bashie, *Machine Learning: Algorithms and Applications*. CRC Press, 2016. doi: 10.1201/9781315371658.

[35] I. H. Sarker, "Machine learning: Algorithms, real-world applications and research directions," *SN Computer Science*, vol. 2, no. 3, 2021. doi: 10.1007/s42979-021-00592-x.

[36] O. Adeleke, S. Akinlabi, T. C. Jen, and I. Dunmade, "A machine learning approach for investigating the impact of seasonal variation on physical composition of municipal solid waste," *Journal of Reliable Intelligent Environments*, 2022, doi: 10.1007/s40860-021-00168-9.

[37] S. Srivastava, R. N. Shah, C. Teodoriu, and A. Sharma, "Impact of data quality on supervised machine learning: Case study on drilling vibrations," *Journal of Petroleum Science and Engineering*, vol. 219, 2022, doi: 10.1016/j.petrol.2022.111058.

[38] A. B. Haque, A. K. M. N. Islam, and P. Mikalef, "Explainable Artificial Intelligence (XAI) from a user perspective: A synthesis of prior literature and problematizing avenues for future research," *Technological Forecasting and Social Change*, vol. 186, 2023, doi: 10.1016/j.techfore.2022.122120.

[39] M. Langer *et al.*, "What do we want from Explainable Artificial Intelligence (XAI)? - A stakeholder perspective on XAI and a conceptual model guiding interdisciplinary XAI research," *Artificial Intelligence*, vol. 296, 2021, doi: 10.1016/j.artint.2021.103473.

[40] A. Frini, A. Guitouni, and J. M. Martel, "A general decomposition approach for multi-criteria decision trees," *European Journal of Operational Research*, vol. 220, no. 2, pp. 452–460, 2012, doi: 10.1016/j.ejor.2012.01.032.

[41] A. Isazadeh, F. Mahan, and W. Pedrycz, "MFlexDT: Multi flexible fuzzy decision tree for data stream classification," *Soft computing*, vol. 20, no. 9, pp. 3719–3733, 2016, doi: 10.1007/s00500-015-1733-2.

[42] Matthew Stewart, "The limitations of machine learning," 2019, https://towardsdatascience.com/the-limitations-of-machine-learning-a00e0c3040c6.

[43] J. L. Andrews, "Addressing overfitting and underfitting in Gaussian model-based clustering," *Computational Statistics & Data Analysis*, vol. 127, pp. 160–171, 2018, doi: 10.1016/j.csda.2018.05.015.

[44] O. Adeleke, S. A. Akinlabi, T. Jen, and I. Dunmade, "Application of artificial neural networks for predicting the physical composition of municipal solid waste: An assessment of the impact of seasonal variation," *Waste Management & Research*, vol. 39, no. 8, pp. 1058–1068, 2021, doi: 10.1177/0734242X21991642.

[45] C. C. Paul Fergus, *Applied Deep Learning: Tools, Techniques and Implementation*, 1st edition, France: Springers, 2022.

[46] Iryna Sydorenko, "How to choose the right machine learning algorithm: A pragmatic approach," 2021, https://labelyourdata.com/articles/how-to-choose-a-machine-learning-algorithm.

[47] P. Rattan, D. D. Penrice, and D. A. Simonetto, "Artificial intelligence and machine learning: What you always wanted to know but were afraid to ask," *Gastro Hep Advances*, vol. 1, no. 1, pp. 70–78, 2022, doi: 10.1016/j.gastha.2021.11.001.

[48] K. Maharana, S. Mondal, and B. Nemade, "A review: Data pre-processing and data augmentation techniques," *Global Transitions Proceedings*, vol. 3, no. 1, pp. 91–99, 2022, doi: 10.1016/j.gltp.2022.04.020.

[49] A. Kadhim, "An evaluation of preprocessing techniques for text classification pattern recognition view project improvement text classification using log (TF-IDF) with K-NN algorithm view project." [Onlin]. Available: https://sites.google.com/site/ijcsis/.

[50] N. Wang *et al.*, "Search-based cost-sensitive hypergraph learning for anomaly detection," *Information Sciences*, vol. 617, pp. 451–463, 2022, doi: 10.1016/j.ins.2022.07.029.

[51] T. Wu, Y. Hao, B. Yang, and L. Peng, "ECM-EFS: An ensemble feature selection based on enhanced co-association matrix," *Pattern Recognition*, p. 109449, 2023, doi: 10.1016/j.patcog.2023.109449.

[52] Christian Kästner, "Scaling data storage and data processing and machine learning in production systems," 2022, https://ckaestne.medium.com/scaling-ml-enabled-systems-b5c6b1527bc.

[53] J. Huawei, Z. Tao, L. Xiao, B. Shan, and R. Chen, "Optimization of inlets and outlets of an ALD chamber with radiant heating," In *2013 IEEE International Symposium on Assembly and Manufacturing (ISAM)*, pp. 170–175, 2013. doi: 10.1109/ISAM.2013.6643519.

[54] G. P. Gakis, H. Vergnes, E. Scheid, C. Vahlas, B. Caussat, and A. G. Boudouvis, "Computational fluid dynamics simulation of the ALD of alumina from TMA and H_2O in a commercial reactor," *Chemical Engineering Research and Design*, vol. 132, pp. 795–811, 2018, doi: 10.1016/j.cherd.2018.02.031.

[55] J. Huawei, Z. Tao, L. Xiao, B. Shan, and R. Chen, "Optimization of inlets and outlets of an ALD chamber with radiant heating," In *2013 IEEE International Symposium on Assembly and Manufacturing (ISAM)*, pp. 170–175, 2013. doi: 10.1109/ISAM.2013.6643519.

[56] Rukshan Pramoditha, "11 Dimensionality reduction techniques you should know in 2021," 2021, https://towardsdatascience.com/11-dimensionality-reduction-techniques-you-should-know-in-2021-dcb9500d388b.

[57] N. Trendafilov and M. Gallo, "PCA and other dimensionality-reduction techniques," In *International Encyclopedia of Education*, 4th edition, R. J. Tierney, F. Rizvi, and K. Ercikan, Eds. Oxford: Elsevier, pp. 590–599, 2023. doi: 10.1016/B978-0-12-818630-5.10014-4.

[58] F. Anowar, S. Sadaoui, and B. Selim, "Conceptual and empirical comparison of dimensionality reduction algorithms (pca, kpca, lda, mds, svd, lle, isomap, le, ica, t-sne)," *Computer Science Review*, vol. 40, 2021. doi: 10.1016/j.cosrev.2021.100378.

[59] F. Sun and Q. Zhang, "Robust transfer learning of high-dimensional generalized linear model," *Physica A: Statistical Mechanics and its Applications*, p. 128674, 2023, doi: 10.1016/j.physa.2023.128674.

[60] T. Zhao, Y. Zheng, and Z. Wu, "Improving computational efficiency of machine learning modeling of nonlinear processes using sensitivity analysis and active learning," *Digital Chemical Engineering*, vol. 3, p. 100027, 2022, doi: 10.1016/j.dche.2022.100027.

[61] O. Chakir *et al.*, "An empirical assessment of ensemble methods and traditional machine learning techniques for web-based attack detection in industry 5.0," *Journal of King Saud University - Computer and Information Sciences*, vol. 35, no. 3, pp. 103–119, 2023, doi: 10.1016/j.jksuci.2023.02.009.

11 Machine-Learning-Based Predictive Analysis in ALD

11.1 INTRODUCTION

Thin films that are uniform and conformal can be deposited using the very accurate and reproducible atomic layer deposition (ALD) technology with a high-level and sub-nanometer-level precision [1–4]. While the optimization of the deposition procedure and forecast of the characteristics of the deposited thin films remain difficult tasks, machine learning can provide some directions in this regard. By utilizing the power of data and algorithms in machine learning-based predictive analysis, we gain new insights into the intricate behavior of ALD and create novel materials and devices with unparalleled accuracy and speed [5–7]. Experts in this dynamic space are pushing the frontiers of what is conceivable, and the prospect for game-changing discoveries is limitless. In the field of ALD, predictive analysis is becoming more and more critical since it allows precise forecast of the properties of thin films and enhances the optimization of the deposition process [8]. The characteristics of the consequent deposited thin film can be influenced by a variety of factors during the deposition, including the precursor chemistry, temperature, pressure, and exposure duration, while the substrate's characteristics, namely, crystal shape, make-up, and structure can greatly influence the qualities of the final thin film [9]. Predictive analysis is important for ALD because it can facilitate the deposition process and cut down on the number of experimental trials needed to attain the desired qualities by anticipating the properties of thin films before they are deposited [5,10]. In addition to saving time and resources, this also makes it possible to find new materials that weren't previously available through conventional trial-and-error approaches.

11.2 DATA SOURCES FOR PREDICTIVE ANALYSIS IN ALD

Machine learning algorithms use data and statistical models to make accurate predictions. However, each machine learning algorithm's success depends on its data [11]. Poor gathering of data inhibits machine learning models. In other words, feeding a model with bad and unreliable data won't produce any reasonable results, regardless of how brilliant your model is, how talented a data scientist is, or how much effort has been put into a task. High-quality data are critical for machine learning algorithms to be successful in the context of ALD process. The first step in the data collection process is identifying the data's sources. This has therefore

DOI: 10.1201/9781003346234-14

necessitated the need to evaluate the data sources that can be employed in ALD machine learning-based applications. We will briefly study the different data types that can be used to enhance the ALD process, ranging from experimental and sensor data, process monitoring systems data, material characterization and computational simulations data. We categorize these sources into two broad groups, namely, ALD experimental data source and ALD simulation data source.

11.2.1 Data from ALD Experiment

An important data source for building a machine learning-based prediction model in ALD is experimental data. A machine learning system can discover patterns and correlations between input and output variables using these data. The model can predict the outcomes of new input variables after training. The chemical reaction that takes place to allow materials to be deposited on the substrate's surface layer by layer is described experimentally by the following steps [2]:

   i. All pollutants that can hinder the deposition process must be properly removed off the substrate surface. Often, this is accomplished by either plasma cleaning or solvents. The ALD reactor receives the cleaned substrate for amplification.
  ii. The first precursor is initially exposed in the reactor chamber. The chamber is filled with the initial precursor gas, and the substrate is given a set amount of time to breathe it in. In order to ensure that a layer is deposited, the time and temperature are carefully regulated.
 iii. Purge the chamber of any excess precursor gas and reaction by-product using an inert gas, such as nitrogen or argon.
  iv. The chamber is filled with the second precursor gas, which is then released onto the substrate for a predetermined amount of time. This reaction with the first precursor layer produces the required material.
   v. In order to get rid of any extra precursor gas and reaction by-products, the chamber is once more purged with an inert gas.
  vi. Steps i–v is repeated until the material reaches the desired thickness.

A sample experimental set up for ALD process is shown in Figure 11.1. Prominent tools needed for ALD experiments are ALD reactors incorporated with precise controls for gas flow, temperature, and pressure controllers. Mass flow, temperature, and gas flow controllers are further instruments. Figure 11.2 shows some other tools required for ALD experimental procedures.

To obtain the desired film properties in ALD studies, precise optimization of these process parameters is essential. More insights from the growth mechanisms, structure–property correlations, and functionality of thin films produced can be gained by deploying a machine learning-based predictive analysis using the data extracted from the ALD experiments [5–8,10]. Table 11.1 summarizes the independent and dependent variables obtained from the ALD experimental procedures. Some of the measurement techniques used in measuring the deposited film properties listed in

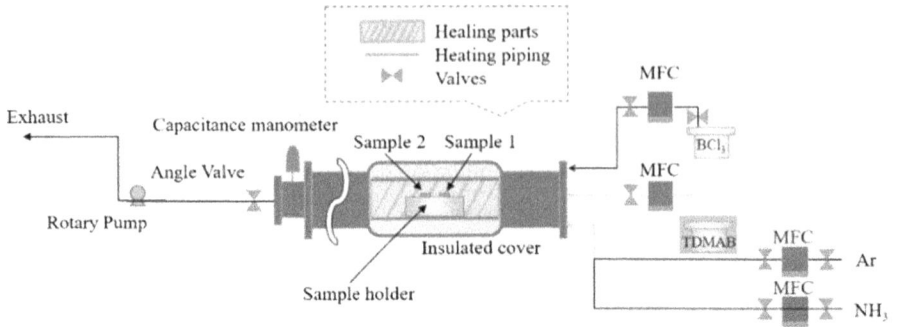

FIGURE 11.1   A sample schematic diagram of the atomic layer deposition (ALD) experimental set up [12].

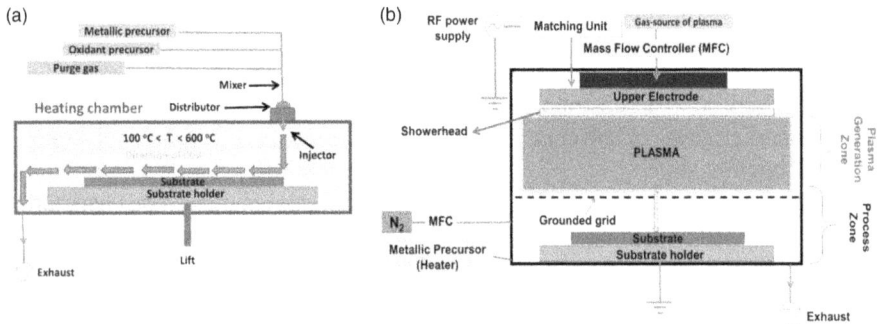

FIGURE 11.2   ALD experimental tools: (a) cross-flow thermal ALD and (b) capacitively linked plasma reactor used in remote plasma mode for plasma-enhanced atomic layer deposition (PEALD) operations [13].

Table 11.1 (process parameters and deposited thin film properties) are but not limited to the following:

   i. Ellipsometry
  ii. Transmission electron microscopy (TEM)
 iii. X-ray photoelectron spectroscopy (XPS)
  iv. Energy-dispersive X-ray spectroscopy (EDS)
   v. Scanning electron microscopy (SEM)
  vi. Atomic force microscopy (AFM).
 vii. Spectroscopy
viii. Profilometry
  ix. Mechanical testing

11.2.2   DATA FROM ALD SIMULATIONS

ALD experiments might be very expensive due to the high cost of instrumentations, materials, substrates, and other equipment required [2,4,14]. When experimental

TABLE 11.1
Typical Input (Independent) and Output (Dependent) Variable in Atomic Layer Deposition (ALD) Experiments

Independent Variable (Process Parameters)	Dependent Variable from Experiments (Deposited Film Properties)
Precursor flow rate	Deposition rate
Exposure time	Thickness of deposited film
Temperature	Composition of deposited film
Pressure	Structure of deposited film (e.g. grain size, texture, and crystallinity)
Volume and type of precursor gases	Morphology of deposited film (e.g. roughness and porosity)
Purging time	Properties of deposited film (e.g. electrical, optical, and mechanical)
Properties of substrate surface	

data are not available, feature extraction from the outcome of computational simulations is a viable source of data for machine-learning-based predictive analysis in ALD. Several studies in literature have demonstrated the feasibility of feature and data extraction from computational simulations for machine learning-based applications. The study by Kimaev et al. [15] trained an artificial neural network model for a nonlinear multiscale predictive control of thin film deposition using full multi-scale numerical model to generate 400,000 data points with a lattice size of 50 comprising substrate temperature and inlet precursor mole fraction as input variable while growth rate and roughness of the deposited film were extracted as the output variable. Ding et al. [16] deployed the data generated from a multiphase computational fluid dynamics (CFD) simulation of plasma-enhanced atomic layer deposition (PEALD) process to develop a recurrent neural network (RNN) and long- and short-term memory (LSTM) model for predictive analysis of transient gas phase profile and dynamic surface profile. A data-driven model based on artificial neuron network (ANN) for predicting the transient deposition rate and surface configuration of the deposition of SiO_2 thin film using data extract from DFT and CFD multiphase simulation was recently applied [17]. Sitapure and Kwon [18] aimed at developing a model predictive controller (MPC) for the chemical deposition of thin film by incorporating a neural network-based model which utilizes data generated from multiphase simulation approaches such as microscopic discrete-element method at the surface level. Bokinala et al. [19] integrate the DFT models and the machine learning models, to extract features from the DFT computational simulation models for machine-learning-based applications. Table 11.2 summarizes some of the important characteristics of the thin film and deposition process which can be made available by different simulation techniques.

11.3 ALD DATA CLEANING AND PRE-PROCESSING

ALD experiments and simulations generate massive amounts of data that must be pre-processed and cleaned before being used for machine-learning-based predictive

TABLE 11.2
Description of Target Variable of the Simulation Approaches [1–4,14,20–25]

Simulation Approaches	Target Film/Deposition Properties
Density function theory (DFT)	i. Atomic structure of the deposited film
	ii. Surface energy of the deposited film
	iii. Bonding characteristics of the deposited film
	iv. Electronic property of the deposited film
	v. Optical property of the deposited films
Molecular dynamics	i. Growth mechanism of the deposited film
	ii. Surface chemistry of the deposition process
	iii. Precursor molecule's behavior
Monte Carlo	i. Gas-phase reactions of the deposition process
	ii. Reaction rates of the deposition process
	iii. The number of collisions between the precursors and the substrate
	iv. Precursor's distribution
Kinetic Monte Carlo	i. Deposition rates
	ii. Thin film growth
	iii. Interface roughness of the deposited film
Finite element method	i. Substrate thermal behavior
	ii. Substrate's thermal stress
	iii. Substrate's temperature distribution and heat transfer
Computational fluid dynamics (CFD)	i. Thickness of the deposited film

modeling. Pre-processing converts raw data into an analysis-ready format, while cleaning removes errors, inconsistencies, and outliers. ALD data must be pre-processed and cleaned for machine-learning-based predictive models. Data quality substantially affects model predictions. Pre-processing and cleaning should be done carefully to ensure the dataset is suitable for machine learning predictive analysis.

11.3.1 FORMATTING

Collecting and properly formatting the data is the first stage in every data-driven endeavor. The information used in ALD experiments and simulations might originate from a variety of places, including experimental measurements, computer simulations, and published sources [26–29]. It is crucial to prepare the ALD data from different sources in a structured way in order to make sure that they are easily accessible for predictive modeling based on machine learning. The processing, cleansing, and analysis of data are made easier by a structured format. Using a spreadsheet or database to format the data is one popular method, with each row designating a distinct sample or deposition experiment. The columns should include pertinent data, such as deposition parameters (such as precursor type, deposition temperature, deposition duration, and number of cycles) and the qualities of the produced material (such as layer thickness, surface roughness, and refractive index).

11.3.2 Missing Data Management

Data analysis frequently encounters the issue of missing data, which can occur for a number of reasons, including measurement flaws and mistakes in entry [30,31]. If handled improperly, it might produce estimates that are skewed and ineffective. Consequently, addressing missing data is a crucial component of data analysis, particularly in scientific research, especially in ALD research. Imputation, which substitutes missing values with conceivable values according to the information available, is a popular method for addressing missing data [31,32]. Mean imputation is the simplest type of imputation, replacing missing values with the mean of the variable's observed values [32]. It is critical to evaluate the accuracy of the imputed data and the likelihood of bias before adopting any imputation technique. Comparing the outcomes of the imputed data to the outcomes of the complete dataset is one strategy. Another strategy is to do sensitivity analysis to evaluate how well the findings hold up to various hypotheses on the missing data.

11.3.3 Dealing with Outliers

ALD predictive analysis data pre-processing requires outlier detection. Outliers are data points that deviate from the rest. Measurement errors, data entry errors, and rare events can cause outliers [33,34]. Outliers can skew model parameter estimations and affect model accuracy, which can greatly affect statistical model results. Before fitting any statistical models to the data, outliers must be found and dealt with [33]. The link between two variables is visualized using scatter plots. This aids tracing out observations that stand out from the rest of the dataset's observations. In addition, the interquartile range (IQR) indicates data distribution. IQR is calculated by subtracting Q3 from Q1. Observations beyond the range (Q1 − 1.5IQR, Q3 + 1.5IQR) may be outliers [35]. There are numerous approaches for handling outliers once they have been discovered. To remove the outliers from the dataset is one option. This strategy, meanwhile, has the potential to cause information loss and has an impact on how the data are distributed. The mean or median of the dataset, for example, might be used to replace the values of the outliers. This is an additional option. This strategy can help to preserve the data's distribution while lessening the impact of outliers.

11.3.4 Feature Selection

The process of creating an accurate predictive analysis in ALD applications begins with feature selection. In order to predict the desired result variable, the most significant features from the existing evidence must be identified and chosen. Feature selection reduces overfitting and improves model generalizability by limiting model variables [36]. Data type and research goal determine the approach. There are several ways to execute feature selection in ALD data analysis vis-a-vis principal component analysis (PCA) [37], and recursive feature elimination (RFE) [38]. In ALD data analysis, feature selection is a crucial phase that reduces the dimensionality of the data, reduces overfitting, and increases model accuracy.

11.3.5 NORMALIZATION

Normalization is another pre-processing step which is vital to the predictive analysis of ALD data using machine learning. It entails rescaling the data to provide each characteristic a common scale and an equal range of values. When developing the model, normalization makes it possible to guarantee that every component is given the same weight and that no one feature reigns supreme over the others [39]. With the min-max scaling approach, the data are scaled to a predetermined range, usually between 0 and 1. It is done by dividing the range after removing the feature's minimum value. This is achieved using Equation 11.1. With the Z-zero normalization technique, the data are scaled to have a mean of 0 and a standard deviation of 1. It is done by using Equation 11.2. The particular qualities of the dataset and the algorithm for machine learning being employed determine the normalization approach to be utilized. In general, normalizing the data before training a model is a recommended approach. By doing this, the model is prevented from favoring particular traits and is better able to capture the underlying patterns in the data:

$$y_{norm} = \frac{x - x_{min}}{x_{max} - x_{min}} \tag{11.1}$$

where y_{norm} = the normalized data, x = the mean of the variable, x_{min} = minimum variable, and x_{max} = maximum variable.

$$z = \frac{(x - \mu)}{\sigma} \tag{11.2}$$

where z = z-score, x is the score, μ is the mean, while σ is the standard deviation.

11.4   PREDICTIVE MODEL DEVELOPMENT AND APPLICATIONS IN ALD

Having examined the different sources of data used for machine learning predictive analysis in ALD, which could be experimental or simulation data, we shall now examine some machine learning techniques and algorithms which have been prominently utilized for predictive analysis for process and thin film control and monitoring. After you have your processed data for ALD, the real challenge of how to create a model that can precisely represent the intricate nonlinear interactions between many variables in ALD arises. Model development, a critical stage in the machine learning pipeline that include choosing the best algorithm, refining its parameters, and verifying the model's performance, is where this comes into play. We will delve into the complexities of model construction and applications of machine learning-based predictive analysis in ALD and study a number of the most potential approaches and techniques utilized by researchers in this space.

Regression analysis seeks to establish a connection between the input and output variables, where the target variable is a continuous value. Regression analysis is an effective statistical method that enables us to model the relationship between input factors (ALD process parameters) and output variables (thin film properties), allowing us to forecast how changes in the former will impact the latter. Using supervised learning techniques on input–output data from ALD experiments or simulations, precise predictive models that may be utilized to enhance ALD procedures and modify thin film properties to meet particular demands have been created. Several previous attempts to optimize the parameters of the deposition process have developed conventional statistical approaches. Response surface methodology (RSM) was used by Ebrahimi et al. [40] to determine the ideal deposition temperature and hydrogen flow to create diamond-like carbon coatings (DLC) with a minimal friction coefficient and great wear resistance. Using Taguchi statistical approaches, Segu et al. [41] presented research on the wear and friction characteristics of MoS_2 coatings on laser-textured surfaces.

While these classical statistical approaches are quite effective at simulating the deposition process, they have certain limitations when it comes to handling a large number of process parameters and creating a reasonable correlation between input and output data [42]. Such challenges were noted primarily as a result of the nonlinear and complex deposition process and parameterization [10]. Sequel to this, a paradigm shift from classical model to intelligent models for handling complex nonlinear ALD data have been noted in contemporary times. Prominent machine learning algorithms which have been used for regression analysis in ALD processes are ANN [15–17,29], support vector machine (SVM) [43], extreme gradient boosting (XGBoost) [44], random forest (RF) [43,45], decision tree (DT) [45], and Gaussian process algorithm (GPR) [46], amongst others.

The ANN has found a wide application in the predictive modeling of ALD process. It is the most preferable machine learning technology in the deposition process owing to its ability to learn from the historic data and use it to more correctly anticipate the response variable [10]. The ability of ANNs to accurately simulate complex correlations between process variables and film properties has made them prominent in ALD applications. In ALD applications, ANNs can achieve high prediction accuracy, which is essential for the creation of trustworthy process models. In surface coating and ALD researches, ANN has demonstrated excellent efficiency in improving the parameters and process modeling. We shall therefore attempt to examine the ANN algorithms, their framework, hyperparameters, and model building approaches for ALD applications. Presented in Figure 11.3 is the methodological framework for predictive analysis in ALD from the data collection to model development and the final output of thin film.

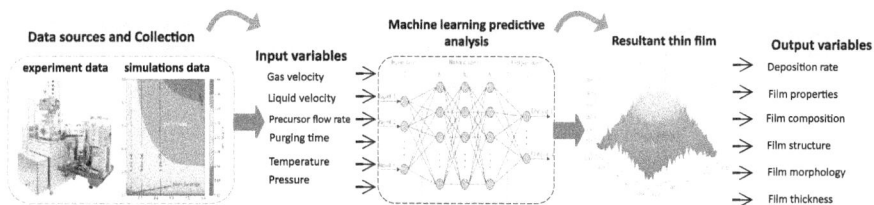

FIGURE 11.3 Methodological framework of machine learning predictive analysis in atomic layer deposition (ALD) variables.

11.4.1 Artificial Neural Network

ANNs are mathematical models of biological brain architecture made up of several processing units (or neurons) connected by weighted synaptic connections, with a self-learning, self-adaptive, and self-organizing approximation function [47]. In order to obtain a precise function mapping between inputs and outputs and to predict future values, ANNs learn from past data. Without understanding the physical connections, ANNs may recognize complex nonlinear interactions between input and output time series data [48]. The initial models had constrained capabilities and needed a lot of training data. The 1980s saw a breakthrough in neural network research as new structures and algorithms were created that made neural network training more effective [49].

The backpropagation neural network, a feed-forward neural network (FFNN) structure that employs an error back propagation algorithm, is now the most prominent neural network design [50]. The two major phases of the backpropagation method are information forward pass and error backpropagation. The output and error are determined after the data are first delivered to the input layer and then forwarded via the network in order to get the hidden and output layer in that sequence. Second, the initial connection weights and thresholds are dynamically changed as the average error is back transmitted via the network to the hidden layer and the input layer in turn. Until the requisite accuracy based on a minimum error is attained, this iterative procedure is repeated [47].

As shown in Figure 11.4, a typical backpropagation neural network comprises three layers, namely, input, hidden, and output layers. Each layer communicates to the layer preceding it, and information transmits through the network in a forward direction via links by the neurons within each layer; however, there isn't a direct link made by the neurons within the same layer [51]. The input data must be changed by

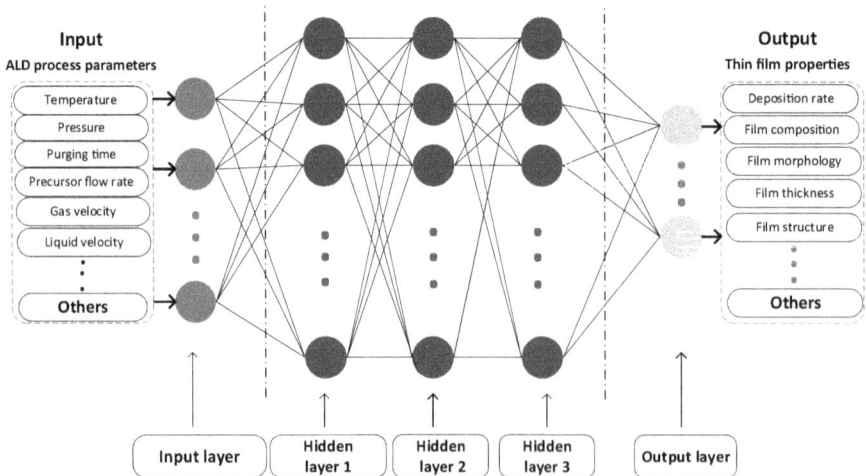

FIGURE 11.4 Artificial neuron network (ANN) architecture for predictive analysis in atomic layer deposition (ALD) process.

the hidden layers into a format that is better suited for the output layer. The network's ultimate output is produced by the output layer. While serving as the network's input interface, an input neuron has no predecessor, likewise output neuron has no successor and as a result acts as the network's output interface. Figure 11.4 illustrates the architectural framework of neural network applied in ALD predictive analysis, with samples variables of the ALD operations at the input layer and samples of output variables vis-à-vis thin film properties at the output layer.

ANN working principle is based on a collection of weights and biases that control how the neurons behave. The biases modify the activation of the neurons, while the weights regulate the strength of the link between neurons. The weights between neurons are adjusted during the training process using the backpropagation method by computing the error between the network's output and the desired output. In order to reduce error and boost network performance, this process is repeated over a number of iterations. For each training set, the biases and weights are updated using Equation 11.3.

$$\theta_{t+1} = \theta_t - \eta.\nabla_{\theta_t} E\left(\theta_t; x^{(i)}; y^{(i)}\right) \tag{11.3}$$

where θ represents the weights and biases, $x^{(i)}$ is the input of the training sample, $y^{(i)}$ is the target label, η is the learning rate, and E is the loss function.

Training is a significant step in building an optimal neural network for several applications. A learning algorithm discovers a decision function that modifies the network's weights [52]. Trying to predict which training algorithm will yield the greatest outcomes is challenging. There are various available training algorithms as presented in Table 11.3. These algorithms are categorized into five groups, namely, Quasi-Newton, Resilient Back propagation, gradient descent, conjugate gradient, and Levenberg–Marquardt. The following training algorithms are commonly used.

Figure 11.5 illustrates the basic working principle of all the training algorithms in neural networks. These groups of training algorithms are briefly described as follows:

i. **Gradient Descent:** Gradient descent operates by figuring out the gradient of the error or loss function with regard to the network's weights and biases. The gradient points in the direction where the function advances the most and represents the steepest ascent of the function. Yet, if we wish to reduce the error or loss function, we must move against the gradient. We then go in the direction indicated by the gradient's negative [52–55]. It is a first-order approach because it needs information from the gradient vector. Assume that $f\left(w^{(i)}\right) = f^{(i)}$ and $\nabla f\left(w^{(i)}\right) = g^{(i)}$ are both true. The process starts at a position called $w^{(0)}$ and proceeds from there until a stopping criterion is fulfilled by moving from $w^{(i)}$ to $w^{(i+1)}$ in the training direction of $d^{(i)} = -g^{(i)}$. As a result, the gradient descent algorithm iterates as follows [56]:

$$w^{(i+1)} = w^{(i)} - g^{(i)}\eta^{(i)}, \text{ for } i = 0,1,..... \tag{11.4}$$

where η is the learning rate. Selecting an appropriate learning rate is essential to comprehending algorithm stability. By picking the right learning rate, you may avoid instability and slow convergence [54].

TABLE 11.3
Different Category of Training Algorithm [52]

S/N	Category	Algorithms	Remark
1	Resilient backpropagation	RP	Resilient backpropagation
2	Conjugate gradient	CGF	Conjugate gradient backpropagation with Fletcher-Reeve starts
		CGB	Conjugate gradient backpropagation with Powel/Beale restarts
		CGP	Conjugate gradient backpropagation with Polak/Ribiere restarts
		SCG	Scale conjugate gradient backpropagation
3	Gradient descent	GDA	Gradient descent with adaptive learning rate backpropagation
		GD	Gradient descent backpropagation
		GDM	Gradient descent with momentum backpropagation
		GDX	Gradient descent with momentum and adaptive learning rate backpropagation
4	Quasi-newton	BFGS	BFGS Quasi-Newton backpropagation
		OSS	One-step secant backpropagation
5	Levenberg–Marquardt	LM	Levenberg–Marquardt backpropagation

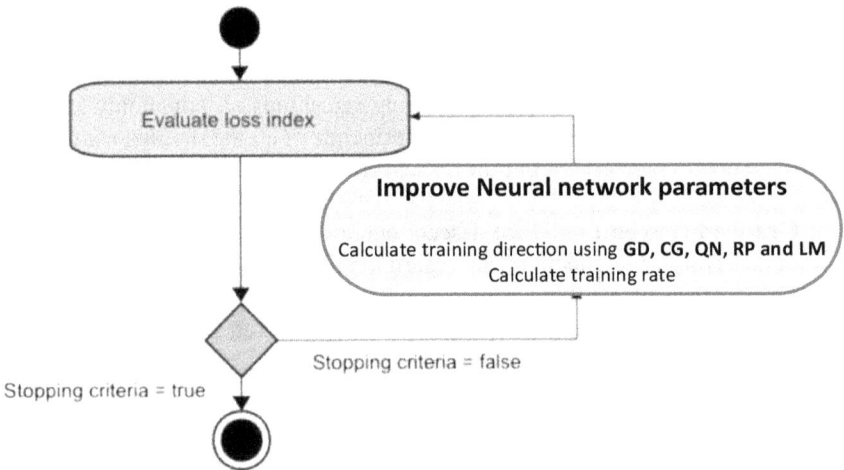

FIGURE 11.5 Basic working principle of training algorithms [56].

ii. **Resilient Backpropagation:** Resilient backpropagation does not employ a learning rate parameter, in contrast to conventional backpropagation algorithms. Instead, it individually modifies the step size for each weight parameter based on the sign of the error's weight-dependent derivative [57]. This makes it converge faster and more effectively than other techniques, especially when the error function's landscape contains a lot of local minima [52]. As illustrated in Equation 11.5, resilient backpropagation is a heuristic learning approach that increased convergence speed by considering only the sign of the derivative rather than the amount of the error function's derivative for the weight update. It decreases the number of adaptive parameters and learning steps, and it easily computes local learning schemes [58]:

$$\Delta x_k = -sign\left(\frac{\Delta E_k}{\Delta x_k}\right)\Delta k \tag{11.5}$$

where Δx_k, ΔE_k, and Δk are used to represent the current weights vector updates, error function E at k, and bias increase, respectively.

iii. **Conjugate Gradient:** Conjugate gradient algorithm, which is a type of optimization technique, is significantly more effective than gradient descent due to its minimal memory requirements and quick convergence [52]. However, it often exhibits instability in large-scale problems [59]. With an initial intuition for a solution, the conjugate gradient algorithm begins by iteratively minimizing a quadratic function. The technique calculates a search direction p that is conjugate to the previous search directions at each iteration and updates the solution by moving in the direction that minimizes the quadratic function along this direction [59,60]. The entire conjugate gradient algorithms operate by looking in the steepest descending direction, which is the opposite of the gradient as described in Equation 11.6 [61].

$$p_o = -g_0 \tag{11.6}$$

The following step involves a series of estimates for a line search as described in Equation 11.7:

$$x_{k+1} = x_k + a_k p_k \tag{11.7}$$

The search direction is denoted with pk in this instance. According to Equation 11.8, the prior search direction determines the choice of the subsequent search direction.

$$p_k = -g_k + \beta_k p_{k-1} \tag{11.8}$$

iv. **Quasi Newton:** Quasi Newton algorithm is comparable to conjugate gradient algorithms in terms of quick optimization, and it can be viewed as the fundamental local approach utilizing second-order information [62]. In comparison to conjugate gradient algorithms, the algorithms' computation costs are higher, denser, and more sophisticated. The quasi-Newton algorithm is a second-order optimization technique that updates the weights by approximating the Hessian matrix. The Newton approach described in equation is used to update the weights [52]. Although it is based on the Newton method, the quasi-Newton (or secant) method does not involve the calculation of second derivatives:

$$x_{k+1} = x_k - H_k^{-1}G_k \tag{11.9}$$

At the present weights and biases, H_k is the Hessian matrix (second derivatives) of the performance index. In Newton algorithms, the new weights x_{k+1} are calculated based on the value of the gradient and the present weight x_k. The matrix of the cost function's second-order partial derivatives with regard to the weights is known as the Hessian matrix. Unfortunately, it is computationally difficult, particularly for large neural networks, to compute the correct Hessian matrix. The quasi-Newton approach, on the other hand, approximates the Hessian matrix by computing it from the gradient of the cost function and the change in the weights.

v. **Levenberg-Marquardt:** Levenberg–Marquardt is acknowledged as a common methodology for resolving nonlinear least squares issues by integrating the Gauss–Newton method and gradient descent. When Levenberg–Marquardt displays adaptive behavior in response to the distance to the solution, it can frequently be certain of the answer [63]. The algorithm is sluggish and far from the solution when backpropagation is gradient descent [64]. On the other hand, if backpropagation is Gauss–Newton, the algorithm is very close to being accurate. For computing the gradient in Levenberg–Marquardt, the Hessian is estimated as in Equation 11.10, while gradient is computed using Equation 11.11 [52].

$$H = J^T J \tag{11.10}$$

$$g = J^T e \tag{11.11}$$

where J and e stand for, respectively, the Jacobian matrix and a vector of network errors. This approach is used by the Levenberg–Marquardt method in the same way as the Newton method as illustrated in Equation 11.12:

$$x_{k+1} = x_k - \left[J^T J + \mu I \right]^{-1} J^T e \tag{11.12}$$

11.4.2 BUILDING AN OPTIMAL NEURAL NETWORK MODEL FOR PREDICTIVE ANALYSIS IN ALD

Neural networks are robust predictive analysis tools owing to their ability to effectively model complex and abstract connections within datasets, hence resulting in superior prediction precision compared to alternative methodologies. However, to build an optimal neural network model for optimal prediction performance, the careful choice of right hyper-parameters, control parameters, topology and general architecture is a vital requisite. Hyper-parameters are parameters that are defined by the user rather than parameters that are learned from the data. Noteworthy is the fact that there are no one-size-fit-all rules or approach in deciding the optimal combination of network architecture and hyperparameter for optimal ANN model building. While it is often based on trial-and-error technique, cross validation and performance metrics may help assist in optimal parameter settings. Some important hyperparameters to think about are listed below:

i. **Number of hidden layers:** This is a critical architectural hyper-parameter of the neural network model which significantly influences the model's complexity and performance. Simple neural network architecture has a single hidden layer while a more complex deep neural network has multiple hidden layers [65].

ii. **Neurons in each layer:** Depending on the designed network architecture and the task at hand, the numbers of neurons in each layer vis-à-vis input, hidden and output varies. Neurons in the input layer signifies the dimensions of the input datasets while neurons in the output layer relates to the output dimensions and the task at hand [65].

iii. **Learning rate:** This is significant to the training dynamics of the neural network as it decides the rate of learning at training phase, hence the weight adjustment. A lower training rate implies a slower training while a higher training rate is determined by a higher learning rate [66].

iv. **Activation function:** The mathematical expression and information flows within neurons of neural network is controlled by the activation function. Its choice is influenced by the problem the neural network is employed for and the type of data set [67]. Typical examples of activation functions used in neural network are sigmoid, tanh, ReLU, and softmax.

v. **Training algorithm:** This is also a vital part of the training process of neural networks which are responsible for the adjustment of key network parameters such as weights and bias. It is an optimization algorithm which minimizes error between the actual and target output called cost function [65,68].

vi. **Batch size:** The batch size decides the quantity of training instances or data points in a single forward or backward pass of training iteration. While a higher batch size could speed up training, it might also call for more memory. Moreso, slower training could be attributed to a smaller batch size [69].

vii. **Epoch number:** The epoch number is another significant neural network parameter whose choice considers the training convergence and model's performance on a validation dataset. It determines how many times the training algorithm will run over the full training dataset [70]. If the epoch number is too high, the model may begin to overfit, which means it starts to memorize the training data and performs poorly on novel data set [71,72].

Neural network model architecture affects learning and prediction. The number of layers, number of neurons in each layer, and kind of activation function affect the model's complexity and ability to identify data patterns [65]. The number of neural network layers determines how many times input data are adjusted before reaching the output layer. Simple neural networks have one hidden layer. Layers depend on issue complexity and data amount [73]. Deeper neural networks can capture more intricate data patterns, but if the dataset is too short, they may overfit [73,74]. ANN captures nonlinearity based on layer neuron count. A neural network with more neurons may be more expressive, but if the number of parameters exceeds the data, it may overfit [75]. Activation functions determine how the neural network's neurons respond to input data. Sigmoid, ReLU, softmax, and hyperbolic tangent activation functions are common. The model data and problem determine the activation function [65]. For a large dataset with complex patterns, a more complex ANN with several hidden layers and many more neurons may be needed [76]. Balance model complexity and data quantity to avoid overfitting and provide accurate predictions. Table 11.4 shows a typical control parameter optimization and hyper-parameter setting for ALD predictive analytic ANN design. Figure 11.6 shows the approach for developing an appropriate neural network for predictive analysis in ALD.

11.4.3 SUPPORT VECTOR MACHINE

Another versatile and relevant but less prominent algorithm for predictive analysis, classification task, and optimization in ALD applications is the SVM. In ALD, SVM has been deployed mainly for regression analysis. The classification potential

TABLE 11.4
A Typical Hyperparameter Setting in an ANN Model

Hyper-Parameters	Values/Range	Description
Number of hidden layers	1–2	An optimal performance can be obtained in one or two hidden layers. A satisfactory model result can be achieved in a single hidden layer architecture while in some applications, two or more hidden layers might be needed
Neurons in hidden layers	1–50	A range of 1–50 neurons can be tested in a single or all hidden layers selected
Activation function in the hidden layer	Tansig, softmax, logsig, and purelin	In each topology, the four available activation functions' optimal combinations in the hidden layers can be examined while the optimization allows to select the most performing function for a particular dataset or a particular task
Activation function in the output layer	Tansig, softmax, logsig, and purelin	In each topology, the four available activation functions' optimal combinations in the output layer can be examined, while the optimization allows to select the most performing function for a particular dataset or a particular task
Training Algorithm	RP, CGF, CGB, CGP. SCG, GDA, GD, GDM, GDX, BFGS, OSS, and LM	Testing each of the 12 will allow the user to determine which algorithm performs best, and use the algorithm to train the final model. In this approach, I obtained the optimal result for my work.
Epoch number	1,000	A higher or lower epoch number depends on the type of problem and the quality of training
Minimum gradient	1e-100	This is the stopping criterion for the training. A minimum error is always selected as the stopping criterion for the training
Training:validation: testing ratio	0.7 : 0.3 : 0.3	This describes how the dataset is divided for training, validation, and testing purposes

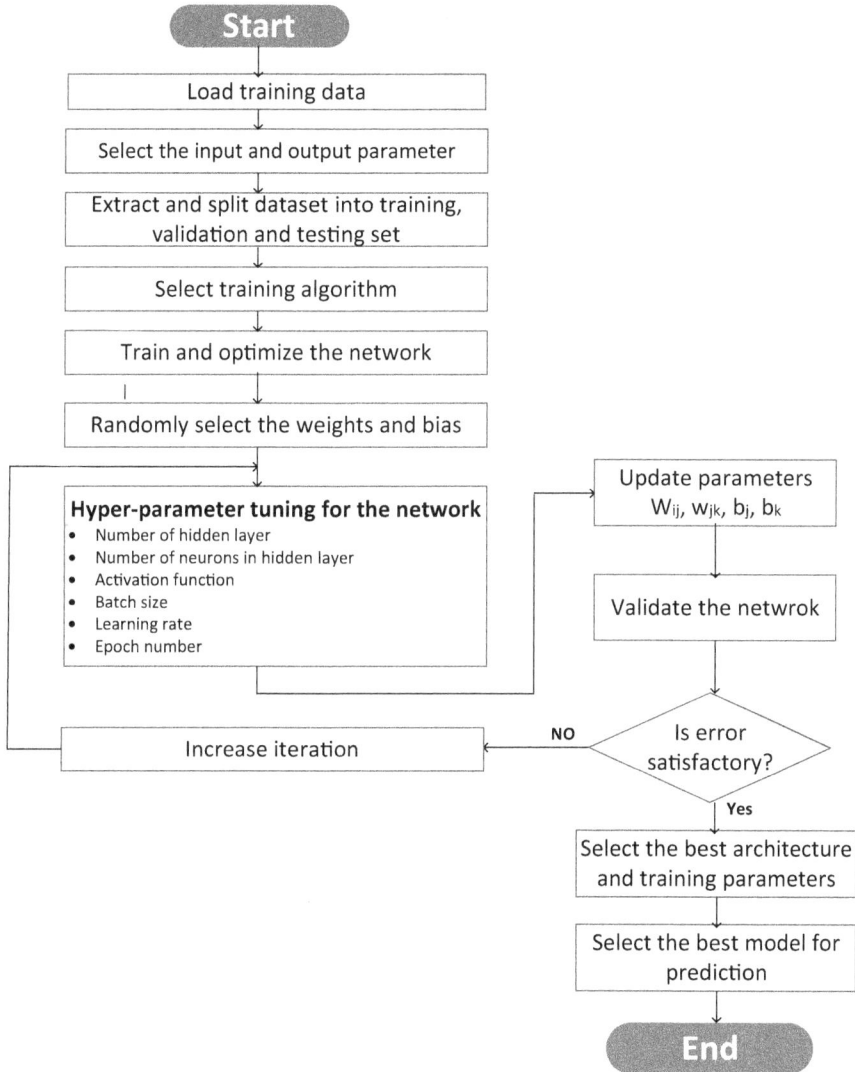

FIGURE 11.6 Process flow diagram for optimal ANN training.

of SVM is a promising technique which has not been fully explored in ALD and thin film deposition process and applications. Based on their characteristics, such as their growth rate, thickness, and film composition, SVM can be used to categorize and classify various types of ALD processes. For instance, it can be used to categorize various ALD processes, such as thermal ALD and plasma-enhanced ALD, based on the process variables temperature, precursor flow rate, and plasma power. The composition of deposited films, such as metal oxides, nitrides, and carbides, can likewise be categorized using SVM. The classification task of SVM was utilized in the study by Arunachalam et al. [45] as a monitoring technique for assessing and controlling thin film thickness in the deposition process of TiO_2 with integration with spectroscopic ellipsometry. The properties of deposited films can also be predicted using SVMs based on the precursor and ALD process variables. The technique can forecast the properties of future films based on the selected precursor and process circumstances by training an SVM on a collection of precursor and process parameters and their corresponding film properties. This can aid in the development of new ALD procedures and the choice of suitable precursors.

11.5 AREA OF APPLICATIONS OF PREDICTIVE CAPABILITY OF MACHINE LEARNING IN ALD

11.5.1 FILM THICKNESS

To ensure the quality of surface modification, thin film thickness evaluation is critical. The successful application of diverse surface engineering is contingent on the thickness of the thin film deposited [77]. As it directly affects permeability, tensile strength, impact resistance, and other properties, film thickness is a crucial component of product packaging as well as other products. The manufacturing needs a dependable system in place for high-value items that will enable them to precisely assess film thickness through every phase of manufacturing. As a result, a number of manufacturing industries have necessitated swift and accurate thin film thickness measurement [77]. Nanomaterials depend on thin layer thickness for electrical, optical, mechanical, and thermal properties. Thus, its thickness must be measured and controlled in various applications [78]. Several methods can measure the thin film's thickness. Factors that play a role in the choice of suitable methods are the transparency of the deposited film in the optical region, and the cost [78]. As a viable alternative technique to overcome some of the limitations of the physical and mechanical techniques of thin film thickness measurement, machine learning models such as ANN and SVM are practicable owing to their ability to handle the hyper nonlinearity in the complex reaction of thin film parameters. ANNs can be used to estimate the thickness of thin films in ALD, which has a number of important consequences for efficiency, product quality, and the discovery of new materials.

Using data extracted from spectroscopic reflectometry in the study by Tabet and Mcgahan [79], an ANN architecture was constructed to predict the thickness and optical parameters of thin films. An attractive ability of the wavelet neural network in extracting features from an experimental data was utilized in the study by Cui et al. [80] to predict the thickness of thin film toward analyzing its field emission properties. The model had a laudable performance with a relative error of 2.98%. Bora [81] demonstrated the measurement of the deposited film thickness measurement using machine learning algorithm vis-à-vis ANN. The study predicted the growth profile of thin film thickness and refractive index based on the data extracted from spectroscopic reflectometry. A study by Wang et al. [77] developed a SVM and ANN model for predicting coating thickness with a time-resolved thermography. The machine learning-assisted thickness estimations were based on surface temperature increment data extracted from the time resolved thermography experiment. This study further compared the predictive performance of both SVM and ANN using a cross-validation approach.

Mariam et al. [82] established the superiority of machine learning models over computational simulations based on accuracy of prediction in the estimation of property of thin film particularly, film thickness. Using data extracted from electrodynamically lubricated (EHL) simulation and full-scale finite element methods, three machine learning algorithms, namely, Gaussian process regression (GPR), ANN, and SVM, were developed for a regression-based task of thin film thickness prediction. The ANN model outperformed other models with a R^2-value of 0.999. By considering a comparison of the machine learning model's result based on international standards instead of the existing analysis method, Lee and Jin [83] developed a novel technique which evaluates the thickness of thin film against an international length standard. The ANN model construction proceeded with data extracted from experimental spectra of certified reference material. Given how well ANN performed in the aforementioned studies, it is clear that ANN has been the primary method used by researchers to accurately forecast thin film thickness. Some other machine learning regression models such as SVM, DT, RF, and k-nearest neighbor are ocassionally used in ALD modeling but are not as quite popular as ANN. The study by Arunachalam [45] developed an integrated framework of spectroscopic ellipsometry and machine learning for an optimal control and monitoring of deposited thin film thickness. Four algorithms developed are SVM, kNN, RF, and DT. The four models gave a laudable outcome in prediction of thin film thickness with accuracy of 83.84%, 88.76%, 84.1%, and 81.32%, respectively.

11.5.2 Deposition Rate

Beyond using computational approaches such as transport model and surface reaction model [84], machine learning models offer a better insight into the modeling and prediction of the deposition rates of different materials in ALD. A regression-based machine learning model can be applied to predict the deposition rate of thin film in ALD based on various process parameters such as temperature, pulse duration, reactor pressure, precursor concentration, and surface morphology, amongst others. To enhance the development process of the ALD, Cognazzo [85] developed a deep learning model to predict the saturation times based on growth rate and dosage time of reagent. The saturation times from the growth rate profile of an ALD reactor was modeled using an ANN model with a stochastic gradient descent (SCG) training algorithm [86]. This was achieved without any prior or additional knowledge of the surface kinetics of the reactor. Ding et al. [17] proposed a multi-scale data-driven model comprising a Bayesian regularized artificial neural network (BRANN) model integrated with computational simulation approaches, namely, kinetic Monte Carlo (KMC), computational fluid dynamics (CFD), and density functional theory (DFT) to predict the transient rate of deposition of SiO_2 based on process parameters such as film coverage, surface heating, and precursor flow rate. The BRANN model was trained using data generated from the KMC, CFD, and DFT models.

11.5.3   MICROSTRUCTURE ANALYSIS

The accuracy of the deposition process is directly correlated with microstructure properties. The most significant characteristics that need to be managed in order to preserve the greater performance of the deposition process are phase content, unmolten particle content, and porosity levels. To improve the efficiency of the deposition process, accurate microstructure feature prediction is required [10]. Machine learning has the capability to bridge the gap between thin film deposition processing parameters and its microstructure either for identification of microstructure [87], cluster analysis of microstructure [88], or defect analysis [89]. Thin film microstructure must be monitored in a multi-parameter space to achieve the best extrinsic qualities, which are typically too difficult to map with a reasonable number of tests. Sequel to this, Banko et al. [90] suggested combining combinatorial experimentation with generative deep learning models to extract synthesis to lower the cost of microstructure design and to comprehend thin film development microstructure complexity. The generative deep learning models developed in their study are able to generate novel data based on a hidden trend in the dataset through techniques such as variation autoencoder (VAE) and generative adversarial network (GAN) [91,92]. It is enticing to create microstructures using GANs. In comparison to experimental methods and physical-based models, GANs can quickly build microstructures of the same quality in considerably larger quantities [93]. In order to determine the relationships between processing-structure and property and to forecast microstructures, Noraas et al. [94] proposed using generative deep learning models for material design. Griffiths and Harris [95] established that the morphological and consequent property changes in the deposited film are caused by alterations in processing parameters during solidification. Thus, their study developed a PCA algorithm for slot die coatings to analyze the microstructures of microparticle distributions. A machine-learning-based model was created by Hashemi et al. [96] to forecast microstructure evolution. Firstly, two-point spatial correlations were used to extracted low-dimensional microstructural elements from the film's microstructures. Afterward, they deployed a Gaussian process vector autoregressive model to forecast how these low-dimensional traits would evolve. Brough et al. [97] have conducted comparable research on the microstructure evolution of polyethylene films. By formalizing the method, they showed that it was three times quicker than numerical simulations [98]. Figure 11.7 represents a deep neural network–integrated framework for analyzing and characterizing the microstructure of materials.

11.5.4   THIN FILM PROPERTIES

Machine learning models could be beneficial in predicting critical properties of deposited thin film such as refractive index, and physical, optical, chemical, and electrical properties based on a number of variables, including the deposition temperature, the number of cycles per cycle, the precursor employed, and the type of substrate during the deposition process. A large dataset of thin films that were created via ALD and for which the deposition parameters and associated thin film properties are known can be used to train ANN and other machine learning models. The ANN may be used to anticipate the properties of new thin films based on their deposition parameters since it can discover the correlation between the input deposition parameters and the properties of deposed thin films. Finding the mechanical properties

(a)

(b)

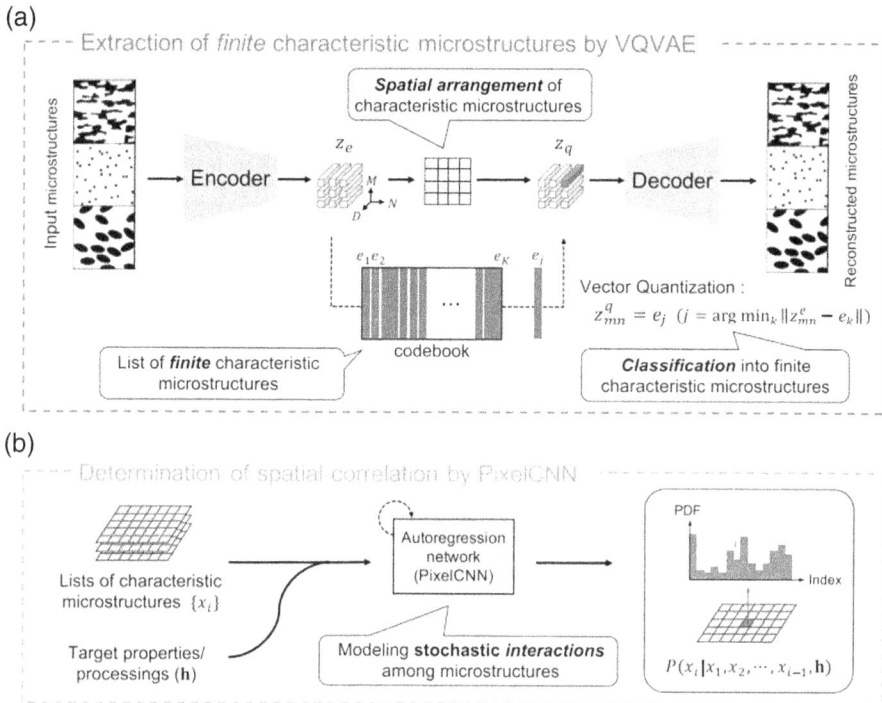

FIGURE 11.7 Machine-learning-integrated framework for microstructural analysis: (a) using VQVAE for finite microstructure extraction and (b) using spatial correlation between extracted microstructural features using Convolutional Neural Network (CNN) [99].

of thin films using conventional physics-based, empirical, or statistical models is challenging. Thus, Long et al. [100] deployed a machine learning algorithm for predicting the mechanical properties of thin films using data from the load-penetration depth curve. Ali et al. [101] proposed an ANN model for predicting the electrical properties of phenol red thin film deposited on silicon. The properties of ZnO thin film via-a-vis its resistivity was predicted using ANN by Sabri et al. [102] based on input parameters, namely, the deposition rate and substrate temperature. Using data generated by molecular dynamics simulations for training and testing a machine learning model, namely, RF algorithms, Wang et al. [103] predicted the mechanical properties of tungsten disulfide (WS_2) based on several input parameters such as temperature, strain rate, chirality, and defect ratio. A regression-based machine learning model was utilized in the study by Vajire et al. [104] to predict the nanoindentation load-deformation profile of Gallium nitride (GaN) thin film deposited on thick substrate used in the manufacturing of sensitive electronic devices. Yang et al. [105] developed an ANN-based prediction for the stress property of deposited Aluminum nitride (AlN) thin film using dataset generated from optical emission spectroscopy after using a PCA algorithm to classify the XRD results into compressive and tensile stress of the thin film. Asafa et al. [106] developed an ANN-integrated model for predicting the intrinsic stress in deposited silicon thin film using deposition parameters such as temperature of substance, pressure of chambers, and dilution ration. The optical properties of thin film have also been modeled using the machine learning model. For instance, an ANN was trained on resilient backpropagation algorithm to model the optical constant of $AS_{30}Se_{70-x}Sn_x$ thin film based on an experimentally generated dataset [107]. A similar study by Kim et al. [108] also applied an ANN model for optical characteristics of perovskite materials.

11.5.5 PROCESS CONTROL AND MONITORING

A comprehensive investigation on the improvement of the protective characteristics of ALD films is difficult due to the large dimensionality of the parameter space. Determining the ideal deposition settings is essential for producing layers free of defects [109]. By successfully overcoming these obstacles and outperforming traditional methods, machine learning approaches are gaining respect in applications in process control and monitoring. There is therefore a demand for effective optimization techniques that can unilaterally choose the processing parameters that would promote the best ALD film growth [110]. To create models that can forecast the ideal process parameters for a given set of film qualities, machine learning algorithms can be trained on past process data. These models can be used to continually monitor the ALD process in real-time and modify the process parameters to obtain the desired thin film properties. This could aid in process optimization and boost the effectiveness and consistency of thin film deposition. They can further aid in uncovering irregularities in the ALD procedure. These models can spot trends that point to anomalous process behavior, like fluctuations in gas flow rates or temperature variations, by continuously monitoring the process data. Further to this, these models can be trained using process data from the past to spot trends indicating the presence of high-quality or low-quality thin films. The study by Dogan et al. [109] used the Bayesian optimization algorithm to find the best deposition parameters which minimize the layer defect density in an ALD-Al_2O_3 passivation layer. Paulson et al. [110] developed three intelligent optimization algorithms, namely, random optimization, expert systems optimization, and Bayesian optimization toward process control and monitoring of the ALD process. Each algorithm discovered optimal timings in 800 ALD cycles or less. Several amounts of measurement noise were used for these optimizations, which have been demonstrated to have a significant impact on the effectiveness of the optimization and the final optimized timings. The random optimization technique is a good starting point and frequently succeeds in the most challenging situations, particularly when the gap between the shortest and longest times is wide. Otsuka et al. [111] also developed a Bayesian optimization algorithm to optimize the ALD process for monitoring the growth of $SrRuO_3$ thin film while the same approach was developed on TiO_2 by Ohkubo et al. [112]. The optimal process condition control and monitoring using intelligent optimization algorithms in ALD is represented in Figure 11.8.

11.5.6 GRAIN SIZE PREDICTION

The thin film's grain size has a significant impact on the way it behaves in terms of its mechanical stability, electrical conductivity, and optical qualities [113]. Understanding the basic mechanics of thin film growth during ALD can also be

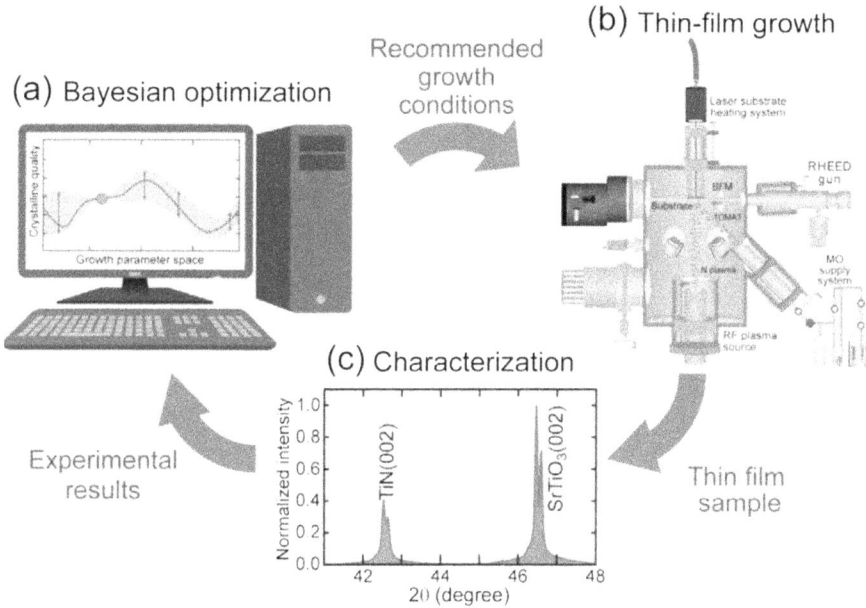

FIGURE 11.8 Machine learning based intelligent optimization for atomic layer deposition (ALD) process control and monitoring [112].

done with the aid of grain size prediction [114]. This can give an insight into how the deposition circumstances affect the film properties, which can be helpful for improving the ALD process and creating novel materials. Predicting the grain size of the thin film that is deposited on the surfaces is crucial for improving the gloss of the surface [10]. ALD-synthesized microelectronic devices can be made more dependable and long lasting by a machine-learning-based predictive analysis of the grain size. Further to this, comprehending how to optimize the grain size can help to increase the efficiency and dependability of the devices because it can have an impact on the mechanical and electrical properties of the thin film. Machine learning techniques have been a viable technique for determining the impact of various parameters on the grain size of the ALD. The grain sizes of titanium nitride (TiN) were predicted by Jaya et al. [115] using the ANN approach. Several neural network models' architecture were created, taking into account the following nodes in the input layer, namely, argon gas pressure (N_2) and turntable speed (TT), while grain size was set as the node in the output layer. In a similar study by Jarrah et al. [116], which examined the deposition of TiN, optimization of the process parameters of the deposition process was carried out using genetic algorithms to control the film grain size. Furthermore, intelligent optimization algorithm, namely, particle swarm optimization was deployed to monitor the grain size of TiN deposition based on optimization of the process parameters [117].

Table 11.5 presents a summary of the application of predictive capabilities of machine learning models in ALD applications, the source of data for the models, materials, and performance metrics.

TABLE 11.5
Summary of Machine Learning-Based Predictive Analysis in Atomic Layer Deposition (ALD) Applications

S/N	Data Source	Algorithm	Summary of Application	Materials/ Substrate	Performance Metrics	Ref.
1	3D multiphase CFD model	LSTM RNN	Integrated data-driven machine learning model for prediction of three outputs, namely, (i) surface precursor partial pressure, (ii) number of Hf physisorption sites, and (iii) number of O physisorption sites using the inlet flow rate as input	Hafnium oxide (HfO$_2$)	Standard error of 2.54%, 1.19%, and 2.85% for pressure prediction at inner, middle, and outer regions, respectively.	[6]
2	Full-scale stochastic multiscale system to generate 400,000 data points with a lattice size of $N=50$	ANN	Data-driven model using ANN was developed to assist the shrinking horizon nonlinear predictive control of chemical vapor deposition with target outputs such as roughness (R) and actual growth rates (G$_r$), while temperature and inlet precursor, mole fraction were considered as input factors	N/A	Mean error of 0.04% and 0.06% for both outputs R and G$_r$, respectively.	[15]
3	Multiscale CFD and kinetic Monte Carlo (KMC) model	ANN	Prediction of deposition rate of the atomic layer deposition of SiO$_2$ based on surface temperature, precursor partial temperature, and transient film coverage	SiO$_2$	Average error of 3.07%	[118]
4	Kinetic Monte Carlo model	ANN	Surface deposition mechanism was predicted using a machine-learning-integrated KMC model	HfO$_2$	6.3% prediction error	[7]
5	Multiscale simulation (discrete-element method (DEM)-based particle aggregation)	ANN	Prediction of film thickness and film roughness in the deposition	Quantum dots (QDs)	$R^2 = 0.99$	[18]

(Continued)

TABLE 11.5 (*Continued*)
Summary of Machine Learning-Based Predictive Analysis in Atomic Layer Deposition (ALD) Applications

S/N	Data Source	Algorithm	Summary of Application	Materials/ Substrate	Performance Metrics	Ref.
6	ALD experiment	PCA	Feature selection and extraction as a method for identifying the contribution of physio-chemical property of material process features on the property of deposited materials	SiO_2	N/A	[119]
7	1D first-order Langmuir kinetics	ANN	Optimization of the ALD process and prediction of optimal saturation time which gives saturation at every reactor location based on input on ALD reactor without prior knowledge of the surface kinetics	N/A	Average error of 3%	[86]
8	Experiment	KNN RF DT SVM	Machine learning ALD monitoring techniques for assessing thin film thickness in the deposition process. Particularly, the spectroscopic ellipsometry was integrated with machine learning to control the thickness of thin films. Machine learning techniques like classification and sample downsizing were deployed with the selected classification models	TiO_2	88.76% (KNN), 81.32% (DT), 83.84% (SVM), and 84.1% (RF)	[45]
9	Experiments with in situ XANES measurement	RF ANN	A screening strategy was developed using supervised machine learning to elucidate the atomic structure of reaction of deposition using simulated X-ray absorption near edge structure (XANES) spectra	ZnS	N/A	[43]

(Continued)

TABLE 11.5 (*Continued*)
Summary of Machine Learning-Based Predictive Analysis in Atomic Layer Deposition (ALD) Applications

S/N	Data Source	Algorithm	Summary of Application	Materials/ Substrate	Performance Metrics	Ref.
10	Simulated data from a stochastic multiphase model (kinetic Monte Carlo)	ANN	Real-time optimization and prediction of the dynamic responses vis-à-vis roughness and growth rate of deposited film with chemical vapor deposition under uncertainty	N/A	−2% relative error estimate of E under uncertainty	[29]
11	Experiment	PCA k-means	Feature extraction and selection from unstructured and unlabeled video data of high-energy electron diffraction from pulsed laser deposition to assist in extracting hidden pattern in the epitaxial film growth	ReSe$_2$ (Metal dichalcogenides)	98.95% variance	[120]
12	TDUS (Two-dimensional Umbrella sampling)	Deep neural network	A machine learning-assisted ab initio modeling of surface reaction using data generated from a TDUS for optimization of the deposition process	Al(Me)$_3$	RMSE of 0.82	[28]
13	Experiment	Bayesian optimization	Optimization of process parameters in ALD toward the effective reduction of defects in the passive layers of deposition	Al$_2$O$_3$	N/A	[27]
14	Experiment	Extreme gradient boosting (XGBoost)	The machine learning algorithm was developed to predict the component ratio of Pt/Al deposited on ∝-Al$_2$O$_3$ based on temperature, stop valve time, precursor pulse time, and reactant pulse time	Nano-film platinum	99.9% accuracy R^2 = 0.99	[44]
15	Experiment	PCA	To understand the chemical dimension of ALD, an unsupervised machine learning algorithm, namely, principal component analysis (PCA) was deployed to extract hidden features. The impact of these features and parameters on the film thickness was further identified and analyzed	SiO$_2$	N/A	[26]

(Continued)

TABLE 11.5 (Continued)
Summary of Machine Learning-Based Predictive Analysis in Atomic Layer Deposition (ALD) Applications

S/N	Data Source	Algorithm	Summary of Application	Materials/ Substrate	Performance Metrics	Ref.
16	Experiment	ANN	Prediction of sheet resistance of indium-doped zinc oxide thin film	Indium-doped zinc oxide (IZO)	$R^2 = 0.795$	[121]
17	Experiment and molecular dynamic simulations	ANN	A feedforward backpropagation neural network with a Bayesian training algorithm based on experimental and simulation data to model film thickness of deposited TiO_2	TiO_2	RMSE = 0.47, SSE = 0.46	[122]
18	Experiment	GA	Optimization of the grain size of the film, based on process parameters, namely, nitrogen gas pressure, argon gas pressure, and turntable speed	TiN	Prediction accuracy of 96.09%	[116]

LSTM, long- and short-term memory; RNN, recurrent neural network; ANN, artificial neuron network; RF, random forest; DT, decision tree; SVM, support vector machine; GA, Genetic Algorithm; KNN, k-Nearest neighbor.

11.6 CONCLUSION

ALD is seen as a technological revolution thanks to machine-learning-based predictive analysis. The ANN is a type of the most prominent machine-learning-based predictive model that offers an accurate and effective method for forecasting key outcomes in ALD. Machine learning algorithms can assess huge and complicated datasets to provide precise predictions on critical factors, such as film thickness, composition, and quality, by utilizing data from many sources, including actual measurement experiments and simulation models. The performance and efficiency of ALD procedures could be greatly enhanced by this technique, which has already shown outstanding results in niche applications like catalyst development and thin-film deposition. We may anticipate seeing many more fascinating uses for machine learning as this technology develops.

REFERENCES

[1] R. W. Johnson, A. Hultqvist, and S. F. Bent, "A brief review of atomic layer deposition: from fundamentals to applications," *Materials Today*, vol. 17, no. 5, pp. 236–246, 2014, doi: 10.1016/j.mattod.2014.04.026.

[2] P. O. Oviroh, R. Akbarzadeh, D. Pan, R. A. M. Coetzee, and T. C. Jen, "New development of atomic layer deposition: processes, methods and applications," *Science and Technology of Advanced Materials*, vol. 20, no. 1, pp. 465–496, 2019. doi: 10.1080/14686996.2019.1599694.

[3] S. Yun, F. Ou, H. Wang, M. Tom, G. Orkoulas, and P. D. Christofides, "Atomistic-mesoscopic modeling of area-selective thermal atomic layer deposition," *Chemical Engineering Research and Design*, vol. 188, pp. 271–286, 2022, doi: 10.1016/j.cherd.2022.09.051.

[4] T. Justin Kunene, L. Kwanda Tartibu, K. Ukoba, and T.-C. Jen, "Review of atomic layer deposition process, application and modeling tools," *Materials Today: Proceedings*, vol. 62, pp. S95–S109, 2022, doi: 10.1016/j.matpr.2022.02.094.

[5] Y. Ding, Y. Zhang, Y. M. Ren, G. Orkoulas, and P. D. Christofides, "Machine learning-based modeling and operation for ALD of SiO_2 thin-films using data from a multiscale CFD simulation," *Chemical Engineering Research and Design*, vol. 151, pp. 131–145, 2019, doi: 10.1016/j.cherd.2019.09.005.

[6] Y. Ding, Y. Zhang, H. Y. Chung, and P. D. Christofides, "Machine learning-based modeling and operation of plasma-enhanced atomic layer deposition of hafnium oxide thin films," *Computers & Chemical Engineering*, vol. 144, 2021, doi: 10.1016/j.compchemeng.2020.107148.

[7] Y. Ding, Y. Zhang, G. Orkoulas, and P. D. Christofides, "Microscopic modeling and optimal operation of plasma enhanced atomic layer deposition," *Chemical Engineering Research and Design*, vol. 159, pp. 439–454, 2020, doi: 10.1016/j.cherd.2020.05.014.

[8] N. Sitapure and J. S. Il Kwon, "Neural network-based model predictive control for thin-film chemical deposition of quantum dots using data from a multiscale simulation," *Chemical Engineering Research and Design*, vol. 183, pp. 595–607, 2022, doi: 10.1016/j.cherd.2022.05.041.

[9] T. Muneshwar, M. Miao, E. R. Borujeny, and K. Cadien, "Chapter 11- Atomic layer deposition: Fundamentals, practice, and challenges," In *Handbook of Thin Film Deposition*, 4th edition, K. Seshan and D. Schepis, Eds., William Andrew Publishing, pp. 359–377, 2018. doi: 10.1016/B978-0-12-812311-9.00011-6.

[10] U. M. R. Paturi, S. Cheruku, and S. R. Geereddy, "Process modeling and parameter optimization of surface coatings using artificial neural networks (ANNs): State-of-the-art review," In *Materials Today: Proceedings*, Elsevier Ltd, pp. 2764–2774, 2020. doi: 10.1016/j.matpr.2020.08.695.

[11] P. Rattan, D. D. Penrice, and D. A. Simonetto, "Artificial intelligence and machine learning: What you always wanted to know but were afraid to ask," *Gastro Hep Advances*, vol. 1, no. 1, pp. 70–78, 2022, doi: 10.1016/j.gastha.2021.11.001.

[12] N. Uene, T. Mabuchi, M. Zaitsu, Y. Jin, S. Yasuhara, and T. Tokumasu, "Growth mechanism study of boron nitride atomic layer deposition by experiment and density functional theory," *Computational Materials Science*, vol. 217, 2023, doi: 10.1016/j.commatsci.2022.111919.

[13] W. Chiappim, M. A. Fraga, H. S. Maciel, and R. S. Pessoa, "An experimental and theoretical study of the impact of the precursor pulse time on the growth per cycle and crystallinity quality of TiO_2 thin films grown by ALD and PEALD technique," *Frontiers in Mechanical Engineering*, vol. 6, 2020, doi: 10.3389/fmech.2020.551085.

[14] M. Kisielewski, A. Maziewski, T. Polyakova, and V. Zablotskii, "Models and simulations for analysis of domain sizes in ultrathin magnetic films," *Journal of Magnetism and Magnetic Materials*, vol. 272–276, pp. E825–E826, 2004, doi: 10.1016/j.jmmm.2003.12.162.

[15] G. Kimaev and L. A. Ricardez-Sandoval, "Nonlinear model predictive control of a multiscale thin film deposition process using artificial neural networks," *Chemical Engineering Science*, vol. 207, pp. 1230–1245, 2019, doi: 10.1016/j.ces.2019.07.044.

[16] Y. Ding, Y. Zhang, H. Y. Chung, and P. D. Christofides, "Machine learning-based modeling and operation of plasma-enhanced atomic layer deposition of hafnium oxide thin films," *Chemical Engineering Science*, vol. 144, 2021, doi: 10.1016/j.compchemeng.2020.107148.

[17] Y. Ding, Y. Zhang, Y. M. Ren, G. Orkoulas, and P. D. Christofides, "Machine learning-based modeling and operation for ALD of SiO_2 thin-films using data from a multiscale CFD simulation," *Chemical Engineering Research and Design*, vol. 151, pp. 131–145, 2019, doi: 10.1016/j.cherd.2019.09.005.

[18] N. Sitapure and J. S. I. Kwon, "Neural network-based model predictive control for thin-film chemical deposition of quantum dots using data from a multiscale simulation," *Chemical Engineering Research and Design*, vol. 183, pp. 595–607, 2022, doi: 10.1016/j.cherd.2022.05.041.

[19] M. A. Bokinala, P. Sinha, P. Halder, and J. K. Singh, "Fusing machine learning strategy with density functional theory to hasten the discovery of 2D MXene based catalysts for hydrogen generation," *Journal of Materials Chemistry*, 2023, doi: 10.1039/D3TA00344B.

[20] A. Yanguas-Gil, J. A. Libera, and J. W. Elam, "Reactor scale simulations of ALD and ALE: Ideal and non-ideal self-limited processes in a cylindrical and a 300 mm wafer cross-flow reactor," *Journal of Vacuum Science & Technology A*, vol. 39, no. 6, p. 062404, 2021, doi: 10.1116/6.0001212.

[21] H. Xia, J. Lu, S. Dabiri, and G. Tryggvason, "Fully resolved numerical simulations of fused deposition modeling. Part I-fluid flow." *Rapid Prototyping Journal*, vol. 24, no. 2, pp. 463–476, 2018.

[22] Y. Abu-Zidan, P. Mendis, and T. Gunawardena, "Optimising the computational domain size in CFD simulations of tall buildings," *Heliyon*, vol. 7, no. 4, p. e06723, 2021, doi: 10.1016/j.heliyon.2021.e06723.

[23] M. Becker and M. Sierka, "Atomistic simulations of plasma-enhanced atomic layer deposition," *Materials*, vol. 12, no. 16, 2019, doi: 10.3390/ma12162605.

[24] J. Huawei, Z. Tao, L. Xiao, B. Shan, and R. Chen, "Optimization of inlets and outlets of an ALD chamber with radiant heating," In *2013 IEEE International*

Symposium on Assembly and Manufacturing (ISAM), pp. 170–175, 2013. doi: 10.1109/ISAM.2013.6643519.

[25] G. P. Gakis, H. Vergnes, E. Scheid, C. Vahlas, B. Caussat, and A. G. Boudouvis, "Computational fluid dynamics simulation of the ALD of alumina from TMA and H_2O in a commercial reactor," *Chemical Engineering Research and Design*, vol. 132, pp. 795–811, 2018, doi: 10.1016/j.cherd.2018.02.031.

[26] C. Kim, T. N. Do, and J. Kim, "Machine learning-based analysis of the physio-chemical properties for the predictive thickness control of atomic layer deposition," *IFAC-PapersOnLine*, pp. 626–631, 2022. doi: 10.1016/j.ifacol.2022.07.513.

[27] G. Dogan et al., "Bayesian machine learning for efficient minimization of defects in ALD passivation layers," *ACS Applied Materials & Interfaces*, vol. 13, no. 45, pp. 54503–54515, 2021, doi: 10.1021/acsami.1c14586.

[28] H. Nakata, M. Filatov(gulak), and C. H. Choi, "Accelerated deep learning dynamics for atomic layer deposition of Al(Me)3and water on OH/Si(111)," *ACS Applied Materials & Interfaces*, vol. 14, no. 22, pp. 26116–26127, 2022, doi: 10.1021/acsami.2c01768.

[29] G. Kimaev and L. A. Ricardez-Sandoval, "Artificial neural network discrimination for parameter estimation and optimal product design of thin films manufactured by chemical vapor deposition," *Journal of Physical Chemistry C*, vol. 124, no. 34, pp. 18615–18627, 2020, doi: 10.1021/acs.jpcc.0c05250.

[30] L. Liu and Y. Liu, "Load image inpainting: An improved U-Net based load missing data recovery method," *Applied Energy*, vol. 327, p. 119988, 2022, doi: 10.1016/j.apenergy.2022.119988.

[31] J. C. Figueroa-García, R. Neruda, and G. Hernandez-Pérez, "A genetic algorithm for multivariate missing data imputation," *Information Sciences*, vol. 619, pp. 947–967, 2023, doi: 10.1016/j.ins.2022.11.037.

[32] H. N. Haliduola, F. Bretz, and U. Mansmann, "Missing data imputation using utility-based regression and sampling approaches," *Computer Methods and Programs in Biomedicine*, vol. 226, p. 107172, 2022, doi: 10.1016/j.cmpb.2022.107172.

[33] T. Zhao, Y. Sun, Z. Chai, and K. Li, "An outlier management framework for building performance data and its application to the power consumption data of building energy systems in non-residential buildings," *Journal of Building Engineering*, vol. 65, p. 105688, 2023, doi: 10.1016/j.jobe.2022.105688.

[34] Q. Hu, Z. Yuan, K. Qin, and J. Zhang, "A novel outlier detection approach based on formal concept analysis," *Knowledge-Based Systems*, vol. 268, p. 110486, 2023, doi: 10.1016/j.knosys.2023.110486.

[35] T. C. Nelsen, "Chapter Five - Outliers," In *Probability and Statistics for Cereals and Grains*, T. C. Nelsen, Ed., Woodhead Publishing, pp. 59–64, 2023. doi: 10.1016/B978-0-323-91724-7.00005-8.

[36] Z. Liu, J. Yang, L. Wang, and Y. Chang, "A novel relation aware wrapper method for feature selection," *Pattern Recognition*, p. 109566, 2023, doi: 10.1016/j.patcog.2023.109566.

[37] A. K. Gárate-Escamila, A. Hajjam El Hassani, and E. Andrès, "Classification models for heart disease prediction using feature selection and PCA," *Informatics in Medicine Unlocked*, vol. 19, p. 100330, 2020, doi: 10.1016/j.imu.2020.100330.

[38] W. Sheikh et al., "Recursive feature elimination (rfe) selects for important features in predicting post-operative cardiac arrest (POCA) in patients undergoing coronary artery bypass grafting (CABG): Insights from the sts registry," *Journal of the American College of Cardiology*, vol. 75, no. 11, Supplement 1, p. 1533, 2020, doi: 10.1016/S0735-1097(20)32160-4.

[39] D. Singh and B. Singh, "Feature wise normalization: An effective way of normalizing data," *Pattern Recognition*, vol. 122, p. 108307, 2022, doi: 10.1016/j.patcog.2021.108307.

[40] M. Ebrahimi, F. Mahboubi, and M. R. Naimi-Jamal, "RSM base study of the effect of deposition temperature and hydrogen flow on the wear behavior of DLC films," *Tribology International*, vol. 91, pp. 23–31, 2015, doi: 10.1016/j.triboint.2015.06.026.

[41] D. Z. Segu, J.-H. Kim, S. G. Choi, Y.-S. Jung, and S.-S. Kim, "Application of Taguchi techniques to study friction and wear properties of MoS2 coatings deposited on laser textured surface," *Surface and Coatings Technology*, vol. 232, pp. 504–514, 2013, doi: 10.1016/j.surfcoat.2013.06.009.

[42] R. Shankar, K. R. Balasubramanian, S. P. Sivapirakasam, and K. Ravikumar, "ANN and RSM models approach for optimization of HVOF coating," *Materials Today: Proceedings*, vol. 46, pp. 9201–9206, 2021, doi: 10.1016/j.matpr.2020.01.211.

[43] O. Trejo *et al.*, "Elucidating the evolving atomic structure in atomic layer deposition reactions with in situ XANES and machine learning," *Chemistry of Materials*, 2019, doi: 10.1021/acs.chemmater.9b03025.

[44] S.-H. Yoon *et al.*, "Extreme gradient boosting to predict atomic layer deposition for platinum nano-film coating," *Langmuir*, 2023, doi: 10.1021/acs.langmuir.2c03465.

[45] A. Arunachalam, S. Novia Berriel, P. Banerjee, and K. Basu, "Machine learning-enhanced efficient spectroscopic ellipsometry modeling," 2022. [Online]. Available: www.aaai.org.

[46] Z. Ma, W. Zhang, Z. Luo, X. Sun, Z. Li, and L. Lin, "Ultrasonic characterization of thermal barrier coatings porosity through BP neural network optimizing Gaussian process regression algorithm," *Ultrasonics*, vol. 100, p. 105981, 2020, doi: 10.1016/j.ultras.2019.105981.

[47] S. Gao, Q. Wu, Z. Zhang, and G. Jiang, "Simulating active layer temperature based on weather factors on the Qinghai-Tibetan Plateau using ANN and wavelet-ANN models," *Cold Regions Science and Technology*, vol. 177, 2020, doi: 10.1016/j.coldregions.2020.103118.

[48] J. Adamowski and K. Sun, "Development of a coupled wavelet transform and neural network method for flow forecasting of non-perennial rivers in semi-arid watersheds," *Journal of Hydrology*, vol. 390, no. 1–2, pp. 85–91, 2010, doi: 10.1016/j.jhydrol.2010.06.033.

[49] I. H. Sarker, "Deep learning: A comprehensive overview on techniques, taxonomy, applications and research directions," *SN Computer Science*, vol. 2, no. 6, 2021. doi: 10.1007/s42979-021-00815-1.

[50] V. Nourani, M. Komasi, and A. Mano, "A multivariate ANN-wavelet approach for rainfall-runoff modeling," *Water Resources Management*, vol. 23, no. 14, pp. 2877–2894, 2009, doi: 10.1007/s11269-009-9414-5.

[51] M. J. Alizadeh and M. R. Kavianpour, "Development of wavelet-ANN models to predict water quality parameters in Hilo Bay, Pacific Ocean," *Marine Pollution Bulletin*, vol. 98, no. 1–2, pp. 171–178, 2015, doi: 10.1016/j.marpolbul.2015.06.052.

[52] Z. Cömert and A. F. Kocamaz, "Journal of science and technology a study of artificial neural network training algorithms for classification of cardiotocography signals," *Journal of Science and Technology*, vol. 7, no. 2, pp. 93–103, 2017, [Online]. Available: www.dergipark.ulakbim.gov.tr/beuscitech/.

[53] P. Gou and J. Yu, "A nonlinear ANN equalizer with mini-batch gradient descent in 40Gbaud PAM-8 IM/DD system," *Optical Fiber Technology*, vol. 46, pp. 113–117, 2018, doi: 10.1016/j.yofte.2018.09.015.

[54] F. Barani, A. Savadi, and H. S. Yazdi, "Convergence behavior of diffusion stochastic gradient descent algorithm," *Signal Processing*, vol. 183, 2021, doi: 10.1016/j.sigpro.2021.108014.

[55] I. Chakroun, T. Haber, and T. J. Ashby, "SW-SGD: The sliding window stochastic gradient descent algorithm," In *Procedia Computer Science*, Elsevier B.V., pp. 2318–2322, 2017. doi: 10.1016/j.procs.2017.05.082.

[56] A. Alberto Quesada, "5 algorithms to train a neural network," 2022, https://www.neuraldesigner.com/blog/5_algorithms_to_train_a_neural_network.

[57] M. Almiani, A. Abughazleh, Y. Jararweh, and A. Razaque, "Resilient back propagation neural network security model for containerized cloud computing," *Simulation Modelling Practice and Theory*, vol. 118, p. 102544, 2022, doi: 10.1016/j.simpat.2022.102544.

[58] M. Riedmiller and H. Braun, "A direct adaptive method for faster backpropagation learning: the RPROP algorithm," In *IEEE International Conference on Neural Networks*, pp. 586–591, 1993, vol.1, doi: 10.1109/ICNN.1993.298623.

[59] M. F. Møller, "A scaled conjugate gradient algorithm for fast supervised learning," *Neural Networks*, vol. 6, no. 4, pp. 525–533, 1993, doi: 10.1016/S0893-6080(05)80056-5.

[60] Z. Chen and X. Chen, "Conjugate gradient-based iterative algorithm for solving generalized periodic coupled Sylvester matrix equations," *Journal of the Franklin Institute*, vol. 359, no. 17, pp. 9925–9951, 2022, doi: 10.1016/j.jfranklin.2022.09.049.

[61] M. T. Hagan and M. B. Menhaj, "Training feedforward networks with the marquardt algorithm," *IEEE Transactions on Neural Networks*, vol. 5, no. 6, pp. 989–993, 1994, doi: 10.1109/72.329697.

[62] R. Battiti, "First- and second-order methods for learning: Between steepest descent and newton's method," *Neural Computing*, vol. 4, no. 2, pp. 141–166, 1992, doi: 10.1162/neco.1992.4.2.141.

[63] D. W. Marquardt, "An algorithm for least-squares estimation of nonlinear parameters," *Journal of the Society for Industrial and Applied Mathematics*, vol. 11, no. 2, pp. 431–441, 1963, [Online]. Available: https://www.jstor.org/stable/2098941.

[64] M. T. Hagan and M. B. Menhaj, "Training feedforward networks with the Marquardt algorithm," *IEEE Transactions on Neural Networks*, vol. 5, no. 6, pp. 989–993, 1994, doi: 10.1109/72.329697.

[65] O. Adeleke, S. A. Akinlabi, T. Jen, and I. Dunmade, "Application of artificial neural networks for predicting the physical composition of municipal solid waste : An assessment of the impact of seasonal variation," *Waste Management & Research*, vol. 39, no. 8, pp. 1058–1068, 2021, doi: 10.1177/0734242X21991642.

[66] X. Hu, S. Wen, and H. K. Lam, "Dynamic random distribution learning rate for neural networks training," *Applied Soft Computing*, vol. 124, 2022, doi: 10.1016/j.asoc.2022.109058.

[67] S. Yu, H. Li, X. Chen, and D. Lin, "Multistability analysis of quaternion-valued neural networks with cosine activation functions," *Applied Mathematics and Computation*, vol. 445, p. 127849, 2023, doi: 10.1016/j.amc.2023.127849.

[68] O. Adeleke, S. Akinlabi, T.-C. Jen, and I. Dunmade, "Prediction of the heating value of municipal solid waste: a case study of the city of Johannesburg," *International Journal of Ambient Energy*, pp. 1–12, 2020, doi: 10.1080/01430750.2020.1861088.

[69] I. Kandel and M. Castelli, "The effect of batch size on the generalizability of the convolutional neural networks on a histopathology dataset," *ICT Express*, vol. 6, no. 4, pp. 312–315, 2020, doi: 10.1016/j.icte.2020.04.010.

[70] O. G. Ajayi and J. Ashi, "Effect of varying training epochs of a faster region-based convolutional neural network on the accuracy of an automatic weed classification scheme," *Smart Agricultural Technology*, vol. 3, p. 100128, 2023, doi: 10.1016/j.atech.2022.100128.

[71] K. A. Adegoke, O. Adeleke, M. O. Adesina, R. O. Adegoke, and O. S. Bello, "Clean technology for sequestering rhodamine B dye on modified mango pod using artificial intelligence techniques," *Current Research in Green and Sustainable Chemistry*, vol. 5, 2022, doi: 10.1016/j.crgsc.2022.100275.

[72] O. Sanni, O. Adeleke, K. Ukoba, J. Ren, and T.-C. Jen, "Application of machine learning models to investigate the performance of stainless steel type 904 with agricultural waste," *Journal of Materials Research and Technology*, vol. 20, pp. 4487–4499, 2022, doi: 10.1016/j.jmrt.2022.08.076.

[73] A. Azadeh, M. Saberi, A. Kazem, V. Ebrahimipour, A. Nourmohammadzadeh, and Z. Saberi, "A flexible algorithm for fault diagnosis in a centrifugal pump with corrupted data and noise based on ANN and support vector machine with hyper-parameters optimization," *Applied Soft Computing*, vol. 13, no. 3, pp. 1478–1485, 2013, doi: 10.1016/j.asoc.2012.06.020.

[74] J. Ma, J. Wang, Y. Han, S. Dong, L. Yin, and Y. Xiao, "Towards data-driven modeling for complex contact phenomena via self-optimized artificial neural network methodology," *Mechanism and Machine Theory*, vol. 182, p. 105223, 2023, doi: 10.1016/j.mechmachtheory.2022.105223.

[75] R. Noori, A. Karbassi, and M. S. Sabahi, "Evaluation of PCA and Gamma test techniques on ANN operation for weekly solid waste prediction," *Journal of Environmental Management*, vol. 91, no. 3, pp. 767–771, 2010, doi: 10.1016/j.jenvman.2009.10.007.

[76] R. Noori, A. Khakpour, B. Omidvar, and A. Farokhnia, "Comparison of ANN and principal component analysis-multivariate linear regression models for predicting the river flow based on developed discrepancy ratio statistic," *Expert Systems with Applications*, vol. 37, no. 8, pp. 5856–5862, 2010, doi: 10.1016/j.eswa.2010.02.020.

[77] H. Wang, S. J. Hsieh, B. Peng, and X. Zhou, "Non-metallic coating thickness prediction using artificial neural network and support vector machine with time resolved thermography," *Infrared Physics & Technology*, vol. 77, pp. 316–324, 2016, doi: 10.1016/j.infrared.2016.06.015.

[78] Bhavna Kaveti, "Methods for measuring thin film thickness," 2023, https://www.azonano.com/article.aspx?ArticleID=6372.

[79] M. F. Tabet and W. A. Mcgahan, "Use of arti(r)cial neural networks to predict thickness and optical constants of thin (r)lms from re‾ectance data." [Online]. Available: www.elsevier.com/locate/tsf.

[80] W. Z. Cui, C. C. Zhu, and H. P. Zhao, "Prediction of thin film thickness of field emission using wavelet neural networks," *Thin Solid Films*, vol. 473, no. 2, pp. 224–229, 2005, doi: 10.1016/j.tsf.2004.06.121.

[81] A. Bora, "Application of artificial neural network to determine the thickness profile of thin film," *Materials Today: Proceedings*, vol. 65, pp. 2807–2811, 2022, doi: 10.1016/j.matpr.2022.06.220.

[82] M. Marian, J. Mursak, M. Bartz, F. J. Profito, A. Rosenkranz, and S. Wartzack, "Predicting EHL film thickness parameters by machine learning approaches," *Friction*, vol. 11, no. 6, pp. 992–1013, 2023, doi: 10.1007/s40544-022-0641-6.

[83] J. Lee and J. Jin, "A novel method to design and evaluate artificial neural network for thin film thickness measurement traceable to the length standard," *Scientific Reports*, vol. 12, no. 1, 2022, doi: 10.1038/s41598-022-06247-y.

[84] H. Habuka, T. Nagoya, M. Mayusumi, M. Katayama, M. Shimada, and K. Okuyama, "Model on transport phenomena and epitaxial growth of silicon thin film in $SiHCl_3$ H_2 system under atmospheric pressure," *Journal of Crystal Growth*, vol. 169, no. 1, pp. 61–72, 1996, doi: 10.1016/0022-0248(96)00376-4.

[85] J. Cagnazzo, O. S. Abuomar, A. Yanguas-Gil, and J. W. Elam, "Atomic layer deposition optimization using convolutional neural networks," In *2021 International Conference on Computational Science and Computational Intelligence (CSCI)*, pp. 228–232, 2021. doi: 10.1109/CSCI54926.2021.00110.

[86] A. Yanguas-Gil and J. W. Elam, "Machine learning and atomic layer deposition: Predicting saturation times from reactor growth profiles using artificial neural networks," *Journal of Vacuum Science & Technology A*, vol. 40, no. 6, p. 062408, 2022, doi: 10.1116/6.0001973.

[87] D. S. Bulgarevich, S. Tsukamoto, T. Kasuya, M. Demura, and M. Watanabe, "Pattern recognition with machine learning on optical microscopy images of typical metallurgical microstructures," *Scientific Reports*, vol. 8, no. 1, p. 2078, 2018, doi: 10.1038/s41598-018-20438-6.

[88] B. L. DeCost, T. Francis, and E. A. Holm, "Exploring the microstructure manifold: Image texture representations applied to ultrahigh carbon steel microstructures," *Acta Mater*, vol. 133, pp. 30–40, 2017.

[89] A. Rovinelli, M. D. Sangid, H. Proudhon, and W. Ludwig, "Using machine learning and a data-driven approach to identify the small fatigue crack driving force in polycrystalline materials," *NPJ Computational Materials*, vol. 4, no. 1, p. 35, 2018, doi: 10.1038/s41524-018-0094-7.

[90] L. Banko, Y. Lysogorskiy, D. Grochla, D. Naujoks, R. Drautz, and A. Ludwig, "Predicting structure zone diagrams for thin film synthesis by generative machine learning," *Communications Materials*, vol. 1, no. 1, 2020, doi: 10.1038/s43246-020-0017-2.

[91] D. P. Kingma and M. Welling, "Auto-encoding variational bayes," 2013, [Online]. Available: https://arxiv.org/abs/1312.6114.

[92] R. Salakhutdinov, "Learning deep generative models," *Annual Review of Statistics and Its Application*, vol. 2, no. 1, pp. 361–385, 2015, doi: 10.1146/annurev-statistics-010814-020120.

[93] A. Joshi *et al.*, "InvNet: Encoding geometric and statistical invariances in deep generative models," *Proceedings of the AAAI Conference on Artificial Intelligence*, vol. 34, no. 4, pp. 4377–4384, 2020, doi: 10.1609/aaai.v34i04.5863.

[94] R., S. N., G. M. & O. O. O. Noraas, *AIAA Scitech 2019 Forum*, California: American Institute of Aeronautics and Astronautics, 2019.

[95] P. R. Griffiths and T. A. L. Harris, "Machine learning workflow for microparticle composite thin-film process-structure linkages," *Journal of Coatings Technology and Research*, vol. 19, no. 1, pp. 83–96, 2022, doi: 10.1007/s11998-021-00512-x.

[96] S. Hashemi and S. R. Kalidindi, "A machine learning framework for the temporal evolution of microstructure during static recrystallization of polycrystalline materials simulated by cellular automaton," *Computational Materials Science*, vol. 188, p. 110132, 2021, doi: 10.1016/j.commatsci.2020.110132.

[97] D. B. Brough, A. Kannan, B. Haaland, D. G. Bucknall, and S. R. Kalidindi, "Extraction of process-structure evolution linkages from x-ray scattering measurements using dimensionality reduction and time series analysis," *Integrating Materials and Manufacturing Innovation*, vol. 6, no. 2, pp. 147–159, 2017, doi: 10.1007/s40192-017-0093-4.

[98] D. B. Brough, D. Wheeler, J. A. Warren, S. R. Kalidindi, and G. W. Woodruff, "Microstructure-based knowledge systems for capturing process-structure evolution linkages," *Current Opinion in Solid State and Materials Science*, vol. 21, no. 3, pp. 129–140 2017.

[99] S. Noguchi, H. Wang, and J. Inoue, "Identification of microstructures critically affecting material properties using machine learning framework based on metallurgists' thinking process," *Scientific Report*, vol. 12, no. 1, 2022, doi: 10.1038/s41598-022-17614-0.

[100] X. Long, C. Lu, Z. Shen, and Y. Su, "Identification of mechanical properties of thin-film elastoplastic materials by machine learning," *Acta Mechanica Solida Sinica*, vol. 36, no. 1, pp. 13–21, 2023, doi: 10.1007/s10338-022-00340-5.

[101] H. A. M. Ali, E. F. M. El-Zaidia, and R. A. Mohamed, "Experimental investigation and modeling of electrical properties for phenol red thin film deposited on silicon using back propagation artificial neural network," *Chinese Journal of Physics*, vol. 67, pp. 602–614, 2020, doi: 10.1016/j.cjph.2020.07.018.

[102] N. M. Sabri, N. D. Md Sin, M. Puteh, and M. Rusop Mahmood, "Prediction of nanostructured ZnO thin film properties based on neural network," *Advanced Materials Research*, pp. 266–269, 2014. doi: 10.4028/www.scientific.net/AMR.832.266.

[103] X. Wang, D. Han, Y. Hong, H. Sun, J. Zhang, and J. Zhang, "Machine learning enabled prediction of mechanical properties of tungsten disulfide monolayer," *ACS Omega*, vol. 4, no. 6, pp. 10121–10128, 2019, doi: 10.1021/acsomega.9b01087.

[104] S. Laxmikant Vajire, A. Prashant Singh, D. Kumar Saini, A. Kumar Mukhopadhyay, K. Singh, and D. Mishra, "Novel machine learning-based prediction approach for nanoindentation load-deformation in a thin film: Applications to electronic industries," *Computers & Industrial Engineering*, vol. 174, p. 108824, 2022, doi: 10.1016/j. cie.2022.108824.

[105] Y. P. Yang *et al.*, "Machine learning assisted classification of aluminum nitride thin film stress via in-situ optical emission spectroscopy data," *Materials*, vol. 14, no. 16, 2021, doi: 10.3390/ma14164445.

[106] T. B. Asafa, N. Tabet, and S. A. M. Said, "Taguchi method-ANN integration for predictive model of intrinsic stress in hydrogenated amorphous silicon film deposited by plasma enhanced chemical vapour deposition," *Neurocomputing*, vol. 106, pp. 86–94, 2013, doi: 10.1016/j.neucom.2012.10.019.

[107] A. A. Attia, M. S. El-Bana, D. M. Habashy, S. S. Fouad, and M. Y. El-Bakry, "Optical constants characterization of As30Se70−xSnx thin films using neural networks," *Journal of Applied Research and Technology*, vol. 15, no. 5, pp. 423–429, 2017, doi: 10.1016/j.jart.2017.03.009.

[108] S. M. Kim, S. D. H. Naqvi, M. G. Kang, H. E. Song, and S. Ahn, "Optical characterization and prediction with neural network modeling of various stoichiometries of perovskite materials using a hyperregression method," *Nanomaterials*, vol. 12, no. 6, 2022, doi: 10.3390/nano12060932.

[109] G. Dogan *et al.*, "Bayesian machine learning for efficient minimization of defects in ALD passivation layers," *ACS Applied Materials & Interfaces*, vol. 13, no. 45, pp. 54503–54515, 2021, doi: 10.1021/acsami.1c14586.

[110] N. H. Paulson, A. Yanguas-Gil, O. Y. Abuomar, and J. W. Elam, "Intelligent agents for the optimization of atomic layer deposition," *ACS Applied Materials & Interfaces*, vol. 13, no. 14, pp. 17022–17033, 2021, doi: 10.1021/acsami.1c00649.

[111] Y. K. Wakabayashi, T. Otsuka, Y. Krockenberger, H. Sawada, Y. Taniyasu, and H. Yamamoto, "Machine-learning-assisted thin-film growth: Bayesian optimization in molecular beam epitaxy of SrRuO3 thin films," *APL Materials*, vol. 7, no. 10, 2019, doi: 10.1063/1.5123019.

[112] I. Ohkubo *et al.*, "Realization of closed-loop optimization of epitaxial titanium nitride thin-film growth via machine learning," *Materials Today Physics*, vol. 16, 2021, doi: 10.1016/j.mtphys.2020.100296.

[113] A. Dulmaa, F. G. Cougnon, R. Dedoncker, and D. Depla, "On the grain size-thickness correlation for thin films," *Acta Materialia*, vol. 212, p. 116896, 2021, doi: 10.1016/j. actamat.2021.116896.

[114] B. Jugdersuren *et al.*, "The effect of ultrasmall grain sizes on the thermal conductivity of nanocrystalline silicon thin films," *Communications Physics*, vol. 4, no. 1, p. 169, 2021, doi: 10.1038/s42005-021-00662-9.

[115] M. J. A. Syukor, M. Ibrahim, M. A. Azam, and M. M. Razali, "Prediction Of grain size in the tin coating using artificial neural network," *International Journal of Applied Engineering Research*, vol. 11, no. 19, 9856–9869, 2016.

[116] A. Ibrahim Jarrah, A. Syukor Mohamad Jaya, M. Razali Muhamad, N. Abd Rahman, A. Samad Hasan Basari, and D. Tunggal, "Modeling and optimization of physical vapour deposition coating process parameters for TiN grain size using combined genetic algorithms with response surface methodology," *Journal of Theoretical and Applied Information Technology* vol. 20, no. 2, 2015, [Online]. Available: www.jatit.org.

[117] M. J. A. Syukor, M. Ibrahim, M. A. Azam, and M. M. Razali, "Intelligence integration of particle swarm optimization and physical vapour deposition for tin grain size coating process parameters," *Journal of Theoretical & Applied Information Technology*, vol. 84, no. 3, 2016.

[118] Y. Ding, Y. Zhang, Y. M. Ren, G. Orkoulas, and P. D. Christofides, "Machine learning-based modeling and operation for ALD of SiO_2 thin-films using data from a multiscale CFD simulation," *Chemical Engineering Research and Design*, vol. 151, pp. 131–145, 2019, doi: 10.1016/j.cherd.2019.09.005.

[119] C. Kim, T. N. Do, and J. Kim, "Machine learning-based analysis of the physio-chemical properties for the predictive thickness control of atomic layer deposition," *IFAC-PapersOnLine*, vol. 55, pp. 626–631, 2022. doi: 10.1016/j.ifacol.2022.07.513.

[120] H. J. Kim *et al.*, "Machine-learning-assisted analysis of transition metal dichalcogen-ide thin-film growth," *Nano Convergence*, vol. 10, no. 1, 2023, doi: 10.1186/s40580-023-00359-5.

[121] A. Salimian, A. Aminishahsavarani, and H. Upadhyaya, "Artificial neural networks to predict sheet resistance of indium-doped zinc oxide thin films deposited via plasma deposition," *Coatings*, vol. 12, no. 2, 2022, doi: 10.3390/coatings12020225.

[122] A. Bahramian, "Study on growth rate of TiO2 nanostructured thin films: Simulation by molecular dynamics approach and modeling by artificial neural network," *Surface and Interface Analysis*, vol. 45, no. 11, pp. 1727–1736, 2013, doi: 10.1002/sia.5314.

12 Machine Learning-Based Classification Techniques in ALD

12.1 INTRODUCTION

In the previous chapter, we established that machine learning-based regression analysis has received a lot of attention in atomic layer deposition (ALD) research for a precise prediction of significant properties of the thin film materials such as film thickness, composition, and quality toward the design of a novel material with specific properties. There is also a rising interest in applying machine learning for classification tasks. Machine learning-based classification techniques in ALD have a wide range of potential applications. Researchers can better understand the underlying mechanics and tailor the deposition process for particular applications by correctly identifying ALD samples. Classification can also be used for quality control, allowing automated sample sorting and screening based on predetermined criteria. A classification algorithm's objective is to create a model using labeled data that can correctly predict the class for brand-new, unforeseen data points. When data are classified, they are labeled, which means that each incoming data point has already been categorized into one of a number of predetermined classes. In this chapter, we shall examine the various classification tasks and techniques along with various potential applications of machine learning-based classification techniques in ALD and thin film deposition. In our final section, we'll examine a number of case studies from the literature where the classification techniques of supervised machine learning methods have been successfully used to improve ALD process and thin film deposition and materials. We will illustrate how these strategies have the potential to transform the thin film deposition space and expand our knowledge of ALD procedures through these instances.

12.2 CLASSIFICATION TECHNIQUES AND ALGORITHMS

Classification techniques are used to assign each item in the data collected to one of the predefined sets of classes or groups [1]. Classification technique in machine learning is used to predict a categorical output variable [2]. This suggests that there are two or more classifications or categories, such as black or white, male or female, yes or no, men or women (binary classification). The binary classification is used to predict one of two possible outputs from a given input. The two outcomes are called good and negative. Multi-class classification can occur. Multi-class

DOI: 10.1201/9781003346234-15

categorization divides input into more than two classes. This indicates that there are numerous classifications to which the input may be categorized. Another type of classification technique is the multi-label classification, where each input can be classified under more than one class or label. This implies that different labels may be simultaneously applied to one input [3–5].The goal of classification is to identify a relationship between the input variables and the output variable, which is frequently represented as a decision boundary on a graph. To predict the class label features (categorical in nature), a model or classifier is built in the classification technique. To accurately forecast the target class for each example in the dataset is the primary goal of the classification technique. We can begin using the classification technique after the dataset's class assignments are determined [2]. Common classifier's algorithms used for classification task in machine learning are multilayer perceptron (MLP) or extreme learning machine (ELM), support vector machine (SVM), Naive Bayes (NB), random forest (RF), decision tree (DT), and logistic regression (LRs). An overview of these classification algorithms is presented in Figure 12.1. This figure depicts the decision boundary of the classification algorithms.

These algorithms were discussed extensively in a previous chapter. For a quick recap of the classifier's algorithms, they are briefly discussed as follows.

12.2.1 MULTILAYER PERCEPTRON

The previous chapters have extensively studied the neural network and the multilayer perceptron (MLP). A chapter investigated the predictive analysis part of the neural network, while its classification task is the focus of this chapter. The classification system of neural network can be deployed in s standalone forma or in hybrid with the predictive analysis task. The output of the MLP is estimated as follows [1]:

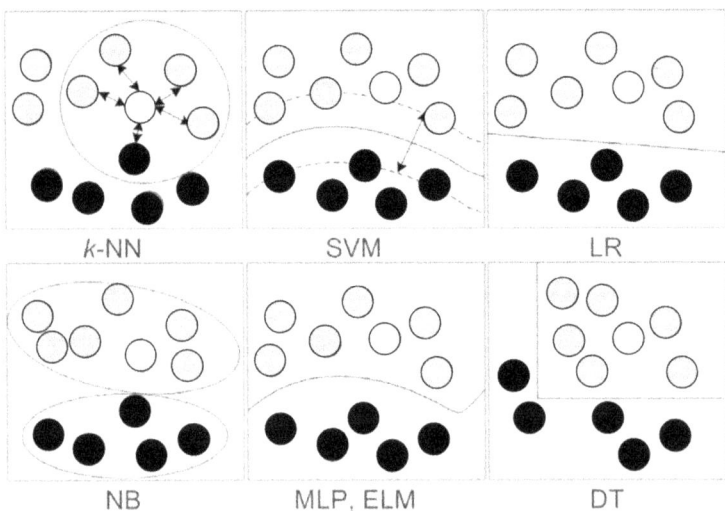

FIGURE 12.1   Decision boundaries of different classifier algorithms [1].

$$net_o \leftarrow \omega + \sum_{j \in Pred(i)} \omega_{ij} a_j \qquad (12.1)$$

$$a_i \leftarrow f_{log}(net_o) \qquad (12.2)$$

where ω denotes the synaptic weights and net_o is the input potential.

The ELM is a feed-forward neural network with a single hidden layer [6]. The output of the ELM can be represented as follows [7]:

$$f(x) = \sum_{i=0}^{M} \beta_i G(a_i, b_i, x) \qquad (12.3)$$

where $G(.)$ denotes the *ith* neuron's activation function in the hidden layer.

12.2.2 k-Nearest Neighbor

kNN is a simple, proximity-based and voting-based classification and regression analysis approach [8]. Particularly for classification task, kNN chooses the new data point's class based on a vote between its k nearest neighbors [9, 10]. This algorithm has been extensively studied in previous chapters. To determine the closest neighbors, the euclidean distance between instances X_1 an X_2 is commonly utilized as in equation 12.4 [11]. Our objective is to minimize the Euclidean distance, and our computation is based on the count of smaller distances. A major demerit of the kNN classification approach is that it exhibits a notable decrease in speed as the number of cases and/or predictors/independent variables increases.

$$dist(X_1, X_2) = \sqrt{\sum_{i=1}^{n}(x_{1i} - x_{2i})^2} \qquad (12.4)$$

The KNN approach was used by Trejo et al. [12] to analyze how the atomic structure changes during ALD reactions. ALD reactions generated in situ X-ray absorption near-edge structure (XANES) spectra were then classified using the KNN algorithm to forecast the ALD process' development cycle.

12.2.3 Naive Bayes

NB is a classification algorithm which employs the probabilistic classification technique known based on the Bayes theorem which estimates the likelihood of an event based on prior knowledge of potential confounding factors [13]. NB has the benefit of being computationally effective and having a short learning curve, even for big datasets [14,15]. Even with the oversimplifying premise of feature independence, it works well in practice. This presumption might, however, not always be true, which could result in projections that are off. When the number of features is high compared to the size of the dataset, NB is frequently employed as a baseline model for text classification tasks. As a probabilistic classifier that relies on the Bayes theorem and assumes that each input attribute is independent (predictors), it is determined by the probability of classes C_i given input qualities X as shown in Equation 12.5 [1]:

$$P(C_i|X) = \frac{P(X \mid C_i)P(C_i)}{P(X)} \qquad (12.5)$$

12.2.4 SUPPORT VECTOR MACHINES

The SVM is a class of classifier algorithms in machine learning used to determine the best boundary between two classes. SVMs, which are based on statistical learning frameworks, are one of the most reliable prediction techniques [16]. Discovering the hyperplane that maximizes the margin between the two classes, or the distance between the closest data points of each class and the hyperplane, is the main goal of the SVM [17]. The distance between the hyperplane and the nearest data points for each class is referred to as the margin. The data points nearest to this hyperplane are referred to as support vectors, and it is known as the optimal boundary. SVM performs well with high-dimensional data and avoids the dimensionality curse by simultaneously maximizing the margin hyperplane and minimizing residuals as follows [18,19]:

$$Y(x) = sign\left(\sum_{i=0}^{n} y_i \propto_i K\left(x_i x_j\right) + b \right) \qquad (12.6)$$

where K represents the kernel function between the $x_i i$ and x_j input vectors.

12.2.5 DECISION TREE

Classification and regression task in supervised machine learning has found a wide use for the DT algorithm. DT predicts the value of a target variable using a variety of input features by segmenting a dataset into successively smaller groups in order to determine the ultimate categorization of a certain observation [20]. The DT method creates a tree-like model of decisions and potential outcomes, with each node in the tree reflecting a decision based on a specific aspect or attribute of the data. Each leaf node represents a class label, and each decision node represents a test on an attribute. With regard to the target variable, a DT aims to divide the data into subsets that are as homogeneous as possible [21]. The DT algorithm's fundamental premise is to divide the data into subsets depending on the values of the attributes, with the goal of having each subset contain data that have the same value for each individual attribute [22]. The process is then repeatedly repeated by the algorithm for each subset up until a halting requirement is satisfied.

12.2.6 LOGISTIC REGRESSION

Logistic regression (LR) is a statistical approach for analyzing the relationship between a dependent class attribute and a number of independent attributes. Logistic regression's key benefit is its capacity to model the relationship between dependent and independent variables, even in the face of intricate and nonlinear correlations. In addition, logistic regression is reasonably simple to comprehend because each independent variable's influence on the projected probability of the positive class can be determined using the independent variable's coefficients. The following relationship is modeled using a logistic function as follows:

$$g(x) = In\left[\frac{\pi(x)}{1-\pi(x)}\right] = In\left[\frac{p(y=1|x)}{p(y=0|x)}\right] = \beta + \sum_{i=1}^{p}\beta_i x_i \qquad (12.7)$$

12.2.7 RANDOM FOREST

A RF is another machine learning classifier which integrates multiple DT algorithms through an ensemble learning technique to increase the model's overall robustness and accuracy. It belongs to the DT class and is a powerful technique for handling binary and multi-class classification and regression issues [13,23]. A random vector of features with a defined probability distribution is used in each DT. By employing the bootstrap aggregating (bagging) technique, it is ensured that base learners are trained on a variety of training data [1].

12.3 POTENTIAL APPLICATIONS OF MACHINE LEARNING-BASED CLASSIFICATION TECHNIQUES IN ALD

12.3.1 MATERIAL CLASSIFICATION

The capability of the machine learning algorithms to analyze huge datasets and draw insightful conclusions from complex data have become an useful approach for materials classification. In order to find materials with particular features that are critical for a variety of applications, machine learning is particularly helpful in the classification of materials. Finding the primary qualities or properties of the materials under study is the first stage in the classification of materials. These characteristics may be structural, like grain size or crystal structure, or functional, like electrical or optical characteristics. The patterns that point to particular material properties and can be detected using a classification algorithm after the pertinent properties have been established. Further to this, the hidden connections between several materials properties can be unveiled to classify the materials. This ability can be particularly demonstrated as a viable alternative to the conventional analytical methods which could be challenging in establishing the connection between crystal structure and electrical conductivity of materials.

The machine learning-based classification technique offers an extra benefit of quality control by categorizing materials according to predetermined criteria, which lowers the need for manual inspection and boosts process efficiency.

12.3.2 OPTIMAL PROCESS PARAMETER SELECTION

The choice of process parameters is a crucial component in the optimization of ALD process optimization. The objective is to get the necessary thin film qualities with the least amount of effort and expense. By categorizing the impact of various process parameters on thin film qualities, machine learning algorithms can aid in this process. Machine learning can help in the selection of the best process conditions for a certain application by highlighting the crucial elements. When trained on a dataset comprising details on the various deposition process parameters and the properties of the produced thin films, the algorithms can be used to forecast the ideal circumstances for a specific set of process parameters. This reduces process optimization trial-and-error time and cost. Machine learning classification approaches can classify how temperature, pulse time, and precursor exposure time affect film thickness, refractive index, and crystallinity. This approach is capable of properly predicting the characteristics of TiO_2 thin films under a variety of processing settings, offering knowledge of the most advantageous processing scenarios for obtaining the desired qualities.

12.3.3 DEFECTS CLASSIFICATION AND DETECTION

While thin films have found wide applications and demonstrated excellent performance in several fields such as electronics, optics, and coatings, defects in thin films can drastically lower their quality and performance, resulting in product failure and wastage. Manual inspection and characterization, which can be time-consuming and arbitrary, are the traditional methods for identifying and defining defects in thin films. An effective alternative is provided by machine learning techniques, which enable automated and data-driven fault detection based on classification techniques with ALD databases. For instance, thin film cracks, pinholes, and delamination can be detected and classified using machine learning algorithms, enabling producers to enhance the caliber and dependability of their goods. Defect identification can also lessen the number of flaws that need to be fixed or thrown away, which reduces waste and boosts productivity. Classifier algorithm can be trained on a dataset of labeled samples of thin film with or without flaws. The algorithm can be used to categorize new samples as either faulty or nondefective when it learns to recognize the aspects or traits of the films that are connected to flaws.

The ability to detect defects in thin films using machine learning may be extremely accurate and dependable. Machine learning algorithms can learn to recognize even tiny flaws that may be challenging for human inspectors to find by employing vast datasets of labeled samples. Defect identification based on machine learning can also be highly automated, which decreases the need for manual inspection and boosts process effectiveness. The ability to identify the causative factors of faults is a significant benefit of machine learning-based defect identification.

12.3.4 DISCOVERY OF MATERIALS

The ability to create novel materials with customized qualities is crucial for the development of these nanotechnologies, yet the conventional method of materials discovery, trial and error, can be time-consuming, expensive, and ineffective. A potent alternative is provided by machine learning techniques, which enable data-driven materials discovery. Machine learning algorithms can be used in ALD to categorize the characteristics of novel materials created by ALD and pinpoint the materials that stand the best chance of being utilized in particular applications. The search space for novel materials synthesized by ALD is enormous, however, and conventional approaches to materials discovery can be prohibitively expensive and time-consuming. In order to find patterns and correlations between the properties of materials and their synthesis circumstances, machine learning-based classification algorithms analyze vast databases of experimental and simulation data. Machine learning algorithms can find the most promising materials for particular applications by learning from these data and help to expedite the discovery of novel materials with desired qualities.

These algorithms can be used to forecast the properties of new materials based on their synthesis circumstances after learning to recognize the traits or characteristics of the materials that are connected with their properties. Further to this, the machine learning-enhanced materials discovery offers the benefit of optimizing the synthesis conditions for novel materials..

12.3.5   ALD QUALITY CONTROL AND MONITORING

To guarantee that the deposited films fulfill the required criteria, thin film production procedures need to tightly manage a number of aspects. This is crucial in fields like semiconductor fabrication, where even minute changes in thin film characteristics can have a big effect on how well a device works. Hence, quality control is a crucial step in the production of thin films, and machine learning can offer a potent tool for identifying departures from predicted behavior. These techniques offer insightful information about the process. A classifier algorithm can be trained on a set of thin films with known attributes as one method of applying machine learning for thin film quality control. The algorithm can then be used to categorize new thin films according to the measured qualities of their composition, thickness, and roughness. The categorization results can then be compared to a list of established parameters to identify deviations from expected behavior. The creation and production of high-quality thin films require real-time quality monitoring and control. Machine learning classifier algorithms can offer insightful information about the process and enable quick response in the event of process deviations or equipment failures. The use of machine learning classifier's algorithms for real-time monitoring and control is anticipated to become more crucial as the field of ALD develops for the creation of new materials and the improvement of the deposition process.

12.4   CASE STUDIES EVALUATION

While there are several potentials and opportunities to enhance an excellent ALD process and achieve a high-quality thin film using the classification techniques of machine learning, the number of researches that particularly use classification algorithms for ALD process analysis and optimization is still limited, despite the growing interest in this area. However, recent studies have indicated that classifiers like kNN and SVM can be used to predict film qualities and optimize ALD process settings. As a result, even while progress is being made, there is still much to discover in this area of machine learning application. Few case studies in literature are discussed as follows.

12.4.1 A CASE STUDY OF MACHINE LEARNING-BASED CLASSIFICATION TECHNIQUES FOR AN ENHANCED EFFICIENT SPECTROSCOPIC ELLIPSOMETRY MODELING

The application of machine learning-based classification algorithms in ALD to improve the effectiveness and accuracy of spectroscopic ellipsometry (SE) modeling was explored in the article titled "Machine Learning-enhanced Efficient Spectroscopic Ellipsometry Modeling" by Arunachalam et al. [25]. The goal of the authors is to create a machine learning-based method to facilitate efficient thin film generation by properly predicting the refractive index of ALD thin films, a crucial factor in the production of microelectronic devices. After giving a general introduction of SE modeling and discussing the difficulties in precisely forecasting the refractive index of ALD thin films, the authors utilized a dataset of SE spectra and associated refractive index values to train a classification algorithm as part of their method for creating a machine learning-based SE model. The authors selected some classification algorithms because they are adept and effective at multi-class classification jobs. The approaches also require little to no prior knowledge of the distribution of input data, are resistant to overfitting during training, and are resilient. The algorithms selected are: support vector machine, k-nearest neighbors, DT, RF, and logistic regression. The ALD process was then optimized for the SE of TiO_2 substrates as a way for the authors to show how their machine learning-based SE model may be used in practice. In order to produce a thin film with refractive index that accurately reflected the desired value, they were capable of identifying the ideal ALD process settings.

This research shows how machine learning-based classification methods have the potential to increase the effectiveness and precision of SE modeling in ALD. The authors were able to create a model that offered more precise predictions than conventional SE modeling approaches by training a classification algorithm on a huge dataset of SE spectra and associated refractive index values. The ALD process may be optimized for the creation of microelectronic devices using this machine learning-based SE model. This study's focus is on a particular variable, the refractive index, which is essential for the creation of microelectronic devices. The authors have significantly advanced the field of ALD by creating a machine learning-based SE model that can correctly forecast this parameter. Also, the authors offer a thorough justification of their methodology, making it understandable to researchers of various levels of experience. The authors were more focused on a particular classification algorithm, namely, the k-nearest neighbor out of all algorithms selected, while its performance was not sufficiently evaluated against that of other algorithms. Some of the significant results of the study are as follows: Figure 12.2 shows the accuracy of all the classification algorithms selected, while changes in the performance accuracy of the kNN classifier for random down sampling with the classes and thickness is shown in Figure 12.3 revealing how level-1 accuracy and level-2 accuracy varied with thickness for the first 300 samples. Figure 12.4 shows the performance accuracy of classification for TiO_2 data for levels 1 and 2.

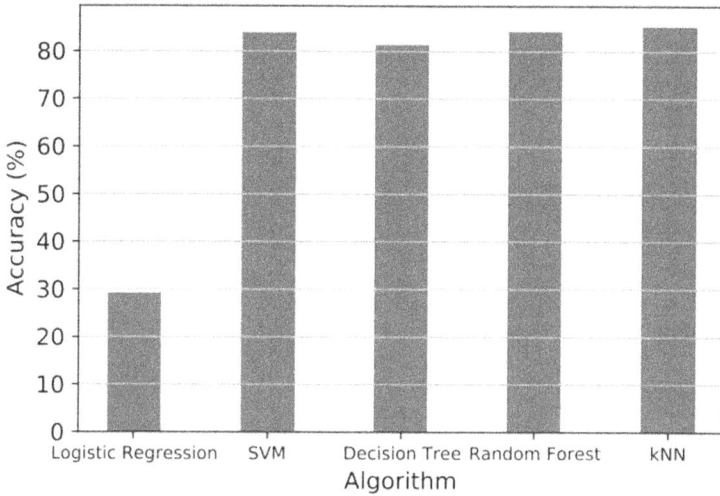

FIGURE 12.2 Performance accuracy of all classifiers selected [25].

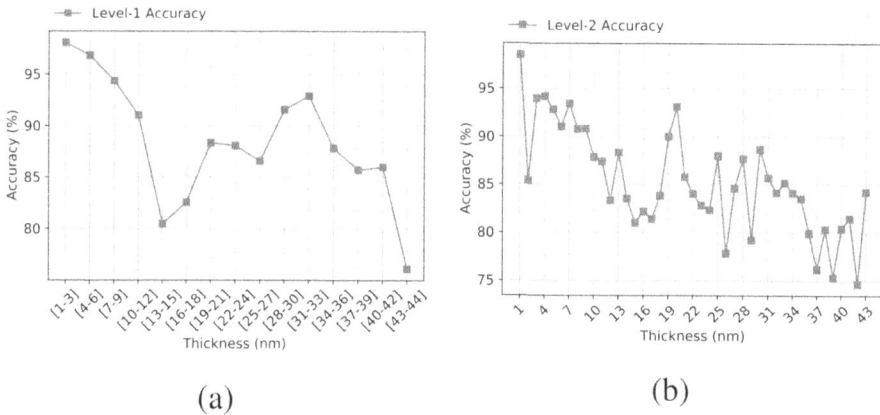

FIGURE 12.3 Changes in the performance accuracy of the kNN classifier for random down sampling with the (b) classes and (b) thickness [25].

12.4.2 A Case Study of Classifier Algorithm for Understanding the Evolving Atomic Structure during the Deposition Process

This research deployed the classification and prediction capabilities of the artificial neural network and RF to understand the evolving atomic structure in ALD reactions with in Situ simulated X-ray absorption near edge structure (XANES) and machine learning. The research was carried out by Trejo et al. [12]. The objective of the study is to improve our knowledge of the ALD reaction processes and the dynamism of atomic structure during the deposition process. The ALD reactions of ZnS thin film and TiO_2 were seen in situ using X-ray absorption near-edge

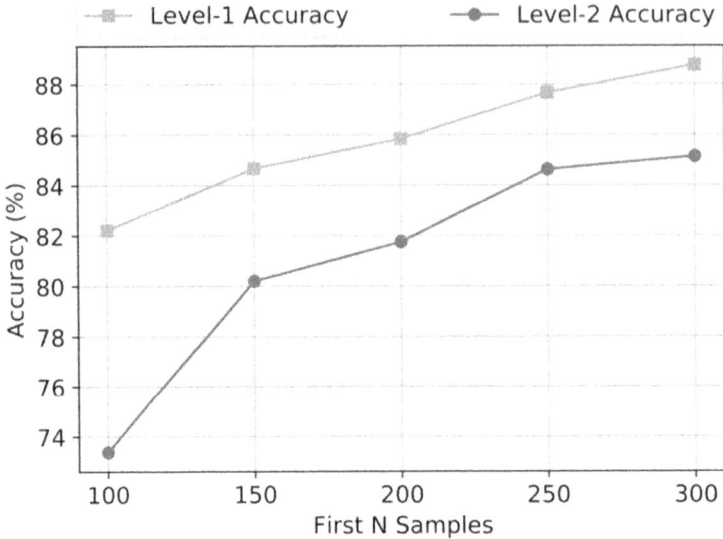

FIGURE 12.4   Performance accuracy of classification for TiO_2 data for levels 1 and 2 [25].

spectroscopy (XANES). The atomic species and their oxidation state can be determined using the spectroscopic fingerprint that XANES offers. Artificial neural network and RF are two machine learning classification techniques that were used to examine the obtained data. The findings demonstrated that the titanium atoms in the formed thin films could be correctly classified into their various oxidation states using the ANN and RF algorithms. In addition, the algorithms were able to spot minor atomic structure changes that occurred throughout the ALD process that were difficult to see with existing analytical methods. The authors also explained how their method may be used to enhance the quality of deposited thin films and to optimize the ALD process parameters. Using in situ XANES and machine learning classifier algorithm, it may be possible to manage the deposition process precisely while also enhancing the reproducibility and homogeneity of the deposited films. The viability, efficiency, and feasibility of machine learning classifier algorithms in ALD research has been established in this research. In situ XANES and machine learning algorithms could work together to create a potent tool for analyzing how the atomic structure changes during ALD reactions and for streamlining the deposition procedure. The strategy outlined in this paper could also be applied to materials and methods other than ALD, offering a promising direction for further study. The study examined two materials, so its findings may not apply to other materials or systems. Algorithms and data quality can also affect the method's success. Yet, the study emphasizes how machine learning algorithms could advance our knowledge of and ability to optimize ALD and other thin film deposition techniques. Figure 12.5 represents the result of atomic schematics, the S k-edge spectra of the structural parameter using the RF.

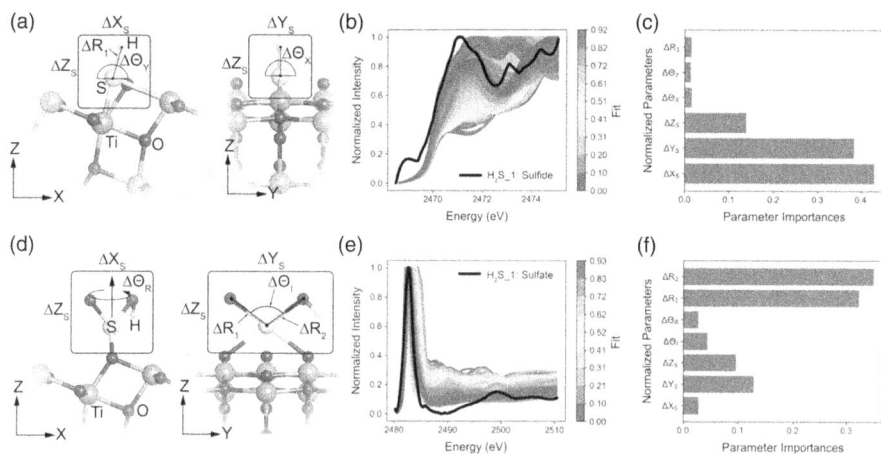

FIGURE 12.5 Result of atomic schematics, the S k-edge spectra of the structural parameter using the random forest [12].

12.4.3 MACHINE LEARNING-ASSISTED ELECTRONIC CLASSIFICATION OF BARCODED PARTICLES FOR MULTIPLEXED DETECTION USING ALD PROCESSES

In this research carried out by Sui et al. [26], nine barcoded particles were created using ALD to create oxide layers of varying thicknesses and different dielectric materials, and then, the precision of particle barcode classification using multi-frequency impedance cytometry in conjunction with supervised machine learning were evaluated. The research proposed a unique machine learning-based approach for multiplexed analysis for barcoded particle identification. The idea of barcoded particles and their potential uses in a variety of industries, including drug development, diagnostics, and environmental monitoring, were initially introduced in the article. The relevance of the ALD technique in creating homogeneous, conformal thin films with exact thickness control is next discussed by the authors. The ALD technique, however, has a low throughput problem that restricts its use in high-throughput screening and analysis. The authors proposed a technique for multiplexed barcoded particle detection utilizing the classification task of supervised machine learning analysis to tackle this problem. The process uses a microfluidic device to produce a stream of barcoded particles that are subsequently identified by an electronic circuit that assesses their electrical characteristics. A machine learning model, namely, SVM which uses the measured data was trained on a Gaussian kernel algorithm to classify the particles based on their barcode and other properties. The accuracy of the SVM classifier, which also shows the impedance difference between various barcoded particles, is assessed by comparing the predicted class with the true class. The authors present interesting experimental findings that show the viability of their method for multiplexed detection of barcoded particles. They demonstrate the outcome of their approach for high-throughput screening and analysis by demonstrating that their machine learning model can obtain a high classification accuracy of up to 91% on a

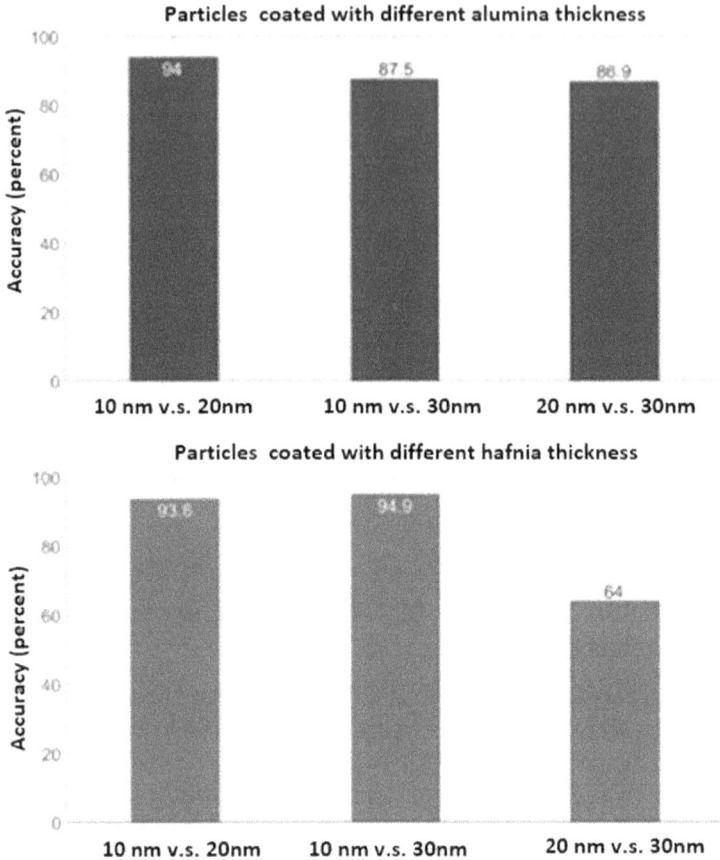

FIGURE 12.6 (A)   The accuracy of the support vector machine (SVM) model for identifying particles with various alumina thicknesses. (b) The accuracy of the SVM model in distinguishing particles with various hafnia thicknesses [26].

test dataset. The result of the SVM classifier accuracy for identifying the thin film with various thicknesses is shown in Figure 12.6.

12.4.4 Classification of Aluminum Nitride Thin Film Stress Using Machine Learning Classification Techniques Using Optical Emission Spectroscopy Data

Large residual stresses are often produced during the formation of the film, which will have a big impact on how well it works. An essential aspect of film strain engineering is the film stress which has an impact on the performance of semiconductor devices and must be controlled and reassessed in connection to process parameters to improve electron mobility in devices [27–29]. This has thus motivated Yang et al. [30] to build a novel approach comprising an unsupervised learning algorithm, namely, principal component analysis (PCA) for data pre-processing and feature extraction and MLP neural network for binary-class classification and prediction of film residual using Optical Emission Spectroscopy (OES) data during the deposition of aluminum nitride (AlN) on Silicon (Si). In situ optical emission spectroscopy (OES) data are used by the authors to determine the stress level in the AlN thin films, a crucial factor in the manufacture of microelectronic devices. The paper starts off by giving a general overview of the significance of identifying, assessing, and controlling the stress level in thin film deposition. The experimental set-up, which consists of an OES system for real-time monitoring of the AlN thin film growth procedure and a specially constructed ALD reactor, is then described by the authors. Figure 12.7 is schematic illustration of the data collection approach from the experimental set-up vis-à-vis the OES data acquisition

After the data are extracted, the PCA algorithm, an unsupervised machine learning algorithm was used for pre-processing the data and feature extraction from the data. Although the data dimension is theoretically lowered, the overall performance of the data is either good or barely different. PCA was deployed to distinguish between data from various principal components (PCs). PC1 and PC2, which are retrieved using PCA, are the directions with the highest and second-largest data disparities, respectively. The PC 1 and PC 2 were then chosen by the authors and entered into the MLP for prediction and classification tasks. The methodological approach for PCA processing, MLP prediction, and classification is illustrated in Figure 12.8. Optimization and hyperparameter settings were carried out to determine the optimal number of PCs that completely represent the data and must be fed into the MLP, while the optimal combination of neurons and hidden layers in the MLP were assessed. After these, the performance of the algorithms was evaluated using metrics such as accuracy and root mean square error (RMSE).

The outcome of the optimal PCA and MLP is shown in Figure 12.9. This final outcome was further validated using a cross-validation method called the confusion matrix. The authors demonstrated that machine learning using PCA and ANNs can more effectively address issues with semiconductor processes. When combined with PCA, the prediction time is substantially preserved with minimal accuracy loss. Also, they showed how well the machine learning algorithms classify the stress

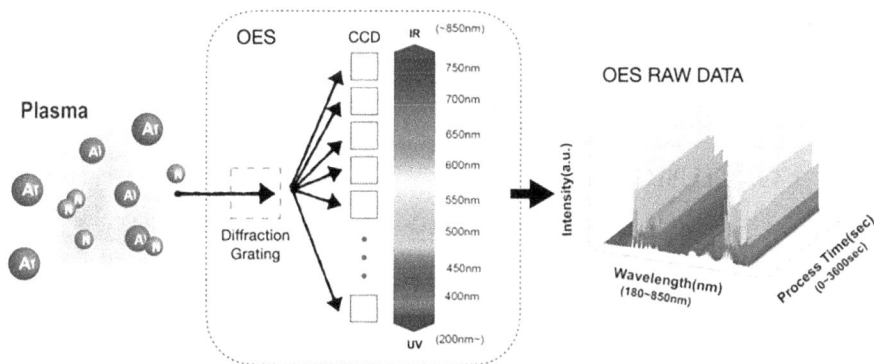

FIGURE 12.7   Schematic flowchart of data acquisition from the spectrometer [30].

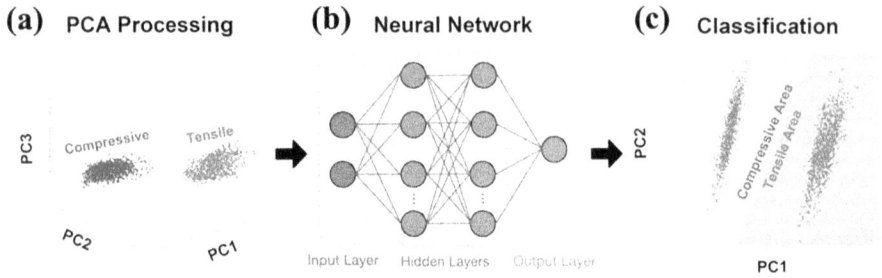

FIGURE 12.8   Methodological framework of the research [30].

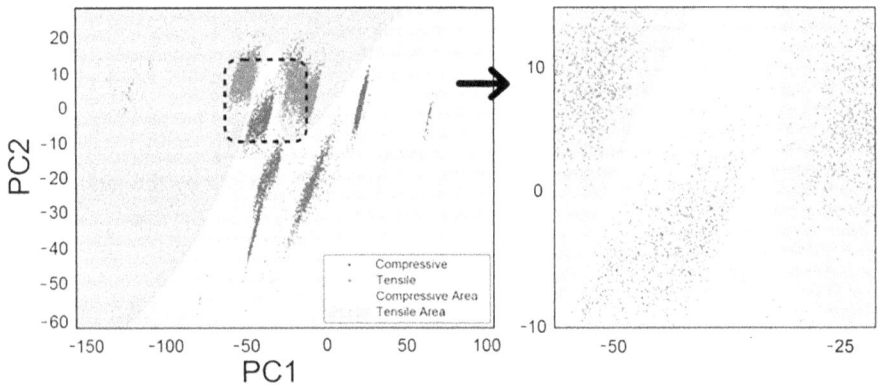

FIGURE 12.9   Outcome of feature extraction and classification task of principal component analysis (PCA) and multilayer perceptron (MLP) using PC1 and PC2 inputs [30].

levels of the AlN thin films. The OES spectra and their relationship to the stress levels in the AlN thin films are also thoroughly examined by the authors. According to the authors, their method can increase the effectiveness and precision of stress control during thin film fabrication, resulting in more dependable microelectronic devices. Finally, the study demonstrated the potential of machine learning classification algorithms to enhance stress control in thin film fabrication, making a significant contribution to the field of ALD. The results show the efficacy of the ANN and PCA algorithms in precisely categorizing stress levels.

12.5   CONCLUSION

Supervised machine learning techniques have demonstrated considerable progress in recent years when used for classification purposes in ALD and thin film deposition. These algorithms have greatly aided in the creation of new materials and procedures because of their capacity to precisely forecast material properties and optimize deposition parameters. Support vector machines and neural network algorithms found wider applications in ALD process and thin film depositions compared to other classifier algorithms in this chapter. Depending on the application, each method has distinct benefits and drawbacks. Aside from that, the case studies examined in this chapter demonstrate how well machine learning-based classification algorithms work in a variety of settings, including thin film stress classification and prediction, machine learning-assisted electronic classification of barcoded particles, understanding atomic structures during deposition, and machine learning-assisted surface characterization, amongst others. These algorithms provide a potent tool for expediting the development and improvement of ALD processes, while more study is required to fully comprehend their limitations and improve their effectiveness.

REFERENCES

[1] M. Z. Abedin, P. Hajek, T. Sharif, M. S. Satu, and M. I. Khan, "Modelling bank customer behaviour using feature engineering and classification techniques," *Research in International Business and Finance*, vol. 65, 2023, doi: 10.1016/j.ribaf.2023.101913.

[2] P. Banerjee, T. Chattopadhyay, and A. K. Chattopadhyay, "Comparison among different clustering and classification techniques: Astronomical data-dependent study," *New Astronomy*, vol. 100, p. 101973, 2023, doi: 10.1016/j.newast.2022.101973.

[3] Y. Liu and F. bin Zheng, "Object-oriented and multi-scale target classification and recognition based on hierarchical ensemble learning," *Computers and Electrical Engineering*, vol. 62, pp. 538–554, 2017, doi: 10.1016/j.compeleceng.2016.12.026.

[4] N. John, R. Surya, R. Ashwini, S. Sachin Kumar, and K. P. Soman, "A low cost implementation of multi-label classification algorithm using mathematica on Raspberry Pi," *Procedia Computer Science*, pp. 306–313, 2015. doi: 10.1016/j.procs.2015.02.025.

[5] A. Huang, R. Xu, Y. Chen, and M. Guo, "Research on multi-label user classification of social media based on ML-KNN algorithm," *Technological Forecasting and Social Change*, vol. 188, p. 122271, 2023, doi: 10.1016/j.techfore.2022.122271.

[6] S. S. Raju and P. Dhandayudam, "Prediction of customer behaviour analysis using classification algorithms," In *AIP Conference Proceedings*, American Institute of Physics Inc., 2018. doi: 10.1063/1.5032060.

[7] A. M. A. Sattar, Ö. F. Ertuğrul, B. Gharabaghi, E. A. McBean, and J. Cao, "Extreme learning machine model for water network management," *Neural Computing and Applications*, vol. 31, no. 1, pp. 157–169, 2019, doi: 10.1007/s00521-017-2987-7.

[8] M. Fopa, M. Gueye, S. Ndiaye, and H. Naacke, "A parameter-free KNN for rating prediction," *Data & Knowledge Engineering*, vol. 142, 2022, doi: 10.1016/j.datak.2022.102095.

[9] A. Belattmania, A. El Arrim, A. Ayouche, G. Charria, K. Hilmi, and B. El Moumni, "K nearest neighbors classification of water masses in the western Alboran Sea using the sigma-pi diagram," *Deep Sea Research Part I: Oceanographic Research Papers*, vol. 196, p. 104024, 2023, doi: 10.1016/j.dsr.2023.104024.

[10] M. M. dos Santos Freitas *et al.*, "KNN algorithm and multivariate analysis to select and classify starch films," *Food Packag Shelf Life*, vol. 34, 2022, doi: 10.1016/j.fpsl.2022.100976.

[11] M. S. Satu, S. Ahamed, F. Hossain, T. Akter, and D. M. Farid, "Mining traffic accident data of N5 national highway in bangladesh employing decision trees," In *2017 IEEE Region 10 Humanitarian Technology Conference (R10-HTC)*, pp. 722–725, 2017. doi: 10.1109/R10-HTC.2017.8289059.

[12] O. Trejo *et al.*, "Elucidating the evolving atomic structure in atomic layer deposition reactions with in situ XANES and machine learning," *Chemistry of Materials*, 2019, doi: 10.1021/acs.chemmater.9b03025.

[13] C. C. Paul Fergus, *Applied Deep Learning: Tools, Techniques and Implementation*, 1st edition, France: Springers, 2022.

[14] X. Zhao and Z. Xia, "Secure outsourced NB: Accurate and efficient privacy-preserving Naive Bayes classification," *Computers & Security*, vol. 124, p. 103011, 2023, doi: 10.1016/j.cose.2022.103011.

[15] L. Li *et al.*, "Naive Bayes classifier based on memristor nonlinear conductance," *Microelectronics Journal*, vol. 129, p. 105574, 2022, doi: 10.1016/j.mejo.2022.105574.

[16] A. Mat Deris, A. Mohd Zain, and R. Sallehuddin, "Overview of support vector machine in modeling machining performances," *Procedia Engineering*, pp. 308–312, 2011. doi: 10.1016/j.proeng.2011.11.2647.

[17] B. Samanta, K. R. Al-Balushi, and S. A. Al-Araimi, "Artificial neural networks and support vector machines with genetic algorithm for bearing fault detection," *Engineering Applications of Artificial Intelligence*, vol. 16, no. 7–8, pp. 657–665, 2003, doi: 10.1016/j.engappai.2003.09.006.

[18] M. Z. Abedin, G. Chi, M. M. Uddin, M. S. Satu, M. I. Khan, and P. Hajek, "Tax default prediction using feature transformation-based machine learning," *IEEE Access*, vol. 9, pp. 19864–19881, 2021, doi: 10.1109/ACCESS.2020.3048018.

[19] F. E. Moula, C. Guotai, and M. Z. Abedin, "Credit default prediction modeling: an application of support vector machine," *Risk Management*, vol. 19, no. 2, pp. 158–187, 2017, doi: 10.1057/s41283-017-0016-x.

[20] X. Han, X. Zhu, W. Pedrycz, and Z. Li, "A three-way classification with fuzzy decision trees," *Applied Soft Computing*, vol. 132, 2023, doi: 10.1016/j.asoc.2022.109788.

[21] F. E. B. Otero, A. A. Freitas, and C. G. Johnson, "Inducing decision trees with an ant colony optimization algorithm," *Applied Soft Computing Journal*, vol. 12, no. 11, pp. 3615–3626, 2012, doi: 10.1016/j.asoc.2012.05.028.

[22] A. Isazadeh, F. Mahan, and W. Pedrycz, "MFlexDT: Multi flexible fuzzy decision tree for data stream classification," *Soft Computing*, vol. 20, no. 9, pp. 3719–3733, 2016, doi: 10.1007/s00500-015-1733-2.

[23] W. Gao, F. Xu, and Z. H. Zhou, "Towards convergence rate analysis of random forests for classification," *Artificial Intelligence*, vol. 313, 2022, doi: 10.1016/j.artint.2022.103788.

[24] E. Castellano-Hernández and G. M. Sacha, "Characterization of thin films by neural networks and analytical approximations," In *2012 12th IEEE International Conference on Nanotechnology (IEEE-NANO)*, pp. 1–5, 2012. doi: 10.1109/NANO.2012.6321943.

[25] A. Arunachalam, S. Novia Berriel, P. Banerjee, and K. Basu, "Machine learning-enhanced efficient spectroscopic ellipsometry modeling," 2022. [Online]. Available: www.aaai.org.

[26] J. Sui, P. Xie, Z. Lin, and M. Javanmard, "Electronic classification of barcoded particles for multiplexed detection using supervised machine learning analysis," *Talanta*, vol. 215, 2020, doi: 10.1016/j.talanta.2020.120791.

[27] G. Abadias *et al.*, "Review article: Stress in thin films and coatings: Current status, challenges, and prospects," *Journal of Vacuum Science & Technology A: Vacuum, Surfaces, and Films*, vol. 36, no. 2, p. 020801, 2018, doi: 10.1116/1.5011790.

[28] H. Uğuz, "A biomedical system based on artificial neural network and principal component analysis for diagnosis of the heart valve diseases," *Journal of medical systems*, vol. 36, no. 1, pp. 61–72, 2012, doi: 10.1007/s10916-010-9446-7.

[29] A. M. Engwall, Z. Rao, and E. Chason, "Origins of residual stress in thin films: Interaction between microstructure and growth kinetics," *Materials & Design*, vol. 110, pp. 616–623, 2016, doi: 10.1016/j.matdes.2016.07.089.

[30] Y. P. Yang *et al.*, "Machine learning assisted classification of aluminum nitride thin film stress via in-situ optical emission spectroscopy data," *Materials*, vol. 14, no. 16, 2021, doi: 10.3390/ma14164445.

13 Deep Learning in Atomic Layer Deposition

13.1 INTRODUCTION

As previously said, deep learning entails teaching neural networks to understand complex data. By gradually abstracting the input data, the network may learn and discover patterns and relationships that conventional machine learning approaches would miss [1]. It is modeled after the composition and operation of the human brain and excels at processing natural language, understanding sounds and images, and making intelligent decisions [2]. To automatically extract and learn features from the input data, deep learning algorithms employ numerous layers of artificial neurons. The deeper the neural network, the more layers it has, therefore the name "deep learning" [3]. Deep learning offers the main benefit of automatically learning features from unprocessed data without the requirement for manually created features. Deep learning is an unprecedented innovation that has transformed artificial intelligence and is propelling significant developments in a wide range of sectors [4].

In the context of atomic layer deposition (ALD), it has the ability to significantly increase the efficiency and accuracy of the process because of its special ability in proffering solutions to complex ALD reactions. With only a few studies proving its efficacy, its current use in ALD is, however, somewhat limited. Deep learning is progressively gaining traction in this space despite the difficulties presented by ALD's complexity. Deep learning has a wide range of possible uses in ALD, including the creation of novel materials, increased process effectiveness, and improved device performance. Deep learning can also help identify and reduce process variations, providing consistent and high-quality deposition. To fully realize the benefits of deep learning integration in ALD, further research and development are necessary. In this chapter, we shall examine an overview of the potential of deep learning for its wider adoption and state-of-the-art in the application of ALD, while emphasizing the challenges associated with its integration.

13.2 COMMON DEEP LEARNING ALGORITHMS AND APPLICATIONS IN ALD

This section explores types, algorithms and applications of the deep learning approach with more emphasis on their distinctive features particularly in the context of ALD.

13.2.1   GENERATIVE ADVERSARIAL NETWORK

Generative models are in two broad categories. The first is the generational algorithms which is based on the conventional machine learning techniques like generative adversarial network (GAN) and variational auto-encoder (VAN) [5–10]. GAN is an intelligent approach to train generative model involving two sub-model vis-a-vis the generator model and the discriminator model as depicted in figure 13.1 [11].

While the training phase encompasses various examples from the training data, the discriminator receives well-established dataset as the first training data while the generator undergoes training based on its fooling capacity to the discriminator. The optimization of key parameters of the ALD process such as exposure time, temperature and precursor flow rates can be achieved using the GAN towards to generate virtual samples of ALD-grown films enhancing the efficiency and quality of the process [12,13]. The ALD process settings and the quality of the produced thin film can be used to train GAN for optimization tasks. Then, GANs can produce novel ALD process parameters that could improve the quality of thin films. Based on the precursor molecules and the ALD process parameters, GANs can also be utilized to forecast the thin film attributes.

13.2.2   CONVOLUTIONAL NEURAL NETWORK

Previous chapters have established the superior characteristics of deep learning techniques such as Convolutional neural networks (CNNs) in handling unstructured data such as visual and image data for task such as image and signal processing, classification, and feature extraction [14–17]. In Figure 13.2, a CNN has an input layer, convolution layers, pooling layers, a fully linked layer, and an output layer, with the convolution and pooling layers connected alternately [18].

The morphology and microstructure of films can be examined using CNNs in ALD. The ALD process parameters, which can be subtly modified to create diverse outcomes, are heavily dependent on the morphology and microstructure of thin films [19,21]. CNNs can be used to examine the characteristics of the thin films, including their thickness, homogeneity, and crystal structure, in order to optimize the ALD process parameters. More detailed applications of CNN in ALD processes are presented in the case studies in subsequent sections.

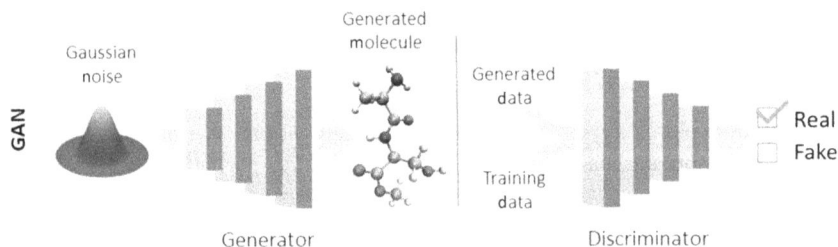

FIGURE 13.1   A typical GAN architecture for molecular discovery [11].

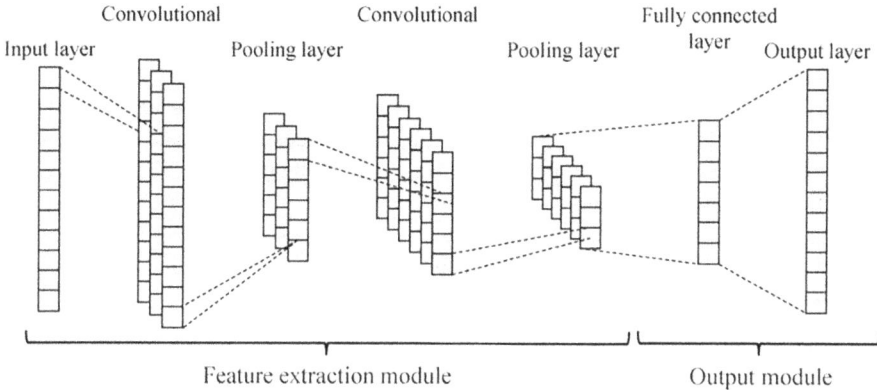

FIGURE 13.2   Fundamental framework of the convolutional neural network (CNN) [18].

13.2.3 RECURRENT NEURAL NETWORK

Recurrent Neural networks (RNNs) are able to process sequences of inputs and keep a "memory" of previous inputs, in contrast to typical feedforward neural networks, which only analyze data in a single pass [23]. They are hence highly suited for tasks like speech recognition and time series prediction [24, 25]. RNNs have a "hidden state" that updates with each input. This concealed state represents the network's "memory" of prior inputs. Training-learned weights and biases update the hidden state. Common RNN architectures are long short-term memory (LSTM) and gated recurrent units (GRUs). Due to their capacity for handling time-series input, they have attracted a lot of interest in the fields of machine learning. Readers can refer to Chapter 8 for more details on RNN.

RNNs can learn the temporal dependencies and forecast the precursor coverage. With greater film quality and deposition rates as a result, this can aid in improving the ALD process parameters. RNNs can also be utilized to simulate the ALD process' growth rate [26]. The reaction kinetics and surface coverage affect how quickly a thin film grows. With a dataset of growth rate under various process parameters, RNNs can be trained. RNNs are able to pick up on temporal dependencies and forecast how quickly the thin film will grow throughout the ALD process. With greater film quality and deposition rates as a result, this can aid in improving the ALD process parameters [27]. RNNs can also be used to simulate the dynamics of the deposition process in real time [28, 29]. The case studies in subsequent sections deals further on the applications of RNN in ALD and thin film deposition process.

13.2.4 IMAGE PROCESSING CAPABILITY OF DEEP LEARNING TECHNIQUES FOR MICROSTRUCTURAL ANALYSIS IN THIN FILM MATERIALS

Achieving the preciseness and accuracy in the thickness and properties of thin film deposited using ALD process is a critical goal in semiconductor industries. To obtain the necessary thin film properties, ALD requires exact control and monitoring of the deposition process. The ability to manage and monitor processes in real time makes image processing for microstructural analysis of materials a crucial tool for deposition analysis [13,29]. With regard to image processing and computer vision tasks, namely, image segmentation, object detection, and classification, deep learning has demonstrated promising outcomes particularly in the microstructural analysis of thin films [13]. Deep learning can be applied to the ALD process to evaluate images of the deposition process and extract pertinent data, such as the thickness of the deposited layer, the regularity of the deposition, and the existence of defects [19,29]. There are various steps involved in the processing of ALD images using deep learning. The system must first take detailed and high-resolution pictures of the deposition procedure, which are afterward pre-processed to get rid of noise and boost contrast. Then, a deep learning algorithm, such as a CNN or GAN, is trained on a dataset of images labeled with the desired output using the pre-processed images [30,31]. The deep learning algorithm gains the ability to recognize patterns and features in the photos that correlate to particular traits of the deposition process during the training phase. After being taught, the algorithm can evaluate fresh images in real time and give the ALD system feedback [32,33].

The capacity of deep learning to adjust to new or changing settings is one of the main benefits of employing it in ALD image processing. Deep learning algorithms can generalize and perform well even when confronted with brand-new, untried images since they can learn from a sizable dataset of images. This adaptability is crucial in ALD since the deposition conditions might vary widely and the system needs to be able to respond quickly to these changes. Deep learning for ALD image processing has the added benefit of processing vast amounts of data quickly. High-speed image analysis by deep learning algorithms can give the ALD system feedback in real time. In the semiconductor sector, where high throughput and efficiency are essential, this capacity is crucial. The precision and effectiveness of the deposition process can be significantly increased by utilizing deep learning in ALD image processing. Deep learning algorithms can rapidly analyze images, detect thickness and homogeneity deviations, and notify the ALD system of corrections. Deep learning algorithms can swiftly process large amounts of data and adapt to changing conditions, making them ideal for the semiconductor industry.

TEM and SEM are used to study thin film microstructure. However, these technologies generate a lot of data that are time-consuming to interpret. Data analysis sometimes involves manual inspection, which is time-consuming and unreliable. A useful technique for the automated investigation of microstructural information of deposited materials is the image processing potential of deep learning [29]. Deep learning algorithms have so far demonstrated tremendous promise in automating the analysis of intricate and high-dimensional datasets, such as those produced by electron microscopy methods. Deep learning–assisted microstructural analysis involves training a deep learning algorithm on a large collection of microstructural data to discover traits and patterns. Once taught, the system can automatically extract significant characteristics from new microstructural data, such as TEM or SEM pictures, and analyze them [34]. By training a deep learning model, we can find and examine anomalies in the microstructure, such as grain boundaries, dislocations, and stacking defects, in thin film materials. In addition, the model may be trained to categorize various grain varieties and determine their gradation. After that, the model may be used to automatically extract data on the texture, distribution, and grain size of the thin film material. The analysis of how microstructures change over time can also be done using deep learning. To monitor the development of thin film materials, and forecast their eventual microstructures based on the original conditions, deep learning models could be helpful. This may aid in enhancing the growing process and enhancing thin film quality [26,35].

13.3 CASE STUDIES EVALUATION OF DEEP LEARNING APPLICATION IN THIN FILM DEPOSITION

Deep learning is becoming more and more popular, and researchers are looking into how this technology might be used to increase the effectiveness and precision of ALD procedures. Researchers can reliably predict the behavior of ALD processes and adjust deposition settings to produce high-quality thin films with an unheard-of level of precision by training deep learning algorithms on vast volumes of data. Thin film deposition researchers now have more options to deep learning. Researchers have used deep learning algorithms to predict critical features of the deposition process and final deposited materials and best deposition conditions through a variety of case studies. The creation of new materials for electronic devices, increasing the effectiveness of solar cells, and other uses in materials science are all significantly impacted by this. This subsection presents an overview of the limitless potential deep learning applications in ALD through some case studies. Although, there is a meager amount of research on deep learning applications in thin film research using ALD and other deposition techniques, the list of case studies presented here is not exhaustive.

13.3.1 Deep Learning–Based Modeling of Plasma-Enhanced Atomic Layer Deposition

In this research carried out by Ding et al. [25], the use of deep learning techniques to enhance the operation of the plasma-enhanced atomic layer deposition (PEALD) of hafnium oxide thin films was explored. The authors initially addressed the difficulty of precisely forecasting the film characteristics and the limitations of managing the PEALD process as well as the hurdles of establishing optimal operating procedures for PEALD with the computational simulation techniques and the shortfalls of the traditional machine learning techniques in capturing the nonlinearity of the complex PEALD process. They then offer a deep learning approach vis-à-vis long-short term memory (LSTM)-based RNN and integrated with a full multi-scale 3D CFD model to enhance the PEALD procedure. DNN is trained on a dataset extracted from a multi-scale CFD model initially carried out to analyze both the transient gas-phase profile development and the dynamic surface profile progression. The RNN and variants were designed for the prediction of the dynamic time sequences of the PEALD process and to capture the complex input–output relationship between operational

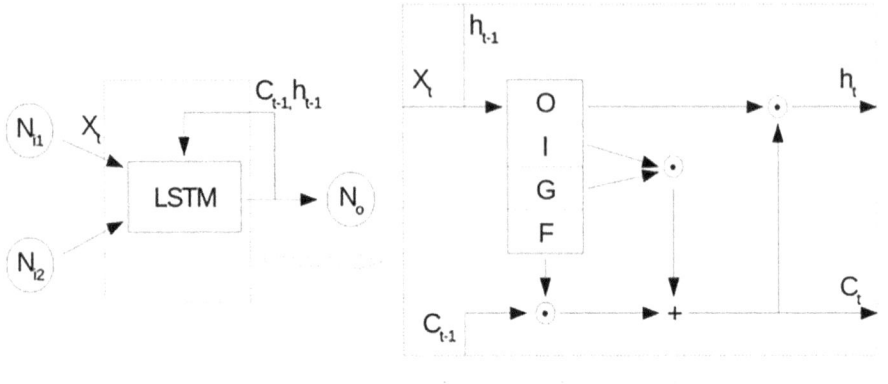

FIGURE 13.3   Methodological framework of the built (long short-term memory) LSTM-ANN [25].

circumstances. The architecture of the LSTM-RNN model built in the research is shown in Figure 13.3. The figure displays a general RNN framework with an LSTM cell as well as the LSTM cell's intricate manifestation, where N_i represents input neurons, N_o represents an output neuron, C_t and C_{t-1} represent the cell memory's state at t and $t-1$ trainings, while h_t and h_{t-1} represent the hidden state at t and $t-1$ trainings.

The first step in ascertaining the performance of the models is to compare their predictions to the output of the original 2D multiscale CFD model. The 3D multiscale CFD model was then contrasted with it using the identical operating conditions and reactor layout. The models that are produced after the training are validated using a set of test conditions. Although there is a strong agreement between the findings of the 2D multiscale CFD model and the RNN-based integrated data-driven model, the authors reported that it would be beneficial to further validate the data-driven model using computations from a full 3D multiscale CFD model. The RNN estimates and CFD results, particularly for O-Cycle, closely resemble one another, as shown in Figure 13.4. The Hf-cycle RNN model's deviation is a little bit larger since Hf-reaction cycle's paths are more intricate. To determine the best operating strategy, the authors deployed the data-driven model's prediction capacity. In order to maximize the production capacity of PEALD processes, cycle times were reduced while maintaining the desired film quality at preferably the most attainable coverage. Sequel to this, the built model is utilized to forecast the system dynamics for a broad range of input flow rates from 2.5×10^6 kg/s to 9.75×10^4 kg/s, being the typical range of operating flow rates, in order to meet both needs. Figure 13.5a–c represents the dynamic profiles for 0-cycles specified range at inner, middle and outer water space, respectively.

Generally, the research offers a thorough analysis of the use of deep learning algorithms to enhance the PEALD method for producing thin films of hafnium oxide. The outcomes show how the suggested method can effectively forecast film qualities and optimize process variables, which have the potential to considerably increase PEALD's efficiency and dependability.

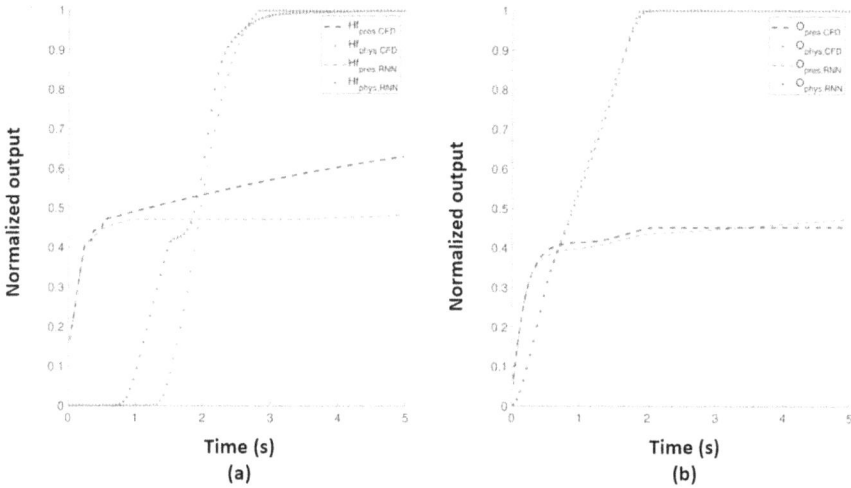

FIGURE 13.4 Comparison of recurrent neural network (RNN) for two half-cycles in the reactor without the showerhead. (a) Hf-Cycle. (b) O-Cycle.

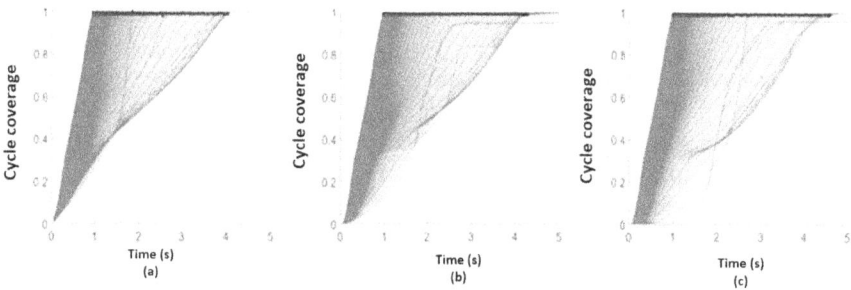

FIGURE 13.5 Dynamic profiles for O-Cycle for flow rates (kg/s) for (a) inner, (b) middle, and (c) outer wafer regions.

13.3.2 STRUCTURAL ZONE DIAGRAM ANALYSIS FOR THE SYNTHESIS OF THIN FILMS

Another dimension of deep learning applications is revealed in the research by Banko et al. [36]. In this research, the authors investigated the use of generative deep learning techniques, namely, variational encoder (VAE) and generative adversarial neural network (GAN) to forecast structure zone diagrams (SZDs) in thin film synthesis, particularly in ALD. This research was motivated by the necessity of microstructure analysis in a multi-parameter zone toward optimal extrinsic attributes which typically requires an excessive number of experiments to test. The authors therefore suggested the integration of combinatorial experimental procedures with generative deep learning models to discover synthesis–composition–microstructure relations in order to master the complexity of thin film processing microstructures and to lower the cost of microstructure design. The research established that generative models offer hitherto novel levels of quality in generated SZDs that can be used to optimize chemical composition and processing conditions in order to achieve the ideal microstructure.

SZDs are an effective instrument for comprehending the connection between a material's composition and its qualities, and they can aid in forecasting the ideal synthesis conditions to achieve desired material characteristics. Given that generative adversarial networks (GANs) are able to learn the underlying distribution of data and produce new data samples that match the distribution, the authors suggest using GANs to predict SZDs in thin films. The study trained a generative model exclusively on experimental data and predicted accurate process–microstructure interactions. Further to this, the developed technique minimized complexity in the following ways: (i) conducting a small number of experiments and effectively creating large training datasets using "processing libraries," (II) processing images of SEM microstructure using the trained deep learning models, (III) visualizing common characteristics between various synthesis routes, and (IV) forecasting microstructures for parameter settings using connections revealed in the training data. Material from metal transition nitrides was considered in this research.

Using the conditional variables, the authors use the GAN model to predict microstructures. Moreover, they needed to define what the model is capable of learning from the experimental data before they could classify the level of prediction. The baseline is created by rebuilding a microstructure from the training set. The figurative comparison of the particle size distribution of experimental and predicted images is as shown in Figure 13.6. By considering two input factors, namely, conditional parameters and a latent sub-space with random noise, the GAN was able to produce these microstructure images. The histograms in Figure 13.6 display the particle counts from 100 patches of both experimental and predicted images. More results from the study are presented in Figure 13.7 which is a further representation of the predicted and experimental microstructural images. The generated SZDs had good agreement with the experimental findings, demonstrating the potential of GANs to

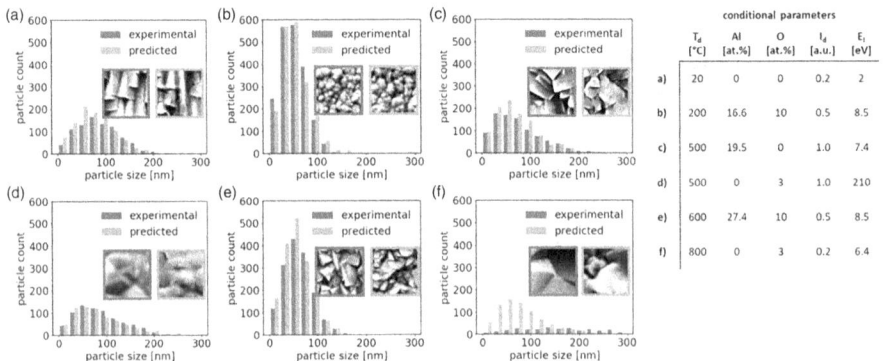

FIGURE 13.6 Comparison of experimental and predicted microstructural images of scanning electron microscopy (SEM) and GAN model, respectively [36].

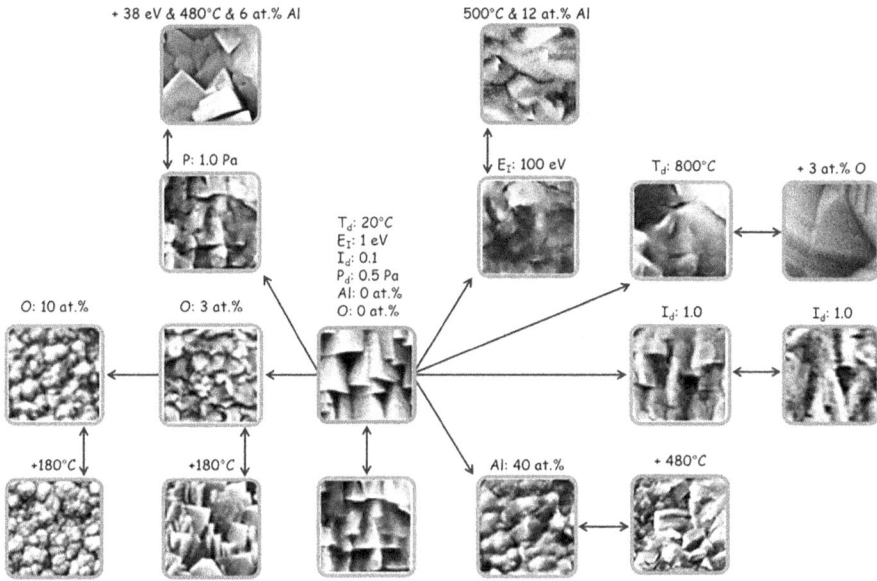

FIGURE 13.7 Comparison of the experimental and predicted microstructure (modified parameters for both experimental and predicted images) [36].

forecast the ideal circumstances for thin film synthesis. The ability to construct SZDs for a variety of materials without prior knowledge of the material's properties or synthesis circumstances is a noteworthy benefit of the suggested approach. This is especially helpful for materials that have not been thoroughly explored because it enables a quick study of the synthesis conditions and material property prediction. The research shows a promising use of deep learning algorithms in thin film synthesis and ALD, particularly in the area of SZD prediction. The development and improvement of novel materials with desired qualities could be greatly accelerated by the suggested method.

13.3.3 ACCELERATED DEEP LEARNING DYNAMICS FOR ATOMIC LAYER DEPOSITION

A case study of the use of deep learning algorithms for speeding up the simulation of ALD processes is presented in the article "Accelerated Deep Learning Dynamics for Atomic Layer Deposition of $Al(Me)_3$ and Water on OH/Si(111)" by Nakata et al. [37]. Having identified some of the current limitations of accurate *ab initio* computational approaches, the study established that computational modeling cannot yet be used to simulate surface processes that occur over a lengthy period of time, such as ALD. Furthermore, the large dataset prerequisite of machine learning models has motivated the authors to develop an iterative approach for optimizing training of the datasets and utilize the machine learning-assisted ab initio calculations for modeling of surface reactions that take place during the $Al(Me)_3/H_2O$ ALD process on the OH-terminated Si (111) surface. The approach developed in the research takes advantage of a newly established low-dimensional projection technique (two-dimensional umbrella sampling method, TDUS), which significantly reduces the quantity of data needed to produce developed deep learning models with high accuracy. The adsorption and reaction kinetics of $Al(Me)_3$ and water on an OH-terminated Si(111) surface, a typical substrate used in ALD procedures, are predicted by the authors using a deep learning model. The DeePMD kit package31, which was interfaced with the TDUS included in LAMMPS software, was used to train the neural network enhanced molecular dynamics simulations. In the architecture, five hidden layers were employed for the deep neural network, with 240, 240, 120, 60, and 30 neurons per hidden layer. Figure 13.8 represents the deep neural network potential at 300 K results in a two-dimensional free energy surface (FES) along reaction r06. The suggested ML-TDUS technique reduces deep learning training time by more than 100. ML-TDUS applied to the ALD of Al (Me)3 and water on the OH/Si (111) surface converges the target free energy RMSE below 1.0 kcal/mol in eight iterations. ML-TDUS free-energy landscapes showed new surface states and reaction paths that static quantum-mechanical computation could not identify. The work reveals how deep learning algorithms may speed up ALD process simulation, which is promising for ALD. The suggested method may greatly speed up the search for and development of new materials with desired qualities, which could have substantial effects on a range of technological applications.

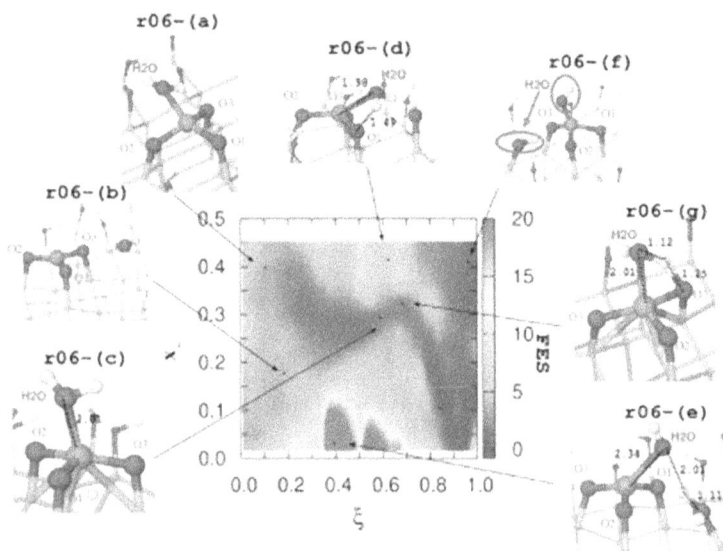

FIGURE 13.8 Free energy surface (FES) in two dimensions along reaction r06 as determined by neural net potential at 300 K [37].

13.3.4   3D CHARACTERIZATION OF ULTRA-THIN EPITAXIAL LAYERS DEPOSITED ON NANOMATERIALS

A cutting-edge technique was developed by Grzonka et al. [34] for the 3D characterization of the nano-system compositions by integrating energy dispersive X-ray spectroscopy-scanning transmission electron tomography (STEM-EDX ET) and the denoising approach of the deep learning techniques. This research demonstrates a case study of how deep learning techniques were used to characterize ultra-thin epitaxial layers deposited on controlled-shape nano-oxides in three dimensions. For the 3D characterization of epitaxial layers, the authors suggest a unique method that combines deep learning and compressed sensing approaches. The suggested method entails collecting a number of 2D photographs of the epitaxial layer using SEM and reconstructing the 3D structure of the layer using a deep learning algorithm. To reduce the number of images needed to precisely recreate the 3D structure, the scientists additionally use compressed sensing approaches. The characterization of epitaxial layers formed on nano-oxides is used by the researchers to show the viability of their suggested methodology. The findings demonstrated that the suggested method can more precisely reconstruct the 3D structure of the epitaxial layer utilizing a great deal lesser SEM image compared with the conventional techniques. The concept put forward in this case study has significant implications for the epitaxial development research space, especially in the context of optimizing cell growth to yield satisfactory material qualities. The suggested method enables fine control of the growth conditions, which can result in the synthesis of novel materials with desired properties by correctly describing the 3D structure of the epitaxial layer. A promising use of deep learning methods is shown, specifically in the 3D characterization of extremely thin epitaxial layers revealed in this research. The suggested method may greatly speed up the search for and development of new materials with desired qualities, which could have substantial effects on a range of technological applications. A comparison of the raw experimental EDX maps with the ones that were denoised using a deep learning algorithm for the projections obtained at 40° and 40° tilts is presented in Figure 13.9. With this procedure, it can be seen that the signal quality is significantly improved compared to the original data. Hence, the structure of the signals is smooth and crisp, while the clouds of scattered points that surrounded the nano-cube were eliminated in the denoised elemental map recorded at 0° tilt. For more details about the result of the integrated deep learning and SEM framework, readers can visit the article "Combining Deep Learning and Compressed Sensing Methods for the 3D Characterization of Ultra-Thin Epitaxial Layers Grown on Controlled-Shape Nano-Oxides" by Grzonka et al. [34].

13.3.5   MICROSCOPIC IMAGE DEBLURRING FOR A NANOMATERIAL

Further in the application of deep learning techniques in thin film generation, optimization, processing, and characterization is the case study of a research by Dong et al. [13] which is focused on a deep learning–based approach for improving the

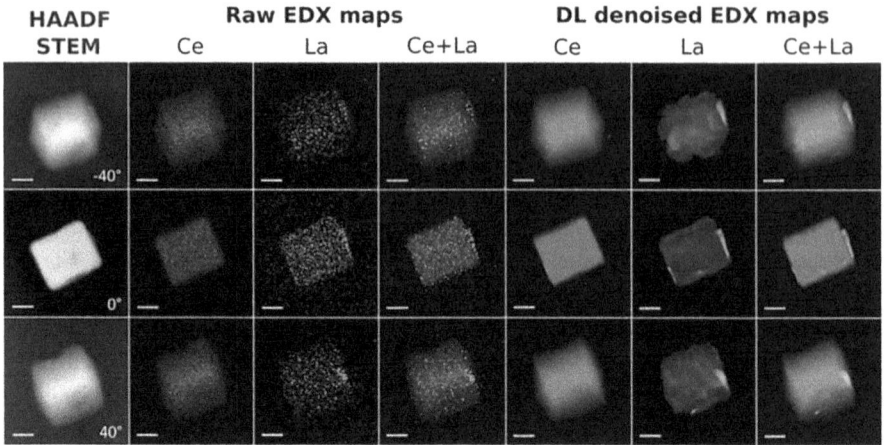

FIGURE 13.9 HAADF-STEM (scanning transmission electron tomography) images of the nanocubes compared with the raw EDX data and deep learning-assisted denoised EDX maps [34].

quality of microscopic images of 2D nanomaterials used in semiconductor production. The authors proposed a generative adversarial network (GAN) to eliminate blurs in the images produced from microscopic observation and concentrate on the use of ALD in the creation of 2D materials. The result of the research was intended to enhance a microscopic 2D characterization of atomic layer of wafer-scale semiconductors. Having firstly established the difficulties of imaging 2D materials with conventional microscopy methods including low contrast and resolution as well as blurring because of the materials' structural thickness, the authors proceeded by demonstrating how they plan to deblur the image using a GAN, which is made up of a generator and a discriminator network. The discriminator network is taught to distinguish between the real and created images, while the generator network is trained on a set of high-quality images to learn the underlying structure of the samples. In this research, a deep learning algorithm, namely, GAN was created to deblur out-of-focus microscopic images of 2D semiconductors using the Pix2Pix architecture and a modified loss function. A sample of MoS_2 with a 270 nm oxidation thickness was measured on a SiO_2/Si surface. Images with varied degrees of defocus were chosen for testing that included both in-focus and out-of-focus photos.

The performance and effectiveness of the GAN-enhanced microscopic imagery deblurring (MID)-generated images were evaluated using structural similarity (SSIM) and peak signal-to-noise ratio (PSNR) assessment metrics. In order to forecast the number distribution maps, the in-focus photos, out-of-focus images, and refocused images produced by MID were sent into a pre-trained segmentation model51 (U-Net). Also, various GAN model loss functions were contrasted in order to evaluate the segmentation and deblurring capabilities. The researchers observed that their GAN-based method was able to greatly improve the quality of the images, producing clearer and more distinct features in the samples when compared to results acquired using a conventional deconvolution algorithm. The authors went ahead to explain the significance of their research for wafer-scale semiconductor characterization, emphasizing how their technique may enhance the effectiveness and precision of quality control in semiconductor manufacturing. In order to provide real-time image processing and analysis, they also examine the feasibility of merging their approach with already-in-use imaging techniques. Figure 13.10 shows the process of deblurring microscopic images by MID and reconstructing distribution maps of species with various atomic layer counts by the U-Net model for 2D nanosheets. Figure 13.10a represents the typical microscope image, while Figure 13.10b shows a new microscopic image. In Figure 13.10c, a microscopic image that has been fixed using the linear preprocessing technique is presented, while in Figure 13.10d–f, the red and green channel values of the monolayer region in Figure 13.10a–c, substrate region, and full FOV image are revealed. Figure 13.10g,h predicts segmentation outcomes for the experimental images' original and corrected versions, and Figure 13.10i presents the reference and experimental microscopic images' sampled pixels' CIE color space. Conclusively, in this research, a novel and promising method for improving the quality of microscopic photographs of 2D nanomaterials utilized in semiconductor production is presented. A GAN-based

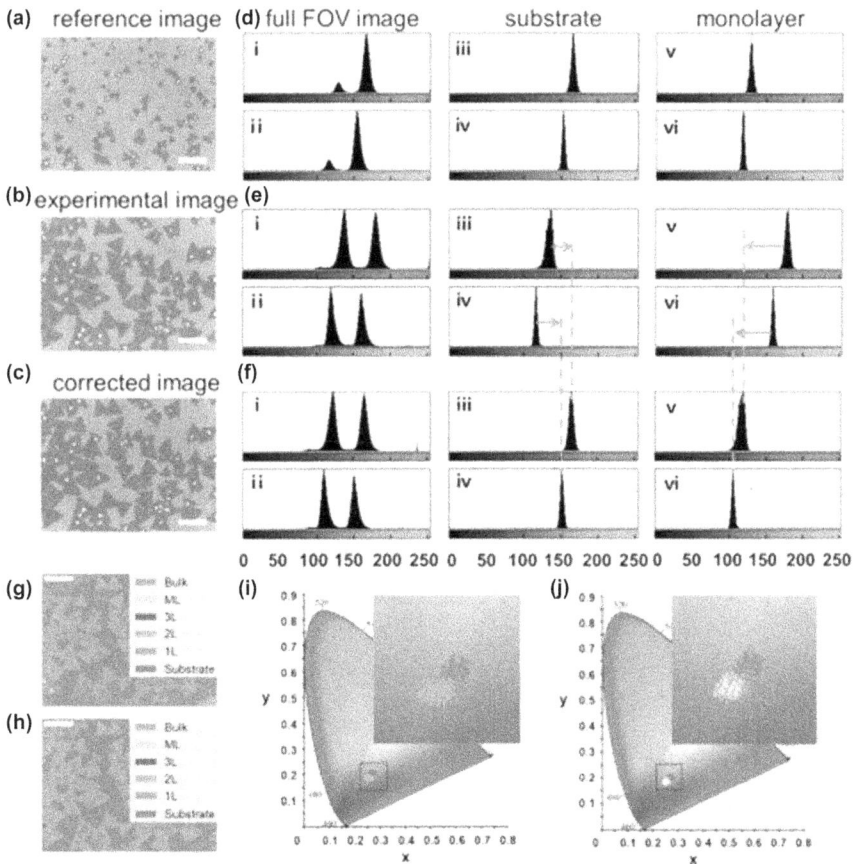

FIGURE 13.10   Deblurring process of the GAN-MID (microscopic imagery deblurring) [13].

technique is used to deblur the images, which is a substantial advancement over conventional deconvolution algorithms and has the potential to increase the effectiveness and precision of semiconductor production operations.

13.3.6 MICROSCOPIC IMAGE PROCESSING FOR A 2D IDENTIFICATION OF DEPOSITED MATERIALS

Recent years have seen the development and implementation of deep learning–based algorithms in a variety of image processing tasks, from simple 2D handwritten digit recognition to object detection in videos by learning from annotated data [1,30]. The three key image processing tasks, namely, classification, segmentation, and detection which are required for the identification of atomic layer numbers based on optical contrast information from the standpoint of computer vision were explored in the research by Dong et al. [29] focused on the development of deep learning algorithms for the identification of 2D semiconductors. The robustness of three deep learning architectures—DenseNet, U-Net, and Mask-region convolutional neural network (RCNN)—were assigned to these tasks while their performance was then assessed based on enhanced 2D microscopic images with various variations in optical contrast. Generally, a deep learning–based classification (single or multilabel) assignment can be used for flakes class prediction [31], while a detection model can classify and localize various 2D flakes using bounding boxes around the objects of interest [33], and a detection model can generate a segmentation map for the presented categories [32]. To identify, categorize, and separate these materials in the images, the authors built a deep learning model using a dataset of microscopic images of 2D semiconductors. To increase the precision of the predictions, they trained the model using a combination of CNN and RNN. Figure 13.11 represents the dataset construction methodology with

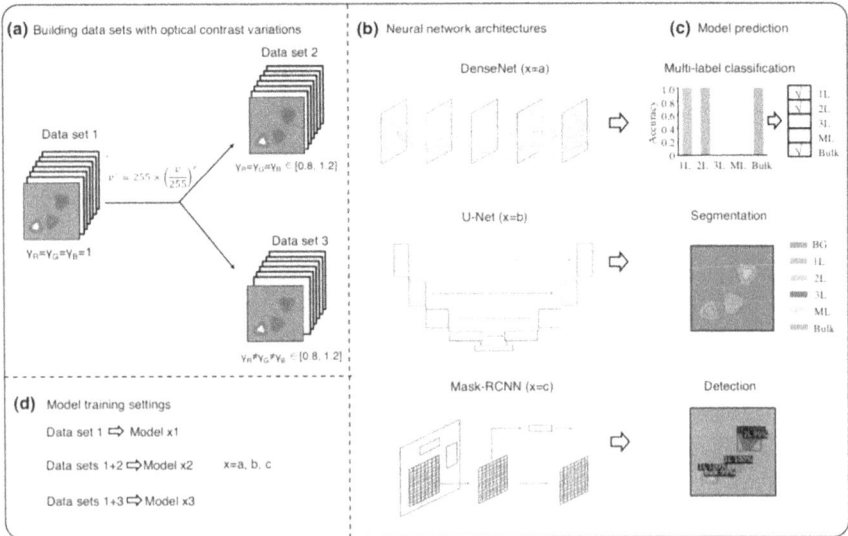

FIGURE 13.11 Methodological framework of the building of the threefold computer vision task in the case study [29].

the three network designs for multilabel segmentation, classification, and detection, respectively. Figure 13.11 also shows the computer vision tasks for processing images of 2D materials, comprising multilabel classification using DenseNet, segmentation using U-Net, and detection using Mask-RCNN while revealing the outputs from three neural network designs that were fed the same. The control parameters of the training and each architecture's dataset are also revealed in Figure 13.11.

The statistics of the training and testing datasets based on RGB optical contrast differences and CIE 1931 color space analyses were used to analyze the neural network models' performances. In the original dataset, lateral categories differ statistically in terms of CIE1931 color-spaced distribution and RGB optical contrast disparities. The distinction between different categories shrank and overlapping regions showed up in the point distribution from the CIE 1931color space study when optical contrast variations were used to enhance the original datasets, which decreased prediction accuracy. Figure 13.12a represents the original and enhanced RGB images that have diverse sampling numbers in the gamma contrast function and the current flake-type

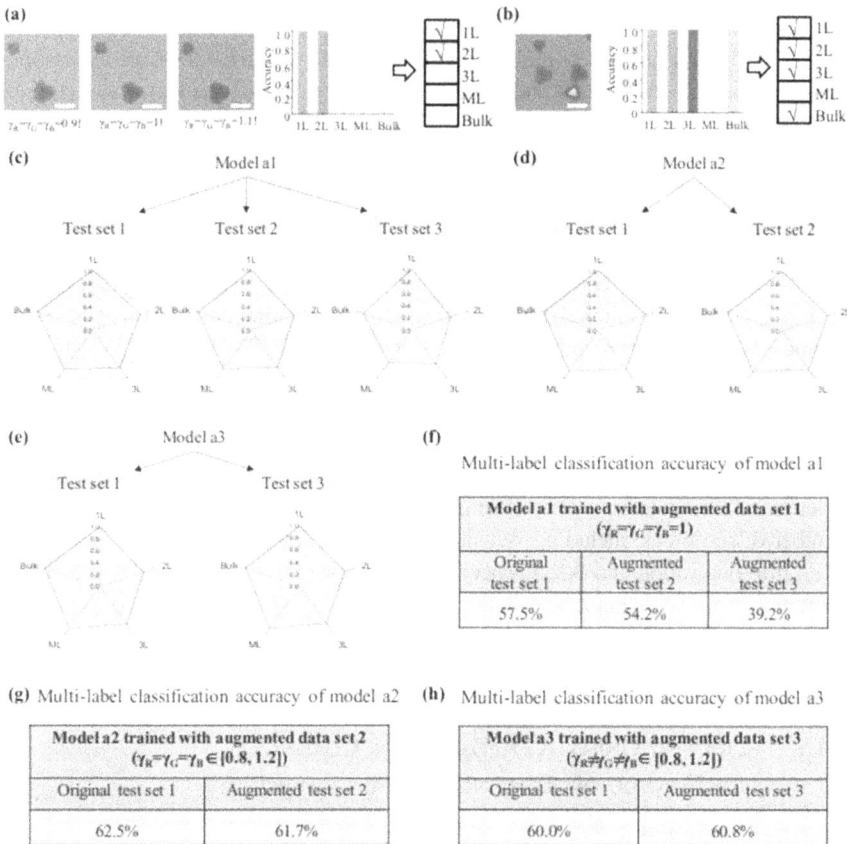

FIGURE 13.12 The raw data and the performance of the classification task of the denseNet model at different test sets [29].

prediction, while Figure 13.12b represents a RGB image with several flake types and a prediction of the ones that will appear. Figure 13.12c–e shows the performance of the denseNet model at different test sets, while Figure 13.12f–h presents the statistical metrics value of the accuracy of the classification with DenseNet models. The identification of 2D semiconductors using deep learning algorithms is demonstrated in the study. This application could have significant ramifications for the creation of cutting-edge electronic gadgets. The suggested method can considerably cut down on the time and work needed to identify these materials, and it may also help find new materials with desirable electrical properties.

13.4   SOME REAL-LIFE CHALLENGES IN THIN FILM DEPOSITION AND MACHINE LEARNING-BASED SOLUTIONS

Electronics, optics, coatings, and the energy sector all heavily rely on the vital and intricate thin film deposition process. However, mass-production of uniform, high-quality films is difficult, and operational errors and failures can make it worse. Process characteristics like temperature, pressure, and gas flow rates, as well as equipment and maintenance concerns, cause these deficits. Machine learning-based solutions that can optimize the thin film deposition process and address these problems can be developed in response to these operational difficulties and pitfalls. The ability of machine learning techniques in proffering solutions to several problems in ALD and thin film synthesis has been studied in some previous chapters. Machine learning algorithms are capable of analyzing data from numerous sensors throughout the deposition process in the context of thin film deposition and making instantaneous modifications to the process parameters. Better and more reliable thin films can be produced over time thanks to these algorithms' capacity to learn from the data. This section discusses a few case studies of real-life operational difficulties and pitfalls encountered during industrial-scale thin film deposition, as well as the machine learning-based solutions created to overcome them. These case studies highlight how crucial it is to comprehend the operational difficulties and traps that thin film deposition presents, as well as the vital role that machine learning can play in minimizing these problems. Researchers and engineers can create more effective and efficient thin film deposition techniques that can satisfy the needs of diverse sectors by understanding these problems and their solutions. The information provided in this chapter was acquired from a number of resources, including websites, reports, bulletins, interviews, and literature on thin film deposition and ALD procedures. The specific identities and origins of this information, however, cannot be revealed due to privacy, confidentiality, and information security restrictions. Privacy, confidentiality, and information security laws prevent the disclosure of this material's origins.

13.4.1   Achieving Uniform Deposition or Coatings in Semi-Conductor, Coating, and Aerospace Industry

13.4.1.1   Challenge

A critical challenge was identified in a semiconductor coating and aerospace industry, which uses deposition of thin films. A crucial part of the semiconductor business is obtaining homogeneous thin film materials across enormous silicon wafers. The coatings business as well encounters a variety of difficulties when trying to apply uniform coatings on surfaces with diverse geometries. Variations in surface morphology or the surface texture and characteristics of a material represent a critical challenge of the coating industry. The final coating quality and adhesion of a coating to a surface are both influenced by its shape. Coatings applied to surfaces with complicated geometries or uneven surfaces may not adhere uniformly, leading to uneven thickness, a rough texture, or even coating failure. Moreso, these varying conditions may lead to irregularities in the deposition process, resulting in nonuniform film thickness and flaws in the electrical devices produced in the semiconductor industry. The industries frequently use time-consuming and expensive trial-and-error techniques to determine the ideal deposition conditions. This identified challenge is typical of all semiconductor industries. The wafer's deposition rate varies, causing nonuniformity. Wafer surface variations, temperature gradients, and other factors may cause this. When these differences take place, the deposition of the material may build up more quickly in some regions of the wafer than in others, resulting in an uneven film thickness.

13.4.1.2 Machine Learning Solution

The literature is replete with machine learning-based approaches for addressing this particular optimization challenge [38,39]. Moreso, we have examined in previous chapters potential applications of machine learning for optimal process optimization and condition monitoring. Recently, experts have also created machine learning algorithms that can forecast the ideal deposition circumstances for producing homogenous films in overcoming this challenge in semi-conductor and coating industries. Throughout the deposition process, the machine learning multi-dimensional optimization algorithms which use data from numerous sensors to optimize the process can make real-time modifications in these particular scenarios. In the coating industry as well, the algorithm can modify in real time to maintain consistent coating thickness and quality across a variety of surfaces with different geometries by taking into account the surface shape and other elements that can influence the coating process. The approach is based on a neural network that can gain knowledge from the information gathered by the sensors and utilize it to forecast the ideal circumstances for deposition. The neural network can recognize patterns and correlations between the input variables and the produced film uniformity since it has been trained on a sizable dataset of deposition conditions and associated film thicknesses. Using the sensor data gathered throughout the procedure, the model may be used to anticipate the ideal conditions for deposition in real time once it has been trained. This technique's capability to manage the intricate relationships between various variables that can affect the deposition process is explored to address this challenge. As opposed to the traditional trial and error method, the machine learning algorithm can swiftly assess massive volumes of data and pinpoint the ideal circumstances for deposition, resulting in a swifter and more reliable production. The difficulty of producing consistent thin films in the semiconductor industry has an optimal solution to this. This deposition approach increases electrical device reliability and performance in numerous applications. Neural networks and real-time sensor data enable this. The operators

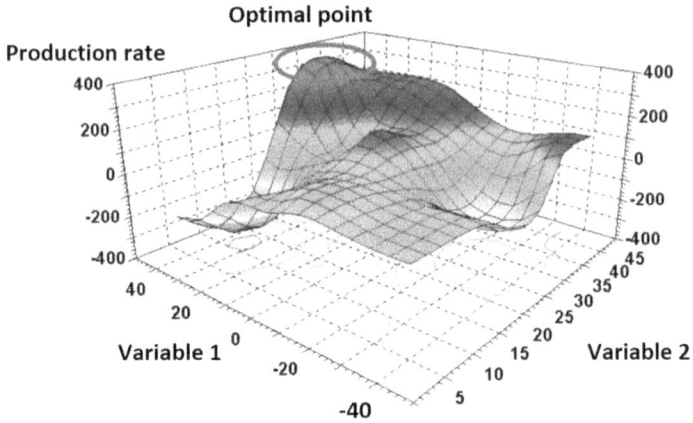

FIGURE 13.13 Multi-dimensional optimization principle of machine learning.

running the process can use this machine learning-based optimization algorithm as a support tool to assist them make smarter choices in order to increase production. These machine learning-based optimization algorithms create a "production-rate" environment that features upward and downward slopes reflecting high and low production scenarios. Further to this, the algorithm searches this environment for the highest peak that corresponds to the highest achievable production rate. Figure 13.13 represents a typical working principle of machine learning-based optimization algorithms that search for optimal production rates based on some variables. The algorithm can provide advice on how to best reach this optimization milestone by navigating this production rate environment.

13.4.2 DEFECTS DETECTION AND CONTROL IN SOLAR CELL MANUFACTURING INDUSTRY

13.4.2.1 Challenge

A solar panel producer which creates a sizable batch of solar cells that will be applied to a solar power facility was identified. Each solar cell is made of thin layers of semiconductor material like silicon and produces a certain quantity of electrical energy when exposed to sunlight. After being formed, the cells have quality issues and don't perform well. On closer inspection, it was found that some of the cells' thin layers have flaws such as pinholes and cracks. These flaws reduce the cells' efficiency, which lowers their capacity to produce energy.

13.4.2.2 Machine Learning Solution

Borrowing from our previous discussion of machine learning applications in thin film deposition, we can proffer a solution to this challenge by using a deep machine learning-based defect detection technique. This approach can be initiated by gathering an extensive archive of thin film images using optical, X-ray, and SEM microscopy. Machine learning algorithms can precisely identify and categorize thin film flaws by being trained on labeled datasets of these images, allowing producers to optimize their manufacturing processes and raise the caliber and dependability of their goods. The CNNs are an ideal algorithm for this particular problem owing to their excellent ability for spotting flaws in thin films because they can automatically learn to recognize patterns in images. A labeled dataset of visual data is needed to train a CNN for defect detection. A range of images with and without faults, representing various defect types and severity levels, would be included in the collection. The CNN is subsequently trained using the labeled dataset, which entails modifying the network's weights to reduce the discrepancy between each image's true label and the predicted label. Once the CNN has been trained, it may be used to analyze new images of thin films and accurately identify flaws. In addition, CNN may be used to categorize defects according to their nature and severity, assisting manufacturers in finding the source of the faults and streamlining their production methods to lower defect density. For the purpose of defect detection in thin films used in solar cells, in addition to CNNs, various other machine learning techniques can be applied. For this reason, decision trees and support vector machines (SVM) have both been employed. The particular application and the type of imaging data being examined will determine which algorithm is used. Solar cell manufacturers may optimize their manufacturing processes and lower production losses by detecting flaws in real time. Furthermore, machine learning-based defect detection may enhance the quality and dependability of products, lowering the possibility of product recalls and boosting the overall satisfaction of customers. Deep learning approaches have been used in a number of studies and research projects to identify cracks and other thin film problems. These researches have demonstrated the ability of machine learning techniques, to precisely identify and categorize thin film flaws, thereby enhancing product quality and minimizing production losses. Manufacturers can enhance deposition conditions and boost the overall performance of thin-film-based technologies by being able to recognize and quantify flaws. Solar cell production is one practical setting in which these results can be put to use. For the conversion of solar energy into electrical energy, solar cells use thin sheets of semiconductor materials. The solar cell's overall performance and efficiency can be negatively impacted by flaws in these thin sheets. Manufacturers can precisely identify and categorize faults in thin films using machine learning-based defect detection approaches, allowing them to improve the deposition conditions and lower the defect density. As a result, the solar cells will operate more effectively and cost-effectively, which is crucial for tackling the global climate crisis. They will also perform better overall.

13.5   CONCLUSION

Deep learning has evolved as a critical tool that has proffered solutions to a wide range of complex problems in different areas of research, including thin film deposition using ALD. Deep learning algorithms' capacity to evaluate huge, complicated datasets has enabled new understandings of thin film technology and ALD. The case studies presented and reviewed in this chapter show how deep learning has the ability to enhance thin film quality, optimize the deposition process, decrease defects, and improve ALD process performance and efficiency. However, deep learning's application in ALD still has a number of drawbacks and difficulties. Lack of adequate data for model training, which can result in overfitting or subpar generalization, is one of the main issues. The requirement for interpretability and transparency presents another difficulty, particularly when deep learning models are applied to crucial applications. To overcome these hurdles, novel strategies for data augmentation and construction of hybrid models that mix deep learning with other techniques such as physics-based modeling must be explored. Moreover, initiatives must be taken to create comprehensible and explainable deep learning models' outcome that can offer perceptions into the algorithmic decision-making process. It is very promising that the use of deep learning in ALD would advance thin film technology and lead to novel insights. Deep learning's difficulties and restrictions can be resolved with new strategies and extensive research. Deep learning has the potential to transform the ALD research space and result in significant discoveries in materials science with further growth and improvement.

REFERENCES

[1] Y. LeCun, Y. Bengio, and G. Hinton, "Deep learning," *Nature*, vol. 521, no. 7553, pp. 436–444, 2015, doi: 10.1038/nature14539.
[2] K. Panwar, S. Kukreja, A. Singh, and K. K. Singh, "Towards deep learning for efficient image encryption," *Procedia Computer Science*, vol. 218, pp. 644–650, 2023, doi: 10.1016/j.procs.2023.01.046.
[3] B. Y. C. A. Goodfellow, *A Deep Learning*, Cambridge, MA: MIT Press, 2016.
[4] C. C. Paul Fergus, *Applied Deep Learning: Tools, Techniques and Implementation*, 1st edition, France: Springers, 2022.
[5] I. Goodfellow *et al.*, "Generative adversarial nets," In *Advances in Neural Information Processing Systems*, Z. Ghahramani, M. Welling, C. Cortes, N. Lawrence, and K. Q. Weinberger, Eds., Curran Associates, Inc., 2014. [Online]. Available: https://proceedings.neurips.cc/paper_files/paper/2014/file/5ca3e9b122f61f8f06494c97b1afccf3-Paper.pdf.
[6] Lawrence R. Rabiner, "A tutorial on hidden Markov models and selected applications in speech recognition," *Proceedings of the IEEE*, vol. 77, no. 2, pp. 267–296, 1989.
[7] Salakhutdinov Ruslan and Geoffrey Hinton, "Deep Boltzmann machines," *Journal of Machine Learning Research*, vol. 5, no. 2, pp. 1967–2006, 2009.
[8] L. Jiang, H. Zhang, and Z. Cai, "A novel bayes model: Hidden Naive Bayes," *IEEE Transactions on Knowledge and Data Engineering*, vol. 21, no. 10, pp. 1361–1371, 2009, doi: 10.1109/TKDE.2008.234.
[9] Y. Pu, Z. Gan, R. Henao, X. Yuan, C. Li, A. Stevens, and L. Carin, "Variational autoencoder for deep learning of images, labels and captions," In *Proceedings of the Conference and Workshop on Neural Information Processing Systems*, pp. 2352–2360, 2016.
[10] L. Jin, F. Tan, and S. Jiang, "Generative adversarial network technologies and applications in computer vision," *Computational Intelligence and Neuroscience*, vol. 2020, 2020. doi: 10.1155/2020/1459107.
[11] C. Bilodeau, W. Jin, T. Jaakkola, R. Barzilay, and K. F. Jensen, "Generative models for molecular discovery: Recent advances and challenges," *Wiley Interdisciplinary Reviews: Computational Molecular Science*, vol. 12, no. 5, 2022. doi: 10.1002/wcms.1608.
[12] A. Carreon, S. Barwey, and V. Raman, "A generative adversarial network (GAN) approach to creating synthetic flame images from experimental data," *Energy and AI*, vol. 13, p. 100238, 2023, doi: https://doi.org/10.1016/j.egyai.2023.100238.

[13] X. Dong *et al.*, "Microscopic image deblurring by a generative adversarial network for 2D nanomaterials: Implications for wafer-scale semiconductor characterization," *ACS Applied Nano Materials*, vol. 5, no. 9, pp. 12855–12864, 2022, doi: 10.1021/acsanm.2c02725.

[14] S. Y. Huang, W. J. An, D. S. Zhang, and N. R. Zhou, "Image classification and adversarial robustness analysis based on hybrid quantum-classical convolutional neural network," *Optics Communications*, vol. 533, 2023, doi: 10.1016/j.optcom.2023.129287.

[15] E. Pintelas, I. E. Livieris, S. Kotsiantis, and P. Pintelas, "A multi-view-CNN framework for deep representation learning in image classification," *Computer Vision and Image Understanding*, vol. 232, 2023, doi: 10.1016/j.cviu.2023.103687.

[16] R. Zenghui and W. Waseem, "Deep convolutional neural networks for image classification: A comprehensive review," *Neural Computing*, vol. 29, pp. 2352–2449, 2017.

[17] W. Fang, L. Ding, B. Zhong, P. E. D. Love, and H. Luo, "Automated detection of workers and heavy equipment on construction sites: A convolutional neural network approach," *Advanced Engineering Informatics*, vol. 37, pp. 139–149, 2018, doi: 10.1016/j.aei.2018.05.003.

[18] Z. Guo, C. Yang, D. Wang, and H. Liu, "A novel deep learning model integrating CNN and GRU to predict particulate matter concentrations," *Process Safety and Environmental Protection*, vol. 173, pp. 604–613, 2023, doi: 10.1016/j.psep.2023.03.052.

[19] J. Cagnazzo, O. S. Abuomar, A. Yanguas-Gil, and J. W. Elam, "Atomic layer deposition optimization using convolutional neural networks," In *2021 International Conference on Computational Science and Computational Intelligence (CSCI)*, pp. 228–232, 2021. doi: 10.1109/CSCI54926.2021.00110.

[20] A.-A. Tulbure, A.-A. Tulbure, and E.-H. Dulf, "A review on modern defect detection models using DCNNs - Deep convolutional neural networks," *Journal of Advanced Research*, vol. 35, pp. 33–48, 2022, doi: https://doi.org/10.1016/j.jare.2021.03.015.

[21] C.-H. D. Tsai and C.-H. Yeh, "Neural network for enhancing microscopic resolution based on images from scanning electron microscope," *Sensors*, vol. 21, no. 6, 2021, doi: 10.3390/s21062139.

[22] C. Perkgöz and M. Z. Angi, "Characterization of artificially generated 2D Materials Using Convolutional Neural Networks," *Eskişehir Technical University Journal of Science and Technology A - Applied Sciences and Engineering*, 2022, doi: 10.18038/estubtda.1149416.

[23] A. Krenker, J. Bešter, and A. Kos, *Introduction to the Artificial Neural Networks, Artificial Neural Networks - Methodological Advances and Biomedical Applications*, ISBN: 978-953-307-243-2, London: InTech, 2011.

[24] R. Khaldi, A. El Afia, R. Chiheb, and S. Tabik, "What is the best RNN-cell structure to forecast each time series behavior?" *Expert Systems with Applications*, vol. 215, p. 119140, 2023, doi: https://doi.org/10.1016/j.eswa.2022.119140.

[25] Y. Ding, Y. Zhang, H. Y. Chung, and P. D. Christofides, "Machine learning-based modeling and operation of plasma-enhanced atomic layer deposition of hafnium oxide thin films," *Computers & Chemical Engineering*, vol. 144, 2021, doi: 10.1016/j.compchemeng.2020.107148.

[26] A. Bahramian, "Study on growth rate of $TiO2$ nanostructured thin films: Simulation by molecular dynamics approach and modeling by artificial neural network," *Surface and Interface Analysis*, vol. 45, no. 11, pp. 1727–1736, 2013, doi: 10.1002/sia.5314.

[27] H. J. Kim *et al.*, "Machine-learning-assisted analysis of transition metal dichalcogenide thin-film growth," *Nano Converg*, vol. 10, no. 1, 2023, doi: 10.1186/s40580-023-00359-5.

[28] B. D. Enrique Ferreira Advisor and B. H. Krogh, *Recurrent Neural Network Models of Multivariable Dynamic Systems: Numerical Methods and Experimental Evaluation*, Carnegie Mellon University's Robotics Institute, 1995, Pittsburgh, Pennsylvania.

[29] X. Dong *et al.*, "Deep-learning-based microscopic imagery classification, segmentation, and detection for the identification of 2D semiconductors," *Advanced Theory and Simulations*, vol. 5, no. 9, 2022, doi: 10.1002/adts.202200140.

[30] A. Garcia-Garcia, S. Orts-Escolano, S. Oprea, V. Villena-Martinez, and J. Garcia-Rodriguez, "A review on deep learning techniques applied to semantic segmentation,", 2017, [Online]. Available: https://arxiv.org/abs/1704.06857.

[31] E. Greplova *et al.*, "Fully automated identification of 2D material samples," 2019, doi: 10.1103/PhysRevApplied.13.064017.

[32] S. Masubuchi *et al.*, "Deep-learning-based image segmentation integrated with optical microscopy for automatically searching for two-dimensional materials," *NPJ 2D Materials and Applications*, vol. 4, no. 1, 2020, doi: 10.1038/s41699-020-0137-z.

[33] B. Han *et al.*, "Deep-learning-enabled fast optical identification and characterization of 2D materials," *Advanced Materials*, vol. 32, no. 29, 2020, doi: 10.1002/adma.202000953.

[34] J. Grzonka, J. Marqueses-Rodríguez, S. Fernández-García, X. Chen, J. J. Calvino, and M. López-Haro, "Combining deep learning and compressed sensing methods for the 3D characterization of ultra-thin epitaxial layers grown on controlled-shape nano-oxides," *Advanced Intelligent Systems*, p. 2200231, 2023, doi: 10.1002/aisy.202200231.

[35] Y. K. Wakabayashi, T. Otsuka, Y. Krockenberger, H. Sawada, Y. Taniyasu, and H. Yamamoto, "Machine-learning-assisted thin-film growth: Bayesian optimization in molecular beam epitaxy of SrRuO3 thin films," *APL Mater*, vol. 7, no. 10, 2019, doi: 10.1063/1.5123019.

[36] L. Banko, Y. Lysogorskiy, D. Grochla, D. Naujoks, R. Drautz, and A. Ludwig, "Predicting structure zone diagrams for thin film synthesis by generative machine learning," *Communications Materials*, vol. 1, no. 1, 2020, doi: 10.1038/s43246-020-0017-2.

[37] H. Nakata, M. Filatov(gulak), and C. H. Choi, "Accelerated deep learning dynamics for atomic layer deposition of Al(Me)3and water on OH/Si(111)," *ACS Applied Materials & Interfaces*, vol. 14, no. 22, pp. 26116–26127, 2022, doi: 10.1021/acsami.2c01768.

[38] D. Weichert, P. Link, A. Stoll, S. Rüping, S. Ihlenfeldt, and S. Wrobel, "A review of machine learning for the optimization of production processes," *International Journal of Advanced Manufacturing Technology*, vol. 104, no. 5–8, pp. 1889–1902, 2019, doi: 10.1007/s00170-019-03988-5.

[39] H. S. Park, D. S. Nguyen, T. Le-Hong, and X. Van Tran, "Machine learning-based optimization of process parameters in selective laser melting for biomedical applications," *Journal of Intelligent Manufacturing*, vol. 33, no. 6, pp. 1843–1858, 2022, doi: 10.1007/s10845-021-01773-4.

14 Feature Engineering in Atomic Layer Deposition

14.1 INTRODUCTION

This chapter examines another significant machine learning application in atomic layer deposition (ALD) namely the feature engineering technique for selecting and modifying raw data in order to create suitable features that may be used effectively by machine learning algorithms [1]. Further to this, the entire spectrum of the impact of feature engineering methodology of extracting significant features from the enormous amounts of data collected during the deposition process in enhancing its precision and efficiency is investigated. Readers are encouraged to visit previous chapters to recall fundamental ideas and theory of feature engineering techniques in machine learning. Feature engineering plays a crucial role in the modification of material characteristics and the optimization of film quality. While deep learning, classification techniques, and predictive analysis studied in previous chapters have demonstrated exciting potentials and functionalities in ALD, their performance is strongly contingent on careful choice and transformations of the pertinent features in ALD data, which is where feature engineering hold significance. In addition, feature engineering is essential to the success of the deposition process, optimizing everything from the film's adherence to the substrate to its mechanical and electrical properties.

14.2 TRADITIONAL TECHNIQUES OF FEATURE ENGINEERING

Despite the growth in the realms of machine learning applications, the traditional approach of feature engineering is still relevant. These techniques involve choosing features based on their individual predictive strength and relevance to the target variable. In these approaches, the significance of features is assessed using statistical tests or heuristics, and a subset of the most instructive characteristics is then chosen for model training. According to Li et al. [2], these techniques could be based on statistics or theoretical information. In the statistics-based method, statistical tests are used to identify the variables that significantly affect the outcome variable [2,3]. These techniques seek to decrease the dataset's dimensionality while keeping the most useful variables, leading to more effective and precise models.

14.2.1 Low Variance Technique

A simple statistics-based classical feature elimination procedure called "Low Variance" is used in feature selection to find variables with low variability or values that are evenly distributed across a dataset. This approach is based on the theory that features with low variance are not informative and hence have little to no influence

DOI: 10.1201/9781003346234-17

on a model's output [2]. This approach eliminates features whose variance falls below a predetermined level. We first determine the variance of each feature in the dataset before using this approach. A feature is deemed to be a low variance feature and is eliminated from the dataset if its variance falls below a predetermined threshold. Either empirical research or domain knowledge can be used to set the threshold value. Because it is impossible to distinguish between instances belonging to various classes when a feature's variance is 0, it should be eliminated for features like these that have the same value for all instances [4]. Given that Boolean features are Bernoulli random variables, and that the dataset solely contains Boolean features, where feature values are either 0 or 1, it is possible to calculate the variance of the dataset as follows [4]:

$$\mathrm{var}\,iance_{score}\left(f_i\right) = p\left(1 - p\right) \tag{14.1}$$

where p is the ratio of instances that have a feature value of 1. The feature with a variance score below a set threshold can be immediately trimmed after the variance of the features has been determined. This approach has the benefit of simplicity and computational efficiency, while it might be helpful when working with high-dimensional datasets where the ratio of features to observations is great [2]. This approach ignores feature correlation and just considers feature variance, which is a big downside. Strongly correlated features may have minimal variation but still provide useful information when coupled with other features.

14.2.2   T-Score Technique

The t-score is the ratio of group variation to mean difference. For feature selection, the t-score evaluates the difference in means between two groups for each attribute. The t-score method is used to solve binary classification problems since most samples have two values for the target variable [2]. Assume that for each feature f_i, the associated standard deviation values are σ_1 and σ_2, while μ_1 and μ_2 are the mean feature values for the instances from the first class and the second class, and n_1 and n_2 are the total number of instances from these two classes. Equation 14.2 can be used to get the t-score for the feature f_i [5]:

$$t_{score(f_i)} = \frac{\left|\mu_1 - \mu_2\right|}{\sqrt{\dfrac{\sigma_1^2}{n_1} + \dfrac{\sigma_2^2}{n_2}}} \tag{14.2}$$

The t-score gives an indication of how important each feature is in separating the two groups. Higher t-score attributes are more significant in separating the two groups and are therefore more likely to be chosen as relevant features. The fundamental goal of the t-score is to determine if a feature can statistically cause the means of two classes to differ from one another. To do this, it compares the variance of two classes to their mean differences. The more significant a trait is, typically, the higher the t-score.

14.2.3 F-Score Technique

The F-score is a statistical indicator that rates a feature's significance according to how well it can distinguish between two or more classes in a given dataset [6]. Recall is the ratio of real positives to all actual positives in the dataset, whereas precision is the ratio of genuine positives to all projected positives. The F-score gives a single score that may be used to rate the significance of various elements in the dataset by striking a balance between precision and recall. The F-score is used in feature selection to determine which features are most pertinent to a given problem. In contrast to the t-score technique which can only handle a binary classification, the f-score can be utilized for a multi-class classification by determining if a feature can effectively separate samples from several classes. The f-score can be calculated using both the within-class variance and the between-class variance as in Equation 14.3 [2]:

$$f\_score(f_i) = \frac{\sum_j \frac{n_j}{c-1}(\mu_j - \mu)^2}{\frac{1}{n-c}\sum_j (n_j - 1)\sigma_j^2} \tag{14.3}$$

where n_j represents the number of instances from class j, μ denotes the mean feature value, while μ_j is the mean feature value on class j, and the standard deviation of feature value on class j is denoted as σ_j, respectively. The more significant a feature is, according to the t-score, the higher the f-score.

14.2.4 Chi-Square Score Technique

The chi-square score technique can be utilized to select the features that are most pertinent or informative for a certain classification or regression task. The chi-square statistic for each feature is first computed, which is a measurement of how much the distribution of that feature's values deviates from what would be anticipated if it were independent of the target variable, in order to execute feature selection using chi-square. To do this, a contingency table that displays the frequency distribution of the feature values and the target variable together is constructed [7]. To determine if a feature is independent of a class label, the chi-square score applies the test of independence. The chi-square score of a given feature can be calculated as follows when the feature has r different feature values [7]:

$$Chi\_square\_score(f_i) = \sum_{j=1}^{r}\sum_{s=1}^{c} \frac{(n_{js} - \mu_{js})^2}{\mu_{js}} \tag{14.4}$$

where n_{js} is the quantity of cases that has feature jth feature value for feature f_i. The relative importance of the feature is indicated by a greater chi-square score.

14.2.5 Mutual Information Maximization

The conventional feature selection method known as Mutual Information Maximization (MIM) tries to choose the most informative subset of features that are pertinent to a particular prediction job [8]. It is founded on the idea of reciprocal information, which quantifies how dependent two random variables are on one another. The importance of each feature in relation to the target variable is assessed in the context of feature selection using mutual information. MIM gauges a feature's significance by examining its relationship to the class label. It is predicated that a feature can aid in achieving good classification performance when there is a substantial association between it and the class label [9]. For a fresh, unselected feature named x_k, the mutual information score is:

$$J_{MIM}(X_k) = I(X_k : Y) \tag{14.5}$$

It is clear that in MIM, each feature's score is evaluated independently from all other features. As a result, only the feature correlation is taken into account in MIM, and the feature redundancy property is completely disregarded. The feature with the highest feature score is picked and added to the group of features that have been selected after it calculates the MIM feature scores for all unselected features [10]. Until the desired number of selected characteristics is obtained, the process is repeated.

14.2.6 Mutual Information Feature Selection

The assumption that features are independent of one another is a drawback of the MIM feature selection criterion [2]. In practice, however, desirable characteristics shouldn't just be substantially connected with class labels but also shouldn't be significantly correlated with one another [11]. In other words, it is best to reduce the correlation between attributes. The feature score for a new unselected feature X_k can be calculated using the Mutual Information Feature Selection (MIFS) criterion, which takes into account both feature relevance and feature redundancy in the feature selection phase. The following formula can be used to calculate the feature score for a new, unselected feature, X_k [11]:

$$J_{MIFS}(X_k) = I(X_k : Y) - \beta \sum_{x_j \in s} I(X_k : X_j) \tag{14.6}$$

The mutual information between each feature and the target variable is computed by MIFS. The most pertinent qualities are those that have the most mutual information. By repeatedly adding features with the highest mutual information until the necessary number of features is reached, the method can be expanded to choose a subset of features.

14.2.7 Minimum Redundancy Maximum Relevance

The Minimum Redundancy Maximum Relevance (MRMR) is used frequently to choose pertinent features from large, dimensional datasets. By decreasing redundancy among them, MRMR seeks to isolate a group of features that are extremely pertinent to the target variable. The MRMR criterion was introduced by Peng et al. [12], in contrast to MIFS, which experimentally sets β to be one:

$$J_{MRMR}(X_k) = I(X_k : Y) - \frac{1}{[S]}\sum_{x_j \in s} I(X_k : X_j)$$

(14.7)

Two criteria are used in the MRMR approach: relevance and redundancy. Relevance is the level of information a characteristic provides about the target variable. When two or more features provide substantially similar information about the target variable, this is referred to as redundancy. The MRMR technique aims to reduce feature redundancy while maximizing the relevance of chosen features.

14.3 OVERVIEW OF MACHINE LEARNING-BASED FEATURE ENGINEERING TECHNIQUES

By creating a desirable feature space that accurately depicts the underlying problem to the prediction models, feature engineering tries to maximize machine learning models [1,13]. Because the raw data are not in a format that allows for learning, feature engineering is required [35]. Therefore, every effort to correct flaws, whether through the generation of new features, the filtering of existing features, or the mapping of existing features, should be covered by feature engineering. In general, feature engineering is the process of turning raw data into features that are relevant or useful for solving prediction challenges by using particular techniques of feature processing, such as feature building, selection, and extraction [1]. Although the idea of feature engineering has been known for a while, researchers still have divergent interpretations of what it means and how it might be used [1]. According to some scholars, the primary goal of feature engineering is to ease the curse of dimensionality brought on by the data's rapid evolution [14]. To decrease data dimensionality, remove redundancy, and reduce noise, feature engineering should incorporate feature selection and extraction. Other researchers opined that the primary goal of feature engineering is to improve machine learning by optimizing the feature space representation [15]. Therefore, feature engineering should also involve feature construction, which tries to give the prediction model useful features alongside feature selection and extraction. It is the last phase of data preparation in the process flow chart of the machine learning task, as illustrated in Figure 14.1. Feature engineering is the in-depth processing of raw data that concentrates on addressing data irrelevance, redundancy, and incompatibility as opposed to data preprocessing, which addresses issues such as missing data, errors, and inconsistencies. It provides a solution to a query of which characteristics have the biggest impact on the reliability and

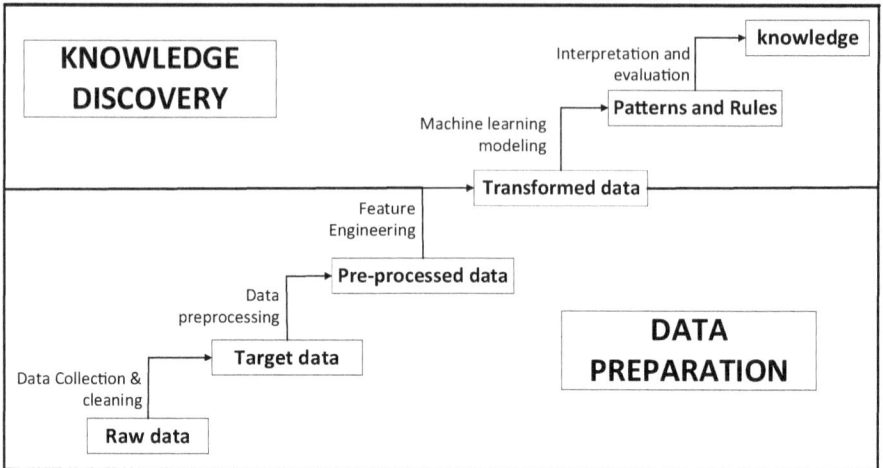

FIGURE 14.1   Workflow diagram of basic machine learning task [1].

accuracy of data-driven forecasts [16]. Feature engineering strengthens the importance of the input information to the forecast parameters (the output data) while decreasing their redundancy and noise to enhance the effectiveness, accuracy, and robustness of machine learning algorithms [17]. This is done by the filtering and reconstruction of the raw data.

Whilst this chapter gives a brief overview of the basic ideas and ALD uses of feature engineering in machine learning-based learning, it is crucial to recognize that the subject is wide and constantly evolving. As such, it is very important to understand that this chapter does not attempt to provide exhaustive coverage of the subject. For more intensive and comprehensive information and understanding of the theory, practices, and most recent developments in feature engineering, readers are recommended to explore additional and specialized textbooks. Once raw data have been collected, cleaned, and pre-processed, future engineering involves the following three important processes:

    i. Future selection
   ii. Feature transformation
  iii. Feature extraction

14.3.1   FEATURE SELECTION

High-dimensional data can be effectively and efficiently prepared for data mining and machine learning challenges using feature selection as a data preparation approach. Building easier-to-understand models, enhancing data mining performance, and creating clear, understandable data are all goals of feature selection [2]. Big data's recent explosion has created both significant obstacles and opportunities for feature selection algorithms. Larger processing and storage requirements are associated with high-dimensional data. Furthermore, noisy, redundant, and irrelevant features

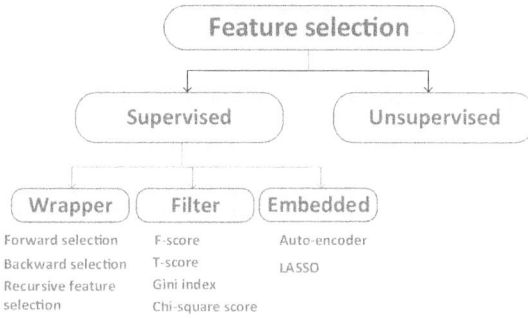

FIGURE 14.2 Categories of feature selection techniques.

in these data contribute to the classification and regression algorithm's poor performance. The removal of redundant, irrelevant, and noisy attributes from the initial feature space makes feature selection necessary in order to pick an effective feature subset, reduce the dimensionality of the data, shorten training time, and simplify model interpretation [18]. Two broad categories of feature selection are the supervised and unsupervised approach as Figure 14.2 reveals.

The unsupervised feature selection approach does not involve the target variable while the supervised feature selection uses the target variable [19]. For classification or regression assignment, supervised feature selection is typically used as it seeks to identify a subset of features that may distinguish samples from various classes, while for clustering assignment, unsupervised feature selection is typically used [2]. Unsupervised feature selection methods look for alternate criteria to determine the significance of features, such as data similarity and local discriminative information, since label information is lacking to do so [2]. The filter-based supervised feature selection techniques utilize statistical metrics to evaluate the correlation and reliance of input factors to be filtered in order to select the most pertinent features [19]. The feature selection process requires domain knowledge and an understanding of the problem being solved, as well as careful consideration of factors such as the data distribution and the interactions between features. The categories of feature selection methods in supervised learning are summarized as follows.

14.3.1.1 Wrapper Technique

This approach employs a model to evaluate numerous feature subsets and select the most significant one. Each fresh subset is used to train a model, and a hold-out set is used to evaluate the model's performance. The features subset with the best model performance is selected [20]. To assess the quality of chosen features, wrapper approaches depend on the accuracy of predictions of a predetermined learning algorithm [2]. Wrapper approaches typically offer the best feature set for the particular model type selected, which is a significant advantage. Common feature selection approaches in this category are forward and backward selection and recursive feature elimination. Figure 14.3 shows the process flow diagram for wrapper methods.

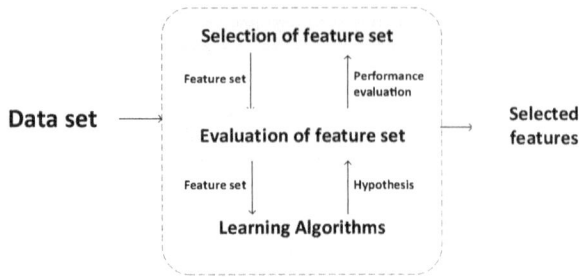

FIGURE 14.3   Process flow diagram of the wrapper feature section technique.

14.3.1.2   Filter Technique

The filter method is a less complex and more direct replacement for the wrapper method. There are no learning algorithms involved with filter approaches. They rely on specific data properties to determine the significance of features [2]. The chosen features, however, might not be the best ones for the target learning algorithms because there wasn't a predefined learning algorithm to direct the feature selection phase. They use statistical metrics like correlation or mutual information to rank each feature in accordance with its statistical significance relative to the model's aim [21,22]. Filter approaches are not only faster than wrappers, but they are also more general because they are model-agnostic and do not overfit to any particular algorithm. They are also quite easy to comprehend: a feature is deleted if it has no statistical significance to the purpose [20]. The chi-square score, t-score, f-score, Gini index, Pearson's correlation, Spearman's rank correlation, and Kendall rank correlation are popular feature selection techniques in this category.

14.3.1.3   Embedded Technique

Filter approaches are highly computationally effective since they choose features without using any learning algorithms. Since, they do not account for the bias of the learning algorithms, the chosen features might not be the best choice for the learning tasks. Wrapper approaches, on the contrary, might produce improved predicted accuracy for that particular learning algorithm by iteratively evaluating the significance of features for the specified learning algorithms. However, when the feature dimension is high, it is computationally difficult in many applications due to the exponential search space [2]. Embedded methods offer a compromise between filter and wrapper techniques that integrate feature selection with model learning, and thus inherit the advantages of both types of methods. The feature selection method is integrated into the model building process using the embedded technique. By combining filter speed with the capacity to find the perfect subset for a specific model, much as from a wrapper, the objective is to attain the best of both ways.

14.3.1.4   Feature Transformation

The technique of modifying data while retaining the information that the data give is known as feature transformation. This kind of data alteration will make it simpler to comprehend machine learning algorithms and produce better outcomes [23].

The objective of feature transformation, which is a mathematical modification, is to apply an equation and then transform the results for additional investigation, and produce a dataset where each format enhances the performance of your models. The ability of machine learning algorithms to comprehend the data they are working with is made possible through feature transformation. Because it frequently contains noise and unimportant details, raw data can be challenging for algorithms to analyze. Most data-driven models assume a normal distribution pattern of variables. However, in practical scenario, it is more probable that variables in actual datasets will have a skewed distribution. Feature transformation is a viable method for implementing the transformation of the skewed variables into normal distribution [24, 25]. The features in the dataset can be transformed to their fullest potential, which will ultimately assist the model's application over time [26]. Since scaling reduces variability, the feature transformation technique lessens the impact of outliers. The nonlinear relationship between the independent feature and the target feature performs better in the model [27]. Feature transformation techniques are but not limited to the following.

14.3.1.4.1 Scaling

Scalability is a critical component of machine learning which informs the decision between local and global feature expansion. Scaling is a method used to normalize the independent variables within a dataset to a consistent range. Data pre-processing includes a procedure to address the presence of significantly disparate magnitudes, values, or units [28]. Furthermore, scaling establishes the standardization of features to a comparable dimension with identical ranges in a process called feature normalization. This holds significance in the impact that of feature size on spectrum of machine learning performance. Examples of algorithms that can be affected by the scale of the features include those that use Euclidean distance metrics, like k-nearest neighbors or linear regression. Scaling is frequently advised in these situations.

14.3.1.4.2 Normalization

A prominent feature transformation method in data preprocessing is normalization, which scales feature values to fall inside a given range or distribution. It is another critical component of feature engineering which ensures the suitability of the data for modelling. It lowers the influence of variations in the magnitude of the input characteristics by normalization to increase the effectiveness and reliability of machine learning algorithms [29]. The data are scaled to a preset range using the min-max scaling approach, which is typically between 0 and 1. After removing the minimum value of the feature, it is accomplished by dividing by the range as in Equation 14.8. The data are scaled to have a mean of 0 and a standard deviation of 1 using the Z-zero normalization method as Equation 14.9 depicts.

$$y_{norm} = \frac{x - x_{min}}{x_{max} - x_{min}} \tag{14.8}$$

where y_{norm} = the normalized data, x = the mean of the variable, x_{min} = minimum variable, and x_{max} = maximum variable:

$$z = \frac{(x - \mu)}{\sigma} \tag{14.9}$$

where $z = z$-sore, x is the score, μ is the mean, while σ is the standard deviation.

14.3.1.4.3 Binning

Data from real-life scenarios are ambiguous and delineated with significant level of noise, outliers and extraneous redundant information and a high level of variability. Data binning, is a pre-processing and transformation method for data that help to lessen the impact of small observational mistakes and noise. It converts numerical data into category representations. The original data values are split up into discrete bins, and then they are substituted with a generic value computed with respect to the bin, thus giving a categorical variable features [30, 31]. When there is a nonlinear relationship between a variable and the target variable, binning is helpful because it can help to reveal relationships that might not be obvious when looking at the continuous variable alone. Furthermore, the number of bins and the method of binning that is selected might have a big impact on the analysis's outcomes. Because of this, it's crucial to carefully assess the best binning approach based on the particulars of the problem. Beyond this realm of applications, binning can also serve as a discretization technique for transforming continuous characteristics, features, or variables into discrete or nominal qualities, features, or intervals. Binning techniques are as follows:

i. **Equal-width (or distance) Binning:** This approach divides the continuous variable's range into bins of equal width. The algorithm splits the continuous variable into a number of groups with bins or ranges that are all the same width. Let x represent the number of groups, and let max and min represent the highest and lowest values in the relevant column, the width is computed as follows:

$$w = \left[\frac{max - min}{x} \right] \tag{14.10}$$

While the categories are computed as follows:

$$categories : \left[min, min + w - 1\right], \left[min + w, min + 2 * w - 1\right],$$
$$\left[min + 2 * w, min * 3 + w - 1 - 1\right].....\left[min + (x - 1) * w, \ max\right] \tag{14.11}$$

ii. **Equal depth (or frequency) Binning:** The observations in this method are split into bins with equal frequencies. Assume n to be the number of data points while the anticipated groups be x, then the frequency would be computed as in Equation 14.12. For instance, if we wanted to partition a continuous variable with 100 observations into 4 bins, each bin would have 25 data:

$$freq = \frac{n}{x} \tag{14.12}$$

iii. **Optimal Binning:** The objective of this approach is to identify the ideal binning approach that maximizes the variable's predictive value. The best bin boundaries are often determined using statistical evaluations or machine learning algorithms.

14.3.2 FEATURE EXTRACTION

Working with datasets containing hundreds (or even thousands) of features is getting more and more common in contemporary times. A machine learning model may get overfitted if the number of features matches or exceeds (or even exceeds!) the number of observations contained in a dataset [32]. Feature extraction techniques are deployed to overcome this challenge. A dataset's dimensionality in machine learning is determined by the number of variables that were utilized to represent it. Feature extraction attempts to decrease the number of features, and the dimensionality of the data by generating novel features from the ones that already exist (and then removing the original features) while retaining as much useful information as possible. The majority of the information in the original collection of characteristics should then be able to be summarized by the new, smaller set of features. Through the combination of the original set, a condensed version of the original features can be produced [32]. Common techniques for feature extraction are as follows.

14.3.2.1 Principal Component Analysis

Principal component analysis (PCA) is a statistical technique that may be used to discover the key features in data and minimize their dimensionality. In order for PCA to function, the data must be transformed into a new collection of orthogonal variables that best captures the original data's variance. The basic goal of PCA is to maintain the maximum variance data set in high-dimensional data while simultaneously providing dimension reduction [33]. It decreases the number of dimensions and compresses the data by identifying the general features in the two-dimensional data. It is a given that some characteristics will be lost with dimension reduction, but the idea is that these attributes that go will reveal nothing about the population. Essentially, PCA combines highly linked variables to create a smaller group of artificial variables, known as "principal components," that account for the majority of variation in the data [34].

14.3.2.2 Linear Discriminant Analysis

A supervised learning algorithm called Linear Discriminant Analysis (LDA) is employed for classifying and reducing dimensionality. LDA seeks to maximize class separability while projecting a high-dimensional dataset onto a lower-dimensional space [35]. LDA also seeks to reduce spreading inside the class while maximizing the distance between the means of each class. LDA consequently employs within-class and between-class measurements [32]. This is a wise decision because, when data are projected in a lower-dimensional space, maximizing the distance between the means of each class can produce better classification outcomes (due to the decreased overlap between the several classes). These linear combinations are referred to as components or discriminants. There are fewer discriminants than or equal to the number of classes divided by one [36].

14.3.2.3   t-Distributed Stochastic Neighbor Embedding

A machine learning approach called t-Distributed Stochastic Neighbor Embedding (t-SNE) is used to visualize high-dimensional datasets [37]. The technique attempts to preserve the pairwise distances between the data points while mapping high-dimensional data points to a low-dimensional environment, usually 2D or 3D [38]. By reducing the divergence between a distribution made up of the pairwise probability similarities of the input characteristics in the original high-dimensional space and its equivalent in the reduced low-dimensional space, t-SNE reduces the variance between the distributions [32]. The t-SNE technique has gained popularity in the area of data visualization and has been used in a variety of tasks, including bioinformatics, image processing, and natural language processing. When used to visualize complicated, high-dimensional datasets like word embedding or gene expression data, it has proven to be extremely beneficial.

14.3.2.4   Autoencoders

A specific kind of neural network known as an autoencoder is trained to compress input into a lower-dimensional form and then reconstruct the original data from this compressed representation. An autoencoder attempts to learn a representation of the input data that captures the most crucial information while ignoring noise and redundancy [39]. Autoencoders have encoders and decoders. Encoders compress input data into a lower-dimensional vector. The decoder reconstructs the input data from this compressed representation [40]. Autoencoders project data from high dimensions to low dimensions through nonlinear transformations [32]. During training, the autoencoder reduces reconstruction error, the difference between input and reconstructed data. This usually involves gradient descent optimization. Denoising, convolutional, sparse, and variational autoencoders exist [32].

14.4   SIGNIFICANCE OF FEATURE ENGINEERING IN MACHINE LEARNING-BASED MODELING OF ALD PROCESSES

Choosing and manipulating ALD raw data into relevant features that machine learning algorithms can utilize to produce precise predictions or classifications is a critical role played by the process of feature engineering. Since it directly affects the model's functionality and accuracy, it is a crucial component in machine learning. Due to the special qualities of ALD thin films, feature engineering is particularly crucial in the machine learning modeling of ALD. A self-limiting chemical reaction between consecutive precursors creates ALD films, producing a highly controlled and exact deposition process. The precursor concentration, substrate temperature, exposure period, and other variables all affect how an ALD film behaves. These elements, however, do not instantly evolve into useful attributes that may be used in machine learning modeling. As a result, feature engineering is required to convert unstructured data into meaningful features that can be utilized to precisely forecast the characteristics of ALD thin films. The development of features based on precursor chemistries is one instance of feature engineering in ALD. This entails taking data from the precursor molecules, including, for example, molecular weight, boiling temperature, and vapor pressure. Because the chemical characteristics of precursors have a direct impact on the chemical reactions that occur during the deposition process, these characteristics can be used to precisely predict the properties of ALD thin films.

The development of features dependent on the deposition circumstances is another instance of feature engineering in ALD. This entails converting unstructured data from deposition experiments, such as temperature, duration, and pressure, into pertinent features that represent the impacts of these circumstances on the characteristics of ALD thin films. Since the deposition circumstances directly affect the structure and morphology of the thin films, these traits can be used to predict the properties of ALD thin films with accuracy. The problem of limited data must also be addressed by feature engineering in ALD. Because ALD thin films are intricate systems, it takes a lot of information to fully describe their characteristics. ALD thin film experimental data collection, however, can be time-consuming and expensive, leading to sparse datasets that might not be appropriate for machine learning modeling. By developing synthetic features that capture the underlying physics and chemistry of the ALD thin films, feature engineering can assist in resolving this problem. For instance, the electronegativity equalization principle, which stipulates that atoms in a molecule will exchange electrons equally if they have identical electronegativities, can be used to produce characteristics. Even with small datasets, these attributes can be utilized to precisely forecast the characteristics of ALD thin films.

ALD machine learning models must be optimized through feature engineering in addition to being crucial for accurately predicting the characteristics of ALD thin films. Feature engineering can enhance the performance and effectiveness of machine learning models by choosing and modifying relevant characteristics, which decreases the computational cost and time needed for predictions. In order to solve difficulties of fairness and bias in machine learning models of ALD, feature engineering can also be applied. The resulting machine learning model might not be able to generalize to different precursors or deposition conditions, for instance, if the dataset utilized for modeling is skewed toward a specific precursor or deposition condition. By designing features that capture the underlying physics and chemistry of the ALD thin films and are not biased toward certain precursors or deposition conditions, feature engineering can be employed to overcome this problem.

14.5 POTENTIAL APPLICATIONS OF FEATURE ENGINEERING IN MACHINE LEARNING-BASED MODELING OF THIN FILM AND ALD PROCESS

14.5.1 Identifying Pertinent ALD Features

There exists a significant variability in the impact of different variables of the deposition process on thin film properties. The process of selecting the most pertinent features that have a significant impact on machine learning models' predictive accuracy is known as feature selection. Some process parameters of the ALD process are contingent on some key input parameters. Process variables including precursor concentrations, deposition time, temperature, and pressure that affect the growth rate, thickness, and other properties of thin films are examples of relevant aspects in

ALD. The optimization of the deposition process and the forecasting of the qualities and attributes of deposited films depend on the ability to recognize pertinent features in ALD. An important parameter that influences the thickness and uniformity of formed films is deposition time. Although longer deposition times can result in thicker films, surface reactions and surface saturation may cause the homogeneity of the film to suffer. While some of these parameters play a critical role in the properties of the deposited materials, others might have negligible impacts.

14.5.2 FILTERING IRRELEVANT FEATURES

After identifying the pertinent characteristics, it's critical to eliminate the irrelevant features that don't significantly improve the predictive power of machine learning models. Noise, redundant features, and features that aren't connected with the target variable are examples of irrelevant features and can introduce noise into the data while making them more dimensional. Hence the filtering of these irrelevant features is critical to machine learning models. Statistical techniques like smoothing, filtering, and averaging can be used in this regard. When several features offer the same information, it results in redundant features which provide another source of irrelevant features. Furthermore, techniques like mutual information and Fisher score can be used to identify the features which are not associated with the target variable. While mutual information evaluates the amount of information shared between the features and the target variable, feature importance ranks the features according to how important they are in predicting the target variable. The Fisher score evaluates the features' ability to discriminate and determines the most discriminatory features.

14.5.3 REDUCTION OF DIMENSIONS OF ALD DATA

Dimensionality reduction is a critical step in the feature engineering procedures for machine learning-based modeling of ALD. ALD process is a complex process owing to the several interconnected variables. It thus generates a high dimensional data. The basic concept of dimensionality reduction has been examined in previous chapters. It entails minimizing information loss while reducing the number of characteristics in a dataset. A model that has fewer features is less likely to overfit. This can be addressed using feature extraction and feature selection to lessen the model's computational complexity, making its training simpler and quicker. The model can be trained more quickly and effectively because it uses less memory and processing power because it has fewer features. In the context of ALD machine learning-based modeling, attempts to make the data less dimensional allow the model to concentrate on the key characteristics that improve prediction accuracy.

14.5.4 ENHANCING THE INTERPRETABILITY OF ALD PREDICTIVE MODELS

To gain useful insights into the interconnections between ALD's critical process parameters and thin film properties, interpretability of machine learning modeling is a crucial part. This understanding can help the process to be optimized and the thin films' quality increased. Feature selection techniques offer a number of benefits to the interpretability of machine learning's model. By focusing on the most crucial elements that influence the model's forecast accuracy, we may narrow our attention. Furthermore, we can learn more about the underlying physical processes by comprehending the connection between these process variables and the characteristics of the thin films. This gives a model that is simpler and easier to understand by lowering the number of features. By choosing the most crucial attributes, we can, for instance, reduce the dimensionality of the data in ALD, making it simpler to comprehend how the process parameters relate to the thin film properties.

14.6 FEATURE ENGINEERING ON IMAGE DATA FOR MICROSTRUCTURAL ANALYSIS OF THIN FILM AND ALD PROCESSES

Understanding the thin films' performance and making the most of their qualities require microstructural investigation. A crucial process has been necessitated in this microstructural analysis of thin film, which is feature engineering. It involves choosing and extracting pertinent features from the image data using domain expertise. Features can be taken from high-resolution scanning electron microscopy (SEM) images in the case of microstructural study of thin films created using ALD. The potential features that could be retrieved from the SEM images include the following:

i. **Grain size**: This a significant structural feature of thin film which can be determined using the image segmentation techniques. The grain size can reveal information about the characteristics of the film, such as its electrical conductivity and mechanical strength.

ii. **Porosity**: Porosity of the thin films can be established by a critical examination and location of pores in the microstructural images. The performance of the film, including its capacity to store energy, is significantly influenced by its porosity.

iii. **Surface roughness**: Texture analysis techniques can be used to determine how rough the thin film's surface is. The optical characteristics of the film, such as its reflectance and transmission, can be impacted by how rough the surface is.

iv. **Grain orientation**: Using crystallographic texture analysis, the orientation of the grains in the thin film can be examined. The mechanical and electrical properties of the film may change depending on the grain orientation.

v. **Defects**: Image processing methods can be used to examine defects including voids, cracks, and contaminants. Defects can alter the film's characteristics, including its mechanical and electrical conductivity.

Machine learning models can be trained using these potential features to forecast end-point performance of the thin film. To comprehend the relationships between various features and the characteristics of the thin film, statistical analysis and clustering can also be done using the extracted features. Feature engineering of image data for microstructural study of thin films created using ALD has a significant potential to offer insightful knowledge into the properties of the films. These methods entail locating and extracting pertinent information and filtering redundant features from the visual data that may be used to comprehend the characteristics and performance of the thin film. Image segmentation, which entails dividing the microstructural image into various sections according to those regions' attributes, such as brightness or texture, can be used in thin film analysis to distinguish between several components, including the substrate, film, and flaws. The characteristics and performance of the film can be deduced from the size, shape, and distribution of these phases. Analyzing a texture entail looking at the spatial relationships between the pixels in the image. In thin film analysis, the surface abrasiveness and porosity of the film can be examined using texture analysis. These characteristics, such as the capacity to store energy, are essential for determining the performance of the film. Data dimensionality can be decreased using the statistical method known as PCA. The most important microstructural features that capture the most important variance in the image data can be found using PCA in thin film analysis. Important traits that might be overlooked by other techniques can be found utilizing this methodology. Wavelet transformation which is a signal processing method can be used to examine the frequency content of image data. This approach can be employed in thin film analysis to examine the surface abrasiveness and porosity of the film at various scales. Different levels of detail can be gleaned from the performance of the movie using this method.

14.7   CONCLUSION

Feature engineering is a crucial component in the ongoing advancement of thin film deposition and ALD process as it enhances and regulates the formation of thin films with the desired level of precision and efficiency. The accuracy and dependability of ALD models can be significantly increased by having the capacity to extract significant features from unprocessed data and translate them into helpful inputs for machine learning algorithms. This chapter examined the diverse techniques of machine-learning based feature engineering and its significance in reducing the dimensions of data, identify and extract irrelevant information and select the most appropriate variables which influences thin film qualities. This strategic manipulation of key process variables via the feature engineering enables the development of novel techniques capable of producing materials with improved characteristics and functions. As feature engineering methods continue to progress and are integrated with machine learning-based modeling, the field of ALD stands to gain significantly. Within the dynamic realm of materials science and nanotechnology, the practice of feature engineering will persist as an indispensable element in propelling innovation and unleashing the complete capabilities of thin film deposition and ALD processes.

REFERENCES

[1] Z. Wang, L. Xia, H. Yuan, R. S. Srinivasan, and X. Song, "Principles, research status, and prospects of feature engineering for data-driven building energy prediction: A comprehensive review," *Journal of Building Engineering*, vol. 58, 2022. doi: 10.1016/j.jobe.2022.105028.

[2] J. Li *et al.*, "Feature selection: A data perspective," *ACM Computing Surveys*, vol. 50, no. 6, 2017. doi: 10.1145/3136625.

[3] R. O. Duda, P. E. Hart, and D. G. Stork, *Pattern Classification*. Wiley, 2012. [Online]. Available: https://books.google.co.za/books?id=Br33IRC3PkQC.

[4] F. Pedregosa et al., "Scikit-learn: Machine learning in Python," *Journal of Machine Learning Research*, vol. 12, pp. 2825–2830, 2011. [Online]. Available: https://scikit-learn.sourceforge.net.

[5] John C. Davis and Robert J. Sampson, *Statistics and Data Analysis in Geology*, vol. 646, New York, NY: Wiley, 1986.

[6] S. Wright, "The interpretation of population structure by F-statistics with special regard to systems of mating," *Evolution*, vol. 19, no. 3, pp. 395–420, 1965, doi: 10.1111/j.1558-5646.1965.tb01731.x.

[7] H. Liu and R. Setiono, "Chi2: feature selection and discretization of numeric attributes," In *Proceedings of 7th IEEE International Conference on Tools with Artificial Intelligence*, pp. 388–391, 1995. doi: 10.1109/TAI.1995.479783.

[8] F. Macedo, R. Valadas, E. Carrasquinha, M. R. Oliveira, and A. Pacheco, "Feature selection using decomposed mutual information maximization," *Neurocomputing*, vol. 513, pp. 215–232, 2022, doi: 10.1016/j.neucom.2022.09.101.

[9] D. D. Lewis, *Feature Selection and Feature Extraction for Text Categorization*, New York: ACM digital library.

[10] A. Fang, J. Wu, Y. Li, and R. Qiao, "Infrared and visible image fusion via mutual information maximization," *Computer Vision and Image Understanding*, vol. 231, p. 103683, 2023, doi: 10.1016/j.cviu.2023.103683.

[11] R. Battiti, "Using mutual information for selecting features in supervised neural net learning," *IEEE Transactions on Neural Networks*, vol. 5, no. 4, pp. 537–550, 1994, doi: 10.1109/72.298224.

[12] H. Peng, F. Long, and C. Ding, "Feature selection based on mutual information criteria of max-dependency, max-relevance, and min-redundancy," *IEEE Transactions on Pattern Analysis and Machine Intelligence*, vol. 27, no. 8, pp. 1226–1238, 2005, doi: 10.1109/TPAMI.2005.159.

[13] P. Domingos, "A few useful things to know about machine learning," *Communications of the ACM*, vol. 55, no. 10, pp. 78–87, 2012, doi: 10.1145/2347736.2347755.

[14] G. Hafeez, K. S. Alimgeer, and I. Khan, "Electric load forecasting based on deep learning and optimized by heuristic algorithm in smart grid," *Applied Energy*, vol. 269, p. 114915, 2020, doi: 10.1016/j.apenergy.2020.114915.

[15] Y. Sun, F. Haghighat, and B. C. M. Fung, "A review of the-state-of-the-art in data-driven approaches for building energy prediction," *Energy and Buildings*, vol. 221, p. 110022, 2020, doi: 10.1016/j.enbuild.2020.110022.

[16] C. Zhang, L. Cao, and A. Romagnoli, "On the feature engineering of building energy data mining," *Sustainable Cities and Society*, vol. 39, pp. 508–518, 2018, doi: 10.1016/j.scs.2018.02.016.

[17] L. Zhang et al., "A review of machine learning in building load prediction," *Applied Energy*, vol. 285, p. 116452, 2021, doi: 10.1016/j.apenergy.2021.116452.

[18] T. Wu, Y. Hao, B. Yang, and L. Peng, "ECM-EFS: An ensemble feature selection based on enhanced co-association matrix," *Pattern Recognition*, p. 109449, 2023, doi: 10.1016/j.patcog.2023.109449.

[19] Jason Brownlee, "How to choose a feature selection method for machine learning," 2019, https://machinelearningmastery.com/feature-selection-with-real-and-categorical-data/#:~:text=Feature%20selection%20is%20the%20process,the%20performance%20of%20the%20model.

[20] Michał Oleszak, "Feature selection methods and how to choose them. Neptune AI," 2023, https://neptune.ai/blog/feature-selection-methods.

[21] M. Chemmakha, O. Habibi, and M. Lazaar, "Improving machine learning models for malware detection using embedded feature selection method," *IFAC-PapersOnLine*, pp. 771–776, 2022. doi: 10.1016/j.ifacol.2022.07.406.

[22] K. Robindro, U. B. Clinton, N. Hoque, and D. K. Bhattacharyya, "JoMIC: A joint MI-based filter feature selection method," *Journal of Computational Mathematics and Data Science*, vol. 6, p. 100075, 2023, doi: 10.1016/j.jcmds.2023.100075.

[23] Ranganathan Rajkumar, "Getting started with Feature transformation for machine learning," 2022, https://tvsnext.io/blog/getting-started-with-feature-transformation-for-machine-learning/.

[24] M. Xiao *et al.*, "Traceable automatic feature transformation via cascading actor-critic agents," 2022, [Online]. Available: https://arxiv.org/abs/2212.13402.

[25] Okan Yenigün, "What are the feature transformation techniques?" 2022, https://python.plainenglish.io/what-are-the-feature-transformation-techniques-ba594b523ec4.

[26] M. M. ElMorshedy, R. Fathalla, and Y. El-Sonbaty, "Feature transformation framework for enhancing compactness and separability of data points in feature space for small datasets," *Applied Sciences*, vol. 12, no. 3, 2022, doi: 10.3390/app12031713.

[27] F. Escolano, P. Suau, and B. Bonev, Eds., "Feature Selection and Transformation," In *Information Theory in Computer Vision and Pattern Recognition*, London: Springer London, pp. 211–269, 2009. doi: 10.1007/978-1-84882-297-9_6.

[28] C. Nkikabahizi, W. Cheruiyot, and A. Kibe, "Chaining Zscore and feature scaling methods to improve neural networks for classification," *Applied Soft Computing*, vol. 123, p. 108908, 2022, doi: 10.1016/j.asoc.2022.108908.

[29] D. Singh and B. Singh, "Feature wise normalization: An effective way of normalizing data," *Pattern Recognition*, vol. 122, p. 108307, 2022, doi: 10.1016/j.patcog.2021.108307.

[30] A. Fazekas and G. Kovács, "Optimal binning for a variance based alternative of mutual information in pattern recognition," *Neurocomputing*, vol. 519, pp. 135–147, 2023, doi: 10.1016/j.neucom.2022.11.037.

[31] X. Chai *et al.*, "Combination of peak-picking and binning for NMR based untargeted metabonomics study," *Journal of Magnetic Resonance*, p. 107429, 2023, doi: 10.1016/j.jmr.2023.107429.

[32] Pier Paolo Ippolito, "Feature extraction techniques," https://towardsdatascience.com/feature-extraction-techniques-d619b56e31be.

[33] B. Yesilkaya, E. Sayilgan, Y. K. Yuce, M. Perc, and Y. Isler, "Principal component analysis and manifold learning techniques for the design of brain-computer interfaces based on steady-state visually evoked potentials," *Journal of Computational Science*, vol. 68, p. 102000, 2023, doi: 10.1016/j.jocs.2023.102000.

[34] S. Wold, K. Esbensen, and P. Geladi, "Principal component Analysis." *Chemometrics and Intelligent Laboratory Systems*, vol. 2, no. 1–3, 37–52, 1987.

[35] K. Yu, S. Lin, and G.-D. Guo, "Quantum dimensionality reduction by linear discriminant analysis," *Physica A: Statistical Mechanics and its Applications*, vol. 614, p. 128554, 2023, doi: 10.1016/j.physa.2023.128554.

[36] F. Zhu, J. Gao, J. Yang, and N. Ye, "Neighborhood linear discriminant analysis," *Pattern Recognit*, vol. 123, p. 108422, 2022, doi: 10.1016/j.patcog.2021.108422.

[37] G. E. Hinton and S. Roweis G, *Stochastic Neighbor Embedding, Advances in Neural Information Processing Systems*, Cambridge, MA: The MIT Press, 833, 2002.

[38] H. Liu *et al.*, "Using t-distributed stochastic neighbor embedding (t-SNE) for cluster analysis and spatial zone delineation of groundwater geochemistry data," *Journal of Hydrology*, vol. 597, p. 126146, 2021, doi: 10.1016/j.jhydrol.2021.126146.

[39] P. Li, Y. Pei, and J. Li, "A comprehensive survey on design and application of autoencoder in deep learning," *Applied Soft Computing*, vol. 138, p. 110176, 2023, doi: 10.1016/j.asoc.2023.110176.

[40] H. Fanai and H. Abbasimehr, "A novel combined approach based on deep Autoencoder and deep classifiers for credit card fraud detection," *Expert Systems with Applications*, vol. 217, p. 119562, 2023, doi: 10.1016/j.eswa.2023.119562.

15 Limitations, Opportunities, and Future Directions

15.1 INTRODUCTION

In this book, we have explored the fast-growing applications of machine learning algorithms in thin film development, optimization, and modelling. The wide spectrum of machine learning applications ranging from predictive analysis to classification techniques, feature engineering and deep learning analysis in transforming the synthesis and development of materials for enhancing thin film deposition process have been investigated. Furthermore, previous chapters have provided useful insights into the strength and weaknesses of these machine learning techniques in the specific domain of ALD. Critical challenges in providing machine learning-based solutions to ALD problems were discussed under several chapters in the second and last section of the book. The difficulties associated with machine learning, such as data accessibility, overfittings and underfittings, and interpretability, have also been examined. The scope and limitation of the book, possibilities, future prospects of machine learning-based applications in ALD, and recommendations for future researches will be discussed in this chapter. We hope to motivate for more intensive research in this fascinating and quickly developing area while also advancing ALD and machine learning concepts.

15.2 SCOPE AND LIMITATION OF THIS BOOK

The studies described in this book have drawbacks, mostly because of its limited scope. Despite extensive coverage, machine learning has many unexplored aspects in this book. While this book has initiated a solid background for comprehending machine learning and ALD process, it has also opened the vast space for more intensive investigations. In this book, we are unable to provide a complete and exhaustive coverage of all feasible applications of machine learning in ALD. While the chapters present most important machine learning applications in ALD, there are many more unexplored applications. For further research, machine learning could be employed to govern the ALD process. There are many additional machine learning methods and algorithms that might be employed to solve ALD problems, even though the algorithms covered in this book represent some of the most prominent strategies in this research space. This book also examined only a few materials and procedures. Although this is significant to establishing a cohesive narrative, however, machine learning may examine many more materials. For more investigation, it could be interesting to model the formation of complicated oxide materials using ALD and machine learning, for instance. Despite this book's strong foundation in ALD machine learning, its limitations are significant and noteworthy. Thus, there is much room for further research in this field, and future research may expand on this book's framework to produce more complicated and exact ALD models.

15.3   LIMITATIONS OF MACHINE LEARNING-BASED ALD MODELING AT INDUSTRIAL-SCALE APPLICATION

Industrial scale thin film deposition process could benefit greatly from machine learning modelling methods. However, there are several hurdles that must be overcome to expand its reach and satisfy the needs of industrial applications. While there are extensive research being carried out in this space at a small-scale ALD and thin film deposition processes, there is still so much work to be done in scaling up its impact at large, commercialized, and industrial-scale applications. The necessity to adjust the process parameters to achieve high throughput and reproducibility at industrial scale while keeping high-quality films is one of the fundamental issues. Since industrial-scale applications involves more sophisticated processes, bigger materials and substrates in a single process, substantial scaling up of machine learning-based ALD research to an industrial scale has been necessitated. As a result, numerous process parameters need to be optimized. Machine learning for process improvement in large-scale manufacturing can speed up commercialization by cutting down on the time and expense associated with creating new procedures while raising yields as well as minimizing rejection rates by improving the films' repeatability and quality. Notably, simultaneous optimization of a number of process variables, which will improve process control effectiveness and efficiency, can be made possible. This book therefore calls for an intensive and in-depth study on scaling up the applications of machine learning-based-solutions to industrial-scale ALD and thin film processes.

15.4   FUTURE PROSPECTS IN MACHINE LEARNING-BASED ALD MODELING

Having established the limited scope of this study, and the wide unexplored domain of machine-learning utilization, we seek to highlights some areas of potential future engagement in advancing thin film trajectories. The precision and effectiveness of ALD procedures could be greatly increased by these future prospects, and they could also help us learn more about the underlying physical and chemical processes detailing the deposition process. Some of these future prospects in machine learning-based ALD modeling research space are briefly discussed as follows.

15.4.1   THE INTRODUCTION OF QUANTUM MACHINE LEARNING

An emerging area that could have significant uses in the realm of ALD research is the use of quantum computing in machine learning. There are potential benefits of incorporating quantum machine learning algorithms in future research in this space to accelerate the development of thin films with enhanced characteristics. By offering novel techniques for data analysis and prediction, this emerging research space has the capacity to transform the field of machine learning-based ALD modeling. Quantum computing proffers a lot of computing benefits over the conventional machine learning techniques. It has the capability to address some issues with optimization substantially more quickly, giving more precise projections of the characteristics of thin films produced by ALD, while interpreting data considerably more effectively, thus minimizing the high processing resources prerequisite for machine learning applications.

15.4.2 ENHANCED MULTISCALE MODELING AND SIMULATIONS

A prominent approach in materials science for integrating several scales, from the atomic scale to the macroscopic scale, is multiscale modeling. This approach finds significance in ALD process in achieving materials forecast at different length scale, ranging from the atomic configuration to the macroscopic features that define the entire film. This provides some steps ahead in performance, time and cost compared to the conventional multiscale simulation approach which combines atomic-scale simulations and continuum modeling. Machine learning approach has proffered viable solutions to address this shortcoming with the conventional multiscale modeling methods. For the purpose of determining the properties of thin films at various scales, machine learning models can be trained on data produced by atomic-scale simulations and continuum modeling,

15.4.3 INTERDISCIPLINARY COLLABORATION OF RELEVANT PROFESSIONAL

A proper comprehension of the complex physical and chemical reactions and processes which makes up thin film deposition and ALD is critical to deploying machine learning-based solutions to ALD problems. ALD research cuts across several interrelated disciplines. Scientists and researchers in computational modeling, machine learning, physics, chemistry, materials science, engineering, and several others must work together and share knowledge in order for this discipline to advance significantly. While experts in machine learning may be able to create models that can precisely anticipate the features of these films, materials scientists, chemistry, and physics experts may have an in-depth comprehension of both the chemical and physical properties of thin films produced by ALD. Collaboratively, these professionals can create fresh strategies for streamlining the ALD procedure and creating thin films with enhanced qualities.

15.4.4 ENHANCING THE INTERPRETABILITY OF MACHINE LEARNING OUTCOME THROUGH EXPLAINABLE AI

As discussed earlier, an important experience which limits the general application of machine learning models in the context of ALD is the interpretability of the model's outcome. Techniques that help researchers to better comprehend how machine learning models are producing predictions, namely, explainable AI can help in this regard. Approaches for developing an explainable AI include model interpretation techniques which function based on the underlying workings of machine learning to acquire knowledge of how it generates predictions, visualization technique which assists researchers to visualize and observe connections between various dataset properties and prediction provided by the models. Furthermore, feature engineering approaches such as feature selection and extraction is significant to understanding the outcome of machine learning. More insights regarding the fundamental causes influencing the ALD process and the characteristics of thin films produced by ALD can get gained by recognizing these selected features.

15.4.5 MACHINE LEARNING-BASED REAL-TIME CHARACTERIZATION OF MATERIALS

Characterization methods namely, Transmission electron microscopy (TEM) and X-ray photoelectron spectroscopy (XPS) are critical to the understanding of the intricate nature of materials and thin film deposition process. However, data processing, which is a vital pre-requisite could be laborious, and less cost-effective, and demand trained professionals. Several dimensions of deep neural network applications could be deployed to overcome this challenge. Real-time data analysis using machine learning is possible for in situ characterization approaches, which could increase the precision of process control and result in the creation of thin films with better properties. Further to this, real-time information that can be generated via this approach and nondetectable in the conventional analytical techniques would be beneficial to experts in enhancing the entire deposition process to generate better quality films.

15.5 RECOMMENDATIONS FOR FUTURE RESEARCHES

The domain of machine-learning-based applications in thin film and ALD modeling will be tremendously transformed by adopting some of these suggestions for future research. This book gives the following recommendations:

i. To enhance the adaptability of machine learning applications in thin film deposition process and its attendant data complexities, this study recommends the development of ALD-specific algorithms.

ii. The robustness, accuracy and practicability of ALD-based machine learning applications can be enhanced by Integrating data from several scales vis-à-vis atomic and macroscopic scales.

iii. The capabilities and competence of ALD procedures can be improved on by investigating novel materials to establish an innovative paradigm in the deposition process. The behavior of new materials can be predicted using machine learning, which could hasten the development process.

iv. Beyond thin film deposition, ALD offers a wide range of possible uses, including energy storage, catalysis, and sensing. These revolutionary ALD applications could also benefit from machine learning while new materials and procedures can be created to satisfy their requirements.

   v. A hybrid modelling applications of machine learning with other state-of-the art modelling techniques will enhance the accuracy and effectiveness of ALD models. This innovative approach was introduced in this book.

  vi. Due to the multi-faceted dimensions of ALD process and applications, it is a highly interdisciplinary field. Thus, collaborative research between experts in other disciplines, like chemistry, materials science, and computer science, can assist to advance research, explore machine learning benefits, and create novel approaches to pressing problems.

 vii. The challenge of precisely characterizing ALD-developed materials can be addressed while its modelling performance is improved by creating novel characterization techniques that can offer thorough details about the structure and properties of these materials.

viii. The deep neural network techniques used in deep learning could offer insightful understanding of microstructural analysis. It hereby recommended because of their significant contributions to quality films productions. Considering the nanometric scale of thin films, monitoring their microstructure plays a critical role in ensuring its quality and performance.

Index

For Product Safety Concerns and Information please contact our EU
representative GPSR@taylorandfrancis.com
Taylor & Francis Verlag GmbH, Kaufingerstraße 24, 80331 München, Germany

www.ingramcontent.com/pod-product-compliance
Lightning Source LLC
Chambersburg PA
CBHW060758220326
41598CB00022B/2472

9 7 8 1 0 3 2 3 8 6 7 3 7